# Understanding Electronic Schematics

**JOHN D. LENK**

*Consulting Technical Writer*

**PRENTICE-HALL, INC.,** *Englewood Cliffs, New Jersey 07632*

*Library of Congress Cataloging in Publication Data*

LENK, JOHN D
Understanding electronic schematics.

Includes index.
1. Electronics—Notation. 2. Electronics—Diagrams.
I. Title.
TK7866.L46 621.381'022'1 80-19920
ISBN 0-13-935908-7

Editorial/production supervision
by Daniela Lodes
Interior design by Gary Samartino
Manufacturing buyer: Joyce Levatino



Printed in the United States of America

10 9 8 7 6 5 4 3 2

PRENTICE-HALL INTERNATIONAL, INC., *London*
PRENTICE-HALL OF AUSTRALIA PTY. LIMITED, *Sydney*
PRENTICE-HALL OF CANADA, LTD., *Toronto*
PRENTICE-HALL OF INDIA PRIVATE LIMITED, *New Delhi*
PRENTICE-HALL OF JAPAN, INC., *Tokyo*
PRENTICE-HALL OF SOUTHEAST ASIA PTE. LTD., *Singapore*
WHITEHALL BOOKS LIMITED, WELLINGTON, *New Zealand*

*To IRENE, the "Sandpiper Lady"*
*and to MR. LAMB, the "Magic Bunny"*

# Contents

# Preface

The word schematic means to show the scheme of things, how things work. In electronics the word schematic is applied to diagrams that identify circuits and show how the circuits work, or how they interrelate to other circuits. To understand such diagrams, you must be able to identify the symbols used, relate these symbols to circuit parts, and then identify the electronic circuits formed by the parts (as represented by the symbols).

The obvious test of any book or course on schematic reading is how well you understand schematics after you have read the book. Throughout this book, and particularly in the last chapters, you are given a variety of schematics. These are both full and partial schematics, from the very simple to the complex. You are asked to "read" the schematics, to identify parts and circuits, describe how the circuits operate, and how the circuits relate to other circuits. If you have read and digested all of the material up to that point, you will be able to read each schematic as it is presented. If you have trouble, you can review the material since each question is related to previous paragraphs or sections.

This is not a math or theory book. Of course, math and theory are not completely ignored, since it is impossible to read schematics without any knowledge of electronics. However, math and theory are kept to an absolute minimum, and used only where necessary to help you read and understand schematics.

In Chapter 1, you are introduced to the electronic schematic. The chapter starts by describing the relationship of schematics to basic electronics, and then proceeds with a discussion of typical schematic symbols, and how symbols are arranged to represent circuits. Next, you are given some actual schematics for study. Although you are not expected to read the schematics at this point in the book, you are taken through the schematics, symbol by symbol, to show what information is available, for example, part values, waveforms, test points, and voltage distribution. The subjects of other electronic diagrams (including pictorial, practical wiring, block, and pinout diagrams) are then discussed, stressing their relationship to the schematic.

With basics out of the way, Chapter 2 launches directly into descriptions of common electronic symbols representing passive electronic devices. Many examples are given to show the variations in electronic symbols, as well as preferred and alternate symbol forms. The symbols covered in Chapter 2 include: conductors, wires, resistors, capacitors, inductors, coils, resonant circuits, a-c circuits, transformers, antennas, grounds, attenuators, pads, batteries, earphones, headsets, loudspeakers, microphones, pickups, circuit breakers, fuses, lamps, relays, switches, connectors, terminals, meters, motors, and generators. By the time you have completed Chapter 2, you should be able to look at a schematic and say "that symbol is a capacitor, and its function in a circuit is . . . ."

Chapter 3 provides the same type of coverage for electronic symbols representing active devices. The symbols covered in Chapter 3 include: vacuum tubes, diodes, transistors (two-junction, bipolar, field effect and unijunction), controlled rectifiers, thyristors, light activated devices (photocells), voltage variable capacitors, saturable reactors, magnetic amplifiers, and integrated circuits. By the time you have completed Chapter 3, you should be able to see a schematic symbol and say "that is a diode, and it functions to pass current in one direction, but not in the opposite direction."

Chapter 4 concentrates on reading the schematics of some very simple, but important, basic circuits. Such circuits are made up of the passive devices described in Chapter 2, and the active devices of Chapter 3. The circuits covered in Chapter 4 include: audio amplifiers, radio frequency amplifiers, oscillators, waveshaping circuits, operational amplifiers, power supplies, attenuators, filters, and control circuits. When you have completed Chapter 4, you should be able to see a circuit schematic and say "that is an oscillator, and it operates as follows . . . ."

Chapter 5 concentrates on reading the schematics of complete electronic equipments or sets. Such equipments are made up of the basic circuits discussed in Chapter 4. In Chapter 5 you are taken on a tour through the schematics of a CB set and a TV set, on a stage by stage, circuit by circuit basis. By the time you have finished Chapter 5, you should be able to look at a TV set schematic, point out every stage, and say "that is the sync separator, and it operates as follows . . . ."

A series of questions follows each chapter. The answers to all questions are given following Chapter 5. To get full benefit from this book, you should try to answer the questions before going on with the next chapter. This will test your skill, and make sure that you are ready for the next chapter.

Many professionals have contributed their talent and knowledge to the preparation of this book. The author willingly acknowledges that the tremendous effort to make this book comprehensive and relevant is impossible for one person, and wishes to thank all who have contributed directly and indirectly. The author wishes to give special thanks to Mr. Joseph A. Labok of Los Angeles Valley College.

JOHN D. LENK

# 1

# Getting Started with Schematics

You can learn to read music without becoming a musician. You can also learn to read electronic schematics without becoming an engineer, technician, or any other form of electronic wizard. Of course, if you are making a study of electronics, at any level, it is quite helpful if you learn to interpret schematics. In fact, it is impossible to go very far in any study of electronics without being able to interpret a schematic diagram. This is because electronics people use schematics as a means of communicating with each other. That is, the *symbols* found in schematics are used to show such vital information as: what *parts* (resistors, capacitors, transistors, etc.) are used, how the parts are connected to form *circuits* (amplifiers, oscillators, logic networks, etc.), and how the circuits are connected to form a piece of *equipment* (television set, citizen's band set, etc.) or a *system* (stereo system, radar, etc.).

## 1-1 SCHEMATICS AND BASIC ELECTRONICS

Just as it is impossible to learn electronics without being able to read schematics, it is equally impossible to learn the art of schematic reading without learning something about electronics. In practical terms, if you learn the schematic symbols, how the symbols are connected to represent circuits, and how the circuits are arranged to form a piece of equipment or a system, *you will be learning basic electronics.* Do not let this frighten you! Unlike most books and courses in basic electronics, this book does not dwell on the abstract theory of how a particular part performs its function, or discuss details of the math involved in the process. Instead, we show you the commonly used symbols and *state in simple terms* what function is performed by the part that the symbol represents. You must accept these statements! (If you cannot accept the statements, or if you insist on knowing the theory back of how a part performs its function, there are a number of well-written books on the subject. But theory and math are not vital to an understanding of electronic schematics.) Once we have given the symbol and an explanation of what the corresponding part does, we go on to talk about how this function is used in a circuit or piece of equipment.

## 1-2 A TYPICAL SYMBOL USED IN SCHEMATICS

Let us consider a typical example of this approach to understanding the electronic schematic. Figure 1-1 shows the symbol of a *solid-state diode.* These diodes are generally made of crystal or crystal-like material, such as silicon, germanium, or selenium. Actually, a diode is a *junction* of two materials. A typical book on theory will tell you (possibly in several paragraphs) that the materials are made up of atoms and that the atoms are made up of electrons. It will go on to say that when voltage is applied across a junction, the electrons in one material near the junction will pass to holes in the atoms of the other material on the other side of the junction. This creates holes in the atoms of the first material, causing more electrons to move, thus creating a flow of electrical current, and so on.

In this book, we bypass all of that and simply state that the symbol shown in Fig. 1-1 is that of a solid-state diode. The pointed (or arrow-like) part of the symbol represents the *anode;* the flat part represents the *cathode.* The junction is the point at which the anode touches the cathode. When the anode is made *more positive* than the cathode, from whatever source and for whatever reason,

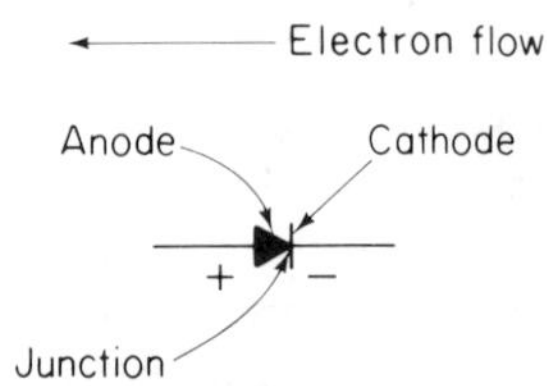


**FIGURE 1-1** Basic solid-state diode symbol

*current will flow through the diode from the cathode to the anode* (against the arrow). When the anode is not more positive than the cathode, no current flows in either direction. You do not have to know how or why, but when you find a diode in a schematic, you must remember that the diode will pass current in *one direction only* and only under one condition (when the anode is more positive than the cathode).

## 1-3 A TYPICAL CIRCUIT USED IN SCHEMATICS

Now that you have learned one basic symbol and the corresponding function of the part represented by that symbol, let us see how the function can be used in an elementary circuit. Figure 1-2 shows the diode connected in a *rectifier* circuit. This very common circuit, usually found in *power supplies* and *detectors*, uses the basic diode function to convert alternating current (ac) into direct current (dc).

There are other symbols in Fig. 1-2. First, there is the circle (containing a sine wave), which represents an *a-c source*. This source could be any alternating current, including normal house power. We could have omitted this symbol and simply stated "a-c power" (as is done on some schematics), but that would have limited understanding to those who read English. Keep in mind that schematic symbols are international in nature (even though there are some

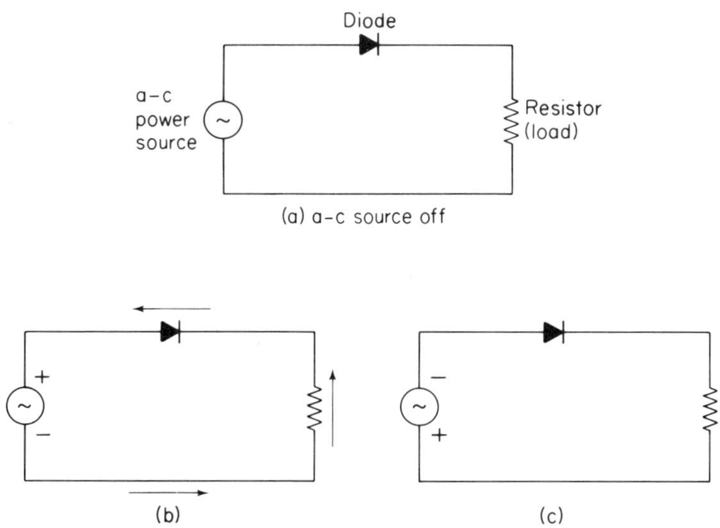


**FIGURE 1-2** Basic diode connected as a half-wave rectifier

variations from country to country). A Russian technician, for example, will instantly recognize the a-c source symbol. Figure 1-2 also shows a *resistor* or resistance element symbol, as well as wires interconnecting the a-c source, diode, and resistor. (In a Russian schematic, the resistor symbol will be a narrow rectangular box instead of the zigzag line.)

Note that the a-c source is shown on the left, with the resistor (labeled the *load*) shown on the right. This is because schematics are normally *read from left to right.* That is, the *signal flow* or *function flow* is shown from left to right. For example, a properly drawn schematic of a stereo system will show the tape head, pre-amplifier, final amplifier, and loudspeaker in that order, from left to right.

Figure 1-2a shows the rectifier circuit when no power is applied (the a-c source is turned off). Figure 1-2b shows operation of the circuit when power is applied and the top line or terminal of the a-c source is positive (with the bottom terminal negative). As you go on with your study of electronics, you will learn that the polarity of alternating current switches regularly. Under the conditions of Fig. 1-2b, the anode of the diode is made positive and the cathode is made negative. Thus, the diode is "turned on" and current flows from the a-c source through the resistor (load) and diode back to the a-c source, as indicated by the arrows.

Figure 1-2c shows the mode of operation when the a-c source has reversed polarity. The anode is now negative, the cathode is positive, and the diode is "turned off." There is no current flow, even though the a-c power source is still available.

When you compare Fig. 1-2b and c, you will see that current flow through the resistor is always in one direction and only on every other polarity change of the source. The circuit of Fig. 1-2 has thus converted alternating current into direct current. As you will learn, one *complete change* of the a-c source polarity (from positive to negative and back to positive again) is called a *cycle* or *wave.* A change from one polarity to the other is called a *half-wave* or *half-cycle.* Thus, the circuit of Fig. 1-2 is generally referred to as a *half-wave rectifier.* We will discuss electronic terms in greater detail later. Here we concentrate on schematics and symbols.

## 1-4 SOME TYPICAL SCHEMATICS

Now let us take a giant leap from the very simple to the real thing—the schematics of actual electronic circuits and equipment. Figures 1-3 through 1-5 show three typical schematic diagrams. Although none

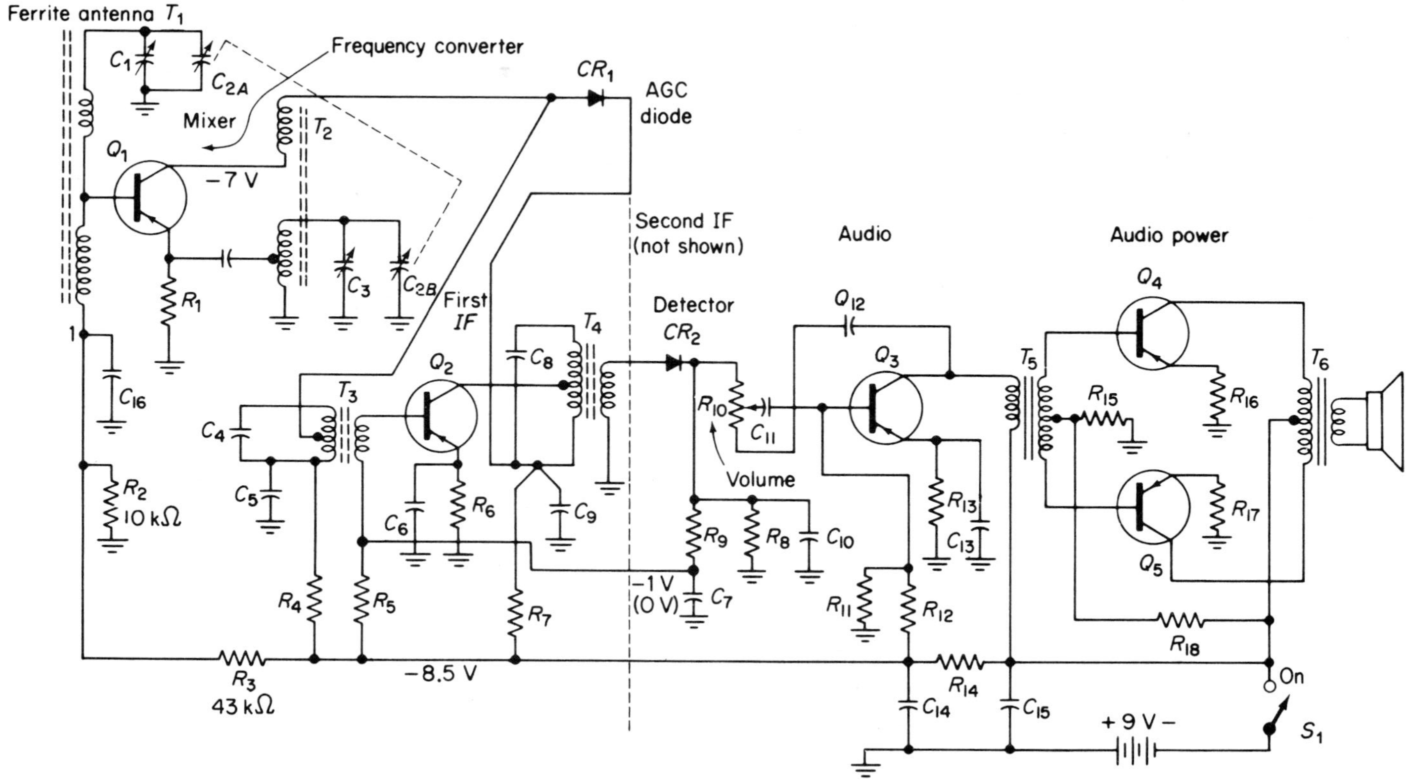


**FIGURE 1-3** Schematic diagram of receiver

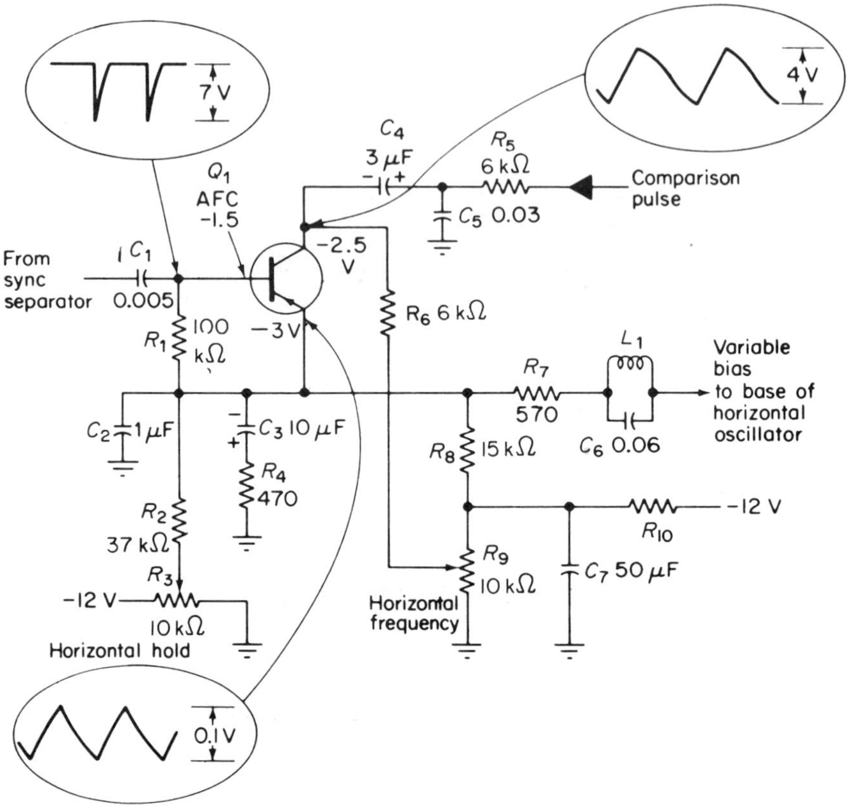


**FIGURE 1-4** Examples of waveform and voltage information found in commercial service literature

of these are really very complex, they probably strike terror in the hearts of readers who are totally unfamiliar with electronics! To such readers, the diagrams are a jumble of symbols, wires, and strange notations. However, when you break down the overall schematic into circuits, and then break down the circuits into a group of symbols, they are not quite as formidable.

As an example, try placing a sheet of paper or your hand over the schematic of Fig. 1-3 so that the paper covers everything to the left of the symbol for transformer T5. [If you cannot find T5, start at the extreme right-hand side (at the loudspeaker symbol) and work your way back to the left.] When you have covered everything to the left of T5, the remaining part of the schematic is a circuit of an *audio power amplifier*, consisting of transistors Q4 and Q5; trans-

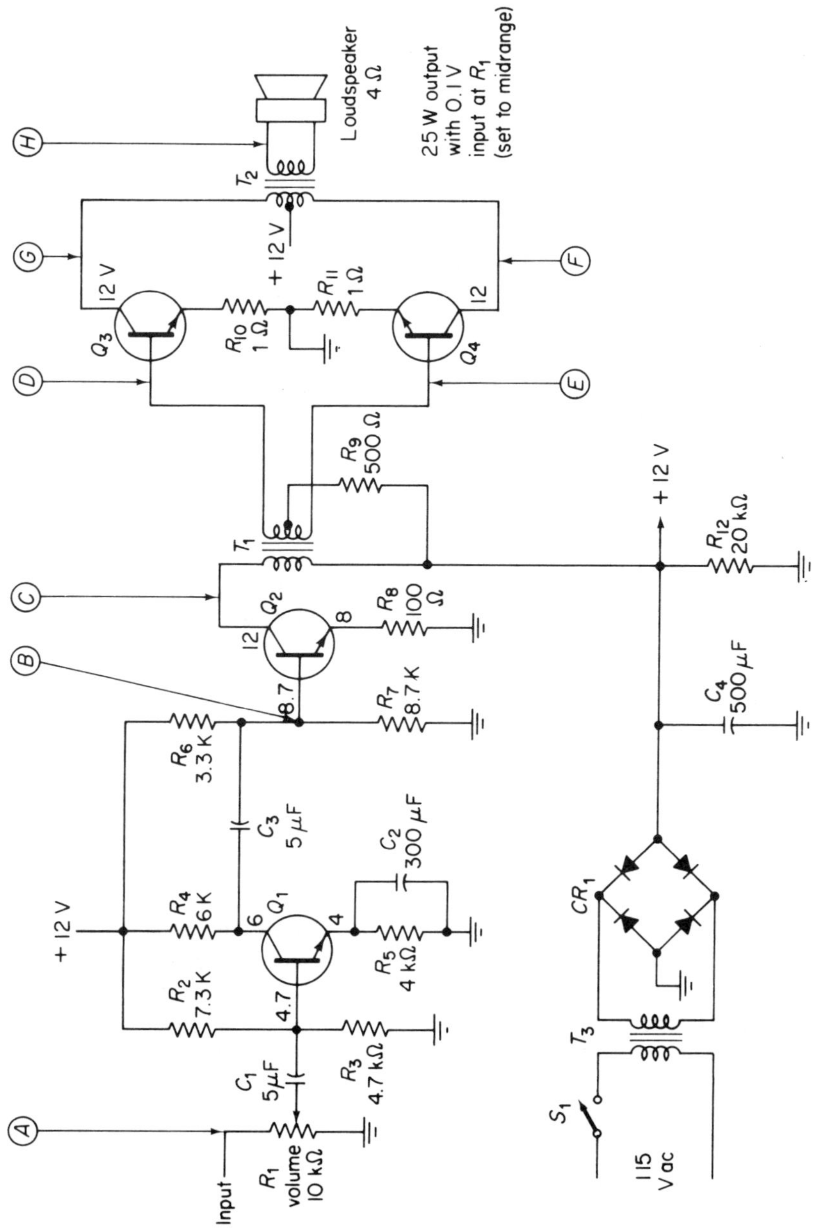


**FIGURE 1-5** Schematic diagram of simple audio amplifier

formers T5 and T6; a loudspeaker; and assorted resistors, each represented by corresponding symbols.

The most complex schematic diagram can be broken down into a number of circuits. In turn, each circuit breaks down into a few parts (represented by corresponding symbols). By the time you have finished reading this book, you will be able to identify the most commonly used symbols and circuits and will understand their functions.

However, before we go on, note well that there are a number of special circuits to be found in electronics. Such circuits, which often result when an electronic designer attempts to solve a difficult problem, may break all the rules of conventional circuit design. Even experienced technicians and engineers may find it difficult even to identify the circuit function, much less to describe how the circuit works! That is why well-prepared service literature includes a "theory of operation" section, which describes all the circuit functions or at least those functions that are "odd." Therefore, do not feel inferior, or that all your time has been wasted, if you can not identify every circuit of a complex schematic or describe the circuit function. But, after reading this book, you should be able to recognize at least 90% of the circuits you encounter.

## 1-5 SCHEMATIC DATA

A well-prepared schematic will show much more than all the parts and circuits involved. Let us examine the schematic of Fig. 1-3. Each part (capacitor, resistor, etc.) is represented by a symbol. In turn, each symbol is identified by combinations of letters and numbers called *reference designations.* The letter is determined by the type of part (C for capacitors, Q for transistors, etc.). The numbers are arbitrarily assigned. In simple equipments, the numbers run from 1 through 100 (higher if necessary). In more complex devices, where there are several groups of circuits, each circuit group is assigned a separate set of reference-designation numbers. For example, in a communications set, the transmitter reference designations may be in the 100 range (C101, R103, T107), with 200s used for the receiver (R201, L201) and the 300s used for the modulator section (T301, Q301).

The letter portion of the reference designation is supposedly consistent on all schematics. In fact, there is a military specification covering reference designations. Unfortunately, not everyone uses this military specification, so you will find many variations in nonmilitary schematics. Transistors and diodes are good examples. The generally accepted letter designation for transistors is Q, but some schematics show TR instead. CR is the accepted designation for diodes, but D is used on some schematics. The letter X is sometimes used to designate miscellaneous or "odd" parts for which there may not be any accepted reference designation. The correct use of an X is to identify the socket for a part. As an example, XQ2 is the socket for transistor Q2.

When a part has more than one section, which is common for filter capacitors and variable capacitors, each section is generally identified by another letter (arbitrarily assigned). For example, if filter capacitor C8 has three sections, the sections can be identified as C8A, C8B, and C8C. This system is used on the schematic of Fig. 1-3 for capacitor C2, which is a two-section variable capacitor. The two sections, C2A and C2B, are geared or "ganged" together, as indicated by a dashed line, the common symbol used to indicate mechanical coupling.

There are other uses for dashed or dotted lines on schematics. In some cases, dashed lines are used to isolate a group of parts or circuits into sections, possibly to show physical relationships or possibly to show a functional relationship. In the schematic of Fig. 1-3, the vertical dashed lines identified with the notation "Second IF (not shown)" indicate that the second IF (intermediate-frequency) stages are now shown on the schematic. Possibly, the IF stages are all contained on a separate circuit parts board or in an integrated circuit (IC). In any event, the parts and circuits for the second IF are not shown on this schematic. (The parts may not be shown on any schematic, as discussed later in the chapter.)

Individual parts are identified by reference designations. Groups of parts that form circuits or sections are identified by a *title*. Generally, these titles or designations are applied to a transistor (or transistors) and the associated circuit parts. For example, transistor Q1 and the associated circuit parts are titled the *mixer* and *frequency converter* (since they have two functions). Transistor Q2 and related parts form the *first IF amplifier.* Transistor Q3 is the *audio amplifier;* transistors Q4 and Q5 form the *audio power amplifier.*

In some cases, an individual part will be identified by a title in addition to the reference designation. For example, diode CR1 is the AGC (automatic-gain-control) diode; diode CR2 is the *detector.* Similarly, variable resistor (or potentiometer) R10 is the *volume* control.

## 1-6 CAPACITOR AND RESISTOR VALUES

It is customary to include the values of capacitors and resistors (and possibly some other parts) on schematic diagrams (but do not count on it!). Note that the schematics of Figs. 1-4 and 1-5 show the values of all resistors and capacitors, whereas the schematic of Fig. 1-3 omits most of the values. To simplify identification, the schematic will often contain a note such as "unless otherwise stated, resistor values are in kilohms and capacitor values are in microfarads." Such notes eliminate the need for a microfarad ($\mu$F) symbol and a kilohm (k$\Omega$) symbol next to each reference designation, as shown in Figs. 1-3 through 1-5.

## 1-7 VOLTAGE VALUES AND WAVEFORMS

It is also customary to include voltage values and waveforms on schematic diagrams, particularly where the diagrams are used for troubleshooting. The amount of voltage and waveform information depends entirely on how well the schematics are prepared. Compare the schematics of Figs. 1-3 through 1-5.

Figure 1-3 shows very few voltage values and no waveforms. The battery is given as 9 V (9 volts), the collector of Q1 is given as −7 V, and the junction of R4-R5-R7 is given as −8.5 V. The junction of R9 and C7 shows two voltages (−1 V and 0 V). Such a notation means that the voltage at this point varies under different circuit conditions. In this particular case, the voltage is −1 V when no signal is being received but changes to 0 V when a signal is present. This is an isolated case but brings up an important point. When reading any schematic, always study all notations and look for any abnormal notations (such as two voltage values at one point). The schematic is trying to tell you something!

Figure 1-4 shows adequate voltage and waveform information, as well as component value data. By "adequate" we mean that the voltages and waveforms are given for each element of each transistor. Transistors are usually the *active devices* for most circuits. Transistor elements are where voltages and waveforms are usually monitored during troubleshooting. For example, if you monitor the collector of Q1 with an oscilloscope, you should find a sawtooth waveform approximately 4 V in amplitude. If you monitor the same collector with a voltmeter, you should find approximately −2.5 V of direct current. If either of these values or measurements is absent or abnormal, the circuit is not functioning normally and you have a starting point for troubleshooting.

When a voltage is given on a schematic without a polarity symbol, it is generally considered to be + (or positive). Properly drawn schematics will show the polarity or give you a note stating something like "all voltage values are positive, unless otherwise indicated."

Voltages and waveforms are also generally measured with respect to ground or a common point on the chassis or circuit board. That is, you must connect one lead of the monitoring device (oscilloscope or meter) to ground or common and the other lead to the point in question, to duplicate the value shown on the schematic. Again, be on the lookout for any notes on the schematic regarding voltage and waveform measurements. One problem often overlooked is that voltages and waveforms may differ from those shown on schematics, depending on how operating controls or adjustment controls are set. A well-prepared schematic will tell you which voltage and waveform values are affected by control settings, and will describe the correct control settings to obtain the values shown on the schematic.

The waveforms shown in Fig. 1-4 are typical of those found in most commercial service literature, such as TV service manuals. They are *approximations* of the actual waveforms. Military-type manuals will usually have actual photographs of the waveforms (or exact duplications of the photos, including the number of horizontal and vertical screen divisions of the oscilloscope).

Many schematics omit all waveforms and show such information on other illustrations. In some cases, the waveforms are omitted simply because waveform information would be meaningless. For example, the schematic of Fig. 1-5 shows an amplifier circuit. The waveforms at any point in such a circuit are amplified versions of whatever waveforms appear at the input.

## 1-8 TEST POINTS AND VOLTAGE DISTRIBUTION

Figure 1-5 shows *test points* at junctions of various parts and circuits. There are many systems used for test points. A schematic drawn in the full military style will have *major test points* (usually numbers in a starred circle) and secondary test points (typically letters in a plain circle), as shown in Fig. 1-5. No matter what system is used, test points usually show the order of signal flow. In the case of Fig. 1-5, test points A through H show signal flow from input to output, in that order. Considering that the equipment shown in Fig. 1-5 is an amplifier, without consultıng any additional information, it would be reasonable to assume that the signals monitored at test points A through H will increase in amplitude from point to point. That is, the signal at B is probably of greater amplitude than the signal at A, and so on.

Two systems of showing *voltage distribution* are illustrated in Fig. 1-5. The power supply, consisting of switch S1, transformer T3, diode rectifier CR1, capacitor C4, and resistor R12, produces a +12-V output. This output is distributed to transformer T1 and resistor R9, as shown by the direct-wire connection. The +12 V is also distributed to transformer T2 and the junction of R2-R4-R6, but without showing the direct-wire connection. The latter system of showing voltage distribution simplifies drawing the schematic and eliminates the clutter caused by the additional lines required to show the direct-wire connections. Both voltage-distribution systems may be used on the same schematic.

As discussed in Sec. 1-5, one reference designation may be used to show several parts when they are assembled and replaced as a package. This is the case with the diode CR1, which acts as the rectifier in the power supply and voltage-distribution circuits of Fig. 1-5. The four diodes are mounted on one circuit board and are interconnected as shown. However, they are never replaced as individual parts but as a group identified as "CR1."

## 1-9 OTHER ELECTRONIC DIAGRAMS

Schematics are not the only diagrams used in electronics. There are *practical wiring diagrams* (sometimes known simply as *wiring diagrams*), *pictorial wiring diagrams*, and *block diagrams*. When ICs are involved, there are also *pinouts* or *pin diagrams*, as discussed in Chapter 2. We do not discuss any of these diagrams to any great extent in this book, since the schematic is still the most commonly used (and best) means of communication among electronics people. However, you should know that other diagrams exist and, more important, their relationship to the schematic diagram. *Practical and pictorial diagrams show physical relationships*, such as placement of parts and point-to-point wiring, whereas *schematics show functional relationships* (such as signal flow) and circuit characteristics (such as voltages and waveforms at various points in the circuits). *Block diagrams are simplified versions of* schematics and show functional relationships but without the clutter of full schematics. *Pinouts* are full or partial schematics, or block diagrams, showing the contents of an IC, with the important points of the circuit shown connected to the IC pins or terminals (arranged as the pins physically exist on the IC).

## 1-10 PRACTICAL AND PICTORIAL WIRING DIAGRAMS

Figure 1-6 shows typical *pictorial wiring diagrams.* Pictorials were used in the early days of electronics and are still in limited use today. Generally, pictorials are used by electronic kit manufacturers and publications aimed at the beginner to show physical relationships and point-to-point wiring. Note that the parts are drawn in three dimensions, with direct interconnecting wiring, on the pictorial of Fig. 1-6. Although this pictorial diagram makes it very easy for someone attempting to assemble the circuit, the system is impractical when there are many parts involved. A pictorial diagram of even a simple receiver such as that shown in Fig. 1-3 would be a hopeless jumble of parts and wires.

The practical wiring diagram of Fig. 1-7 has generally replaced the pictorial when both physical relationships and wiring must be shown. Note that the diagram of Fig. 1-7 shows the practical wiring of transistor Q2 in the amplifier of Fig. 1-5. Figure 1-7 is a flat, two-dimensional view of the circuit parts and wiring as seen from the underside of the amplifier chassis. All the resistors are mounted on a terminal board. Transistor Q2 and transformer T1 are mounted on the chassis, with their terminals extending through the chassis. The wiring is point to point and follows the *approximate path* of the actual wiring. In effect, the practical wiring diagram duplicates approximately what you see when you look at the underside of the

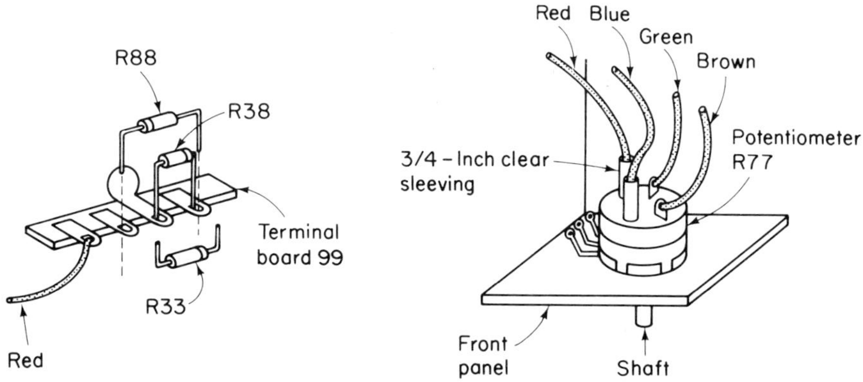


**FIGURE 1-6** Examples of pictorial wiring diagrams

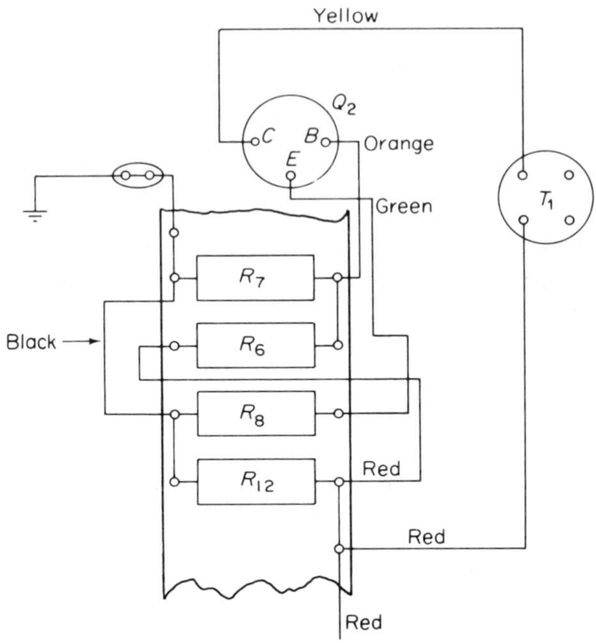


**FIGURE 1-7** Physical relationship of Q2 circuit parts and point-to-point wiring (practical wiring diagram)

chassis but in simplified form (not three-dimensional and not exact wiring as in the pictorial).

Now compare the practical wiring of Fig. 1-7 with the schematic of Fig. 1-5. Both diagrams show that transistor Q2 is connected to transformer T1 and to several resistors. The schematic shows that the Q2 base (or input) is at the left, whereas the Q2 collector (or output) is at the right. Thus, the schematic shows signal flow through the circuit. For example, if you measure the signal at the base and collector (with an oscilloscope or other signal-tracing device), you will probably find that the collector signal is an amplified version of the base signal. (This is not always true in all cases, but it is a logical assumption here.)

The practical wiring of Fig. 1-7 shows no such signal flow or functional relationship. Instead, transistor Q2 is on top, transformer T1 is at the right, and the resistors are all in a row in the middle (as all the parts are physically placed). However, the practical wiring does show that the base of Q2 is connected to resistors R6 and R7 by an orange wire and the collector is connected to transformer T1 by a yellow wire. In turn, T1 is connected to several parts by a red wire. These facts are very helpful if you are trying to locate a defective part, find a broken wire, or find a short between wires.

## 1-11 BLOCK DIAGRAMS

Figure 1-8 shows a block diagram of the same circuits shown on the schematic of Fig. 1-3. In Fig. 1-8, each transistor and the related circuit parts are represented by a block (Q1 mixer, Q2 IF amplifier, etc.). Key parts between each transistor are also shown (detector CR2, transformers T5 and T6, loudspeaker). The block diagram of Fig. 1-8 also includes test points and some simulated waveforms. The obvious advantage of the block is that the signal flow, key test points, and functional relationships can be shown without the clutter of the corresponding full schematic (Fig. 1-3). This makes the block well suited to supplement the "theory of operation" sections in service manuals and as a ready reference for troubleshooting. However, the schematic is still required to show such vital data as circuit wiring, part values, voltages, and so on.

Figure 1-9 is a combination block and schematic diagram. The block at the bottom of the schematic represents the audio amplifier shown in Fig. 1-5. This arrangement might be used where there are

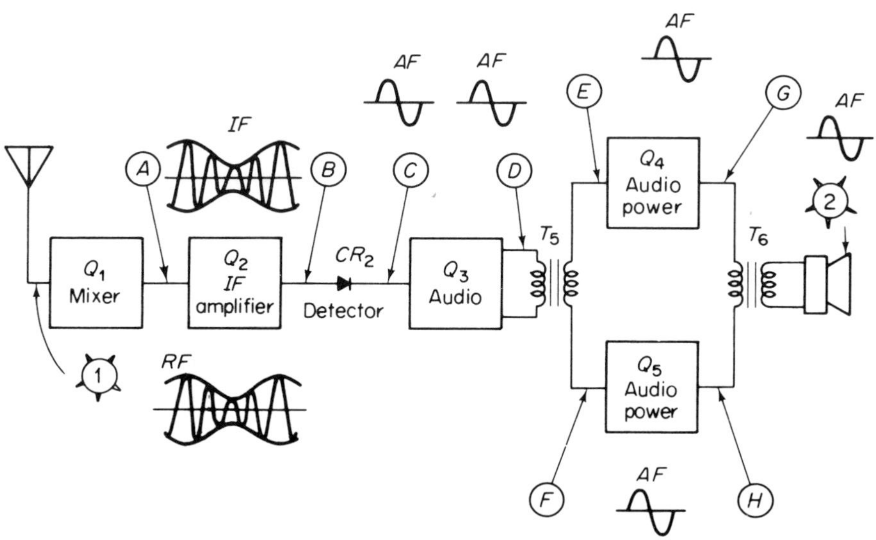


**FIGURE 1-8** Functional block diagram of receiver

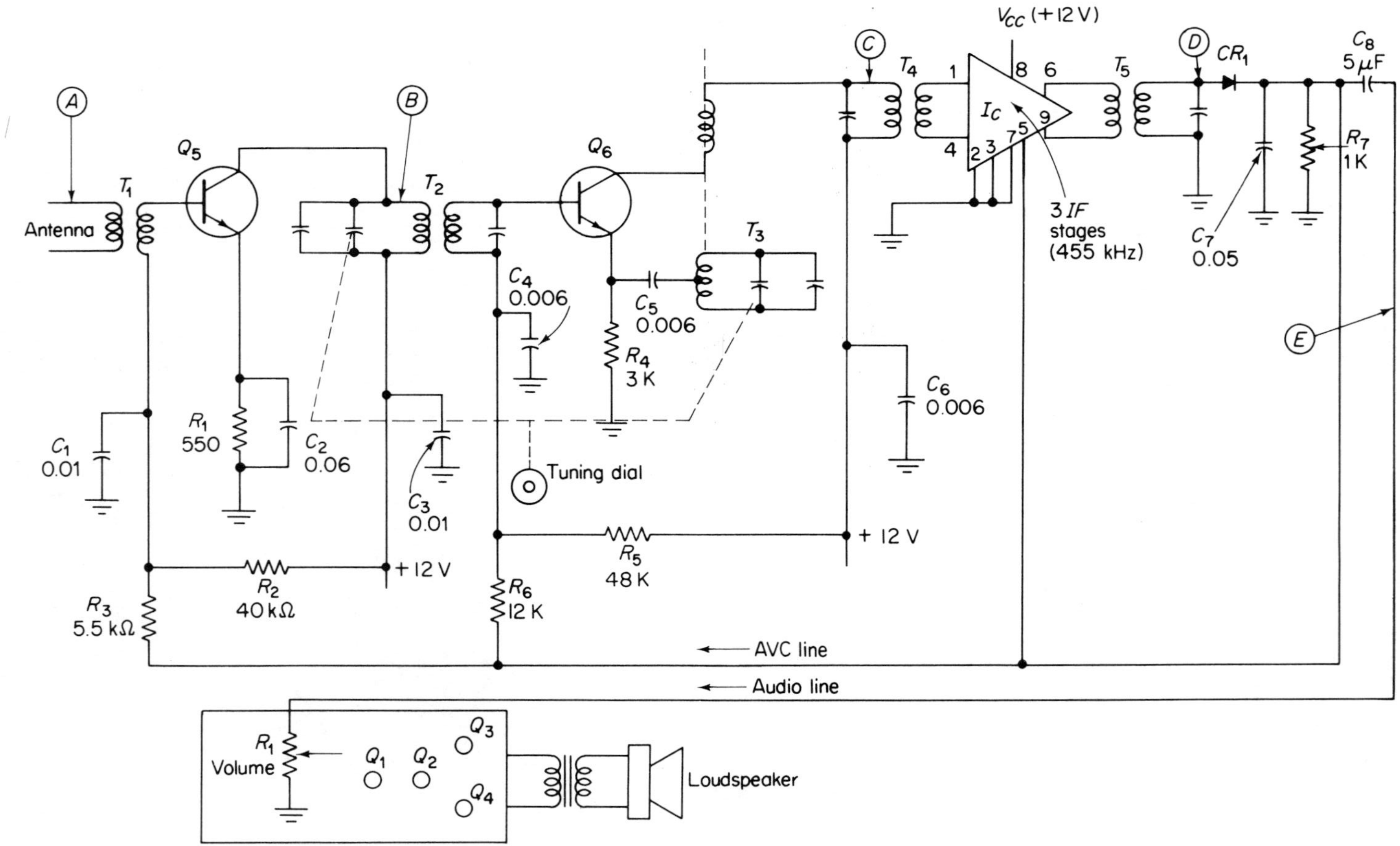


Same as audio-amplifier troubleshooting example

**FIGURE 1-9** Combination schematic/block diagram of AM receiver

two schematics presented in a service manual. An alternative system would be to simply put an arrow on the audio line and include a note stating that the audio section is shown on another schematic.

The drawing of Fig. 1-9 has three additional special features. First, note that the "tuning dial" is connected by a dashed line (indicating mechanical coupling) to capacitors identified with transformers T2 and T3. The capacitors are not identified by reference designations because the capacitors are part of the transformers.

Next, note that the IC is shown between transformers T4 and T5, with a note indicating that the IC contains three IF stages. This means that all parts of the IF stages (except T4 and T5) are contained within the IC and need not be shown on the schematic. Even if the parts were shown, it would be of little value, because the parts are enclosed in a sealed IC and must be replaced as a package.

Finally, note the vertical dashed line from the winding of transformer T3 to another unidentified winding (or coil). This indicates that the unidentified winding is physically part of the transformer (probably wound on the same core) but is shown separately for convenience.

## 1-12 SCHEMATIC VERSUS PRACTICAL AND BLOCK DIAGRAMS

It is obvious from the discussions in Secs. 1-10 and 1-11 that block, practical, and schematic diagrams should be available when attempting to understand, service, or troubleshoot any piece of electronic equipment or any system. This is the case in military electronic manuals but rarely in commercial manuals. At best, commercial manuals provide only the schematic. Of course, this is the logical choice, since the schematic shows more of the most useful information.

Now that you see the value of schematics and understand their relationships to other electronic diagrams, let us proceed with learning how to read schematics, starting with the basic symbols given in the next chapter.

## 1-13 ANSWERING THE QUESTIONS

It is essential that you read the questions at the end of each chapter and try to answer them. Then check your answers against those given at the end of the book. Not only will this test your skill in schematic reading, but it will make certain that you understand what you have learned. The questions and answers are designed to reinforce the text. So before you jump to Chapter 2, answer the following questions.

## EXERCISES

1.1. The parts used in electronic equipment are represented by __________ on schematic diagrams.

1.2. When parts or symbols are connected together on a schematic, they form __________.

1.3. Referring to Fig. 1-1, under what circumstances does current flow from anode to cathode?

1.4. Referring to Fig. 1-2, what electronic part is represented by the circle symbol (containing a sine wave)?

1.5. Referring to Fig. 1-2, what symbols do you find besides the a-c power source and the diode?

1.6. Referring to Fig. 1-2, why is the resistor shown on the right-hand side of the schematic?

1.7. Referring to Fig. 1-2a, why are there no arrows showing current flow?

1.8. Referring to Fig. 1-2c, why are there no arrows showing current flow?

1.9. The circuit of Fig. 1-2 is generally referred to as a __________.

1.10. Referring to Fig. 1-3, what is the function of diode CR2? From what you have read so far, can you describe this function?

1.11. What is the reference designation for the transistor shown at the upper left-hand side of Fig. 1-3?

1.12. Find transistor Q3 in Fig. 1-3. Locate the capacitor directly above Q3. Why is this capacitor given the reference designation Q12?

1.13. What do the dashed lines between C2A and C2B indicate in Fig. 1-3?

1.14. Where is the second IF shown in Fig. 1-3?

1.15. Why is Q1 called a mixer in one place, and a frequency converter in another place in Fig. 1-3?

1.16. What is the value of resistor R2 in Fig. 1-3?

1.17. What is the value of resistor R8 in Fig. 1-3?

1.18. What is the voltage at the junction of R5 and R7 in Fig. 1-3?

1.19. What is the voltage at the junction of R12 and R14 in Fig. 1-3?

1.20. What is the voltage at the junction of R9 and C7 in Fig. 1-3?

1.21. What is the active device in Fig. 1-4?

1.22. What voltage would you expect to find at the emitter of Q1 in Fig. 1-4?

1.23. What signal would you expect to find at the emitter of Q1 in Fig. 1-4?

1.24. What is the polarity of the voltage to which resistor R3 in Fig. 1-4 is connected?

1.25. From what you have read so far, can you describe the function of resistor R3 in Fig. 1-4?

1.26. If the horizontal frequency control resistor R9 in Fig. 1-4 is adjusted, what voltage in Fig. 1-4 would you expect to change?

1.27. The circled letters in Fig. 1-5 are ______________.

1.28. Assume that the circuit of Fig. 1-5 amplifies signals. Would the signal at test point E be larger or smaller than the signal at test point F?

1.29. What are the voltages at the elements of Q1 in Fig. 1-5?

1.30. Are the voltages at the elements of Q1, Q2, and Q3 in Fig. 1-5 positive or negative with respect to ground or common?

1.31. How many parts are represented by the reference designation CR1 in Fig. 1-5? What are the parts?

1.32. What functional relationships are shown by the diagram in Fig. 1-6?

1.33. How is signal flow shown in the diagram of Fig. 1-7?

1.34. What is the color of the wire that connects resistors R7, R8, and R12 in Fig. 1-7 to ground?

1.35. What is the difference between the test points shown as starred numbers and the test points shown as circled letters in Fig. 1-8?

1.36. Describe the signal flow through the circuit shown in Fig. 1-8.

1.37. What part is shown after Q2 in the signal flow path shown in Fig. 1-8?

1.38. What stages are contained in the IC shown in Fig. 1-9?

1.39. What parts are controlled by the tuning dial shown in Fig. 1-9?

1.40. Of all the drawings and diagrams found in electronic literature, which is the most valuable?

# 2

# Some Basic Electronic Schematic Symbols

In this chapter, we concentrate on those symbols most often found on schematics of all types. That is, these symbols can be used in the schematics of such diverse equipment as a portable transistor radio, a citizen's band (CB) set, a TV set, and a stereo system, as well as most other electronic devices. More specialized symbols are discussed in Chapters 3 through 5.

## 2-1 CONDUCTORS AND WIRES

Figure 2-1 shows the symbols for wires, cables, and other conductors. When wires cross and are connected electrically, a dot is placed at the wire junction. When wires cross (for convenience of drawing the schematic) but are not connected electrically, there is no dot at the wire junction. At one time, it was common practice to show crossed wires (without connection) by jumping or looping as shown in the "not recommended" example of Fig. 2-1. This system is still in use on schematics today, but the dot (connection) or no dot (no connection) is preferred. Note that shielded wire and coaxial cable are often confused on schematics because of the similarity of the symbols.

### *2-1.1 FUNCTION OF CONDUCTORS AND WIRES IN CIRCUITS*

The obvious function of conductors and wires in circuits is to carry electrical current (or voltages, signals, pulses, etc.) from one element to another. As discussed, we will not go into how current

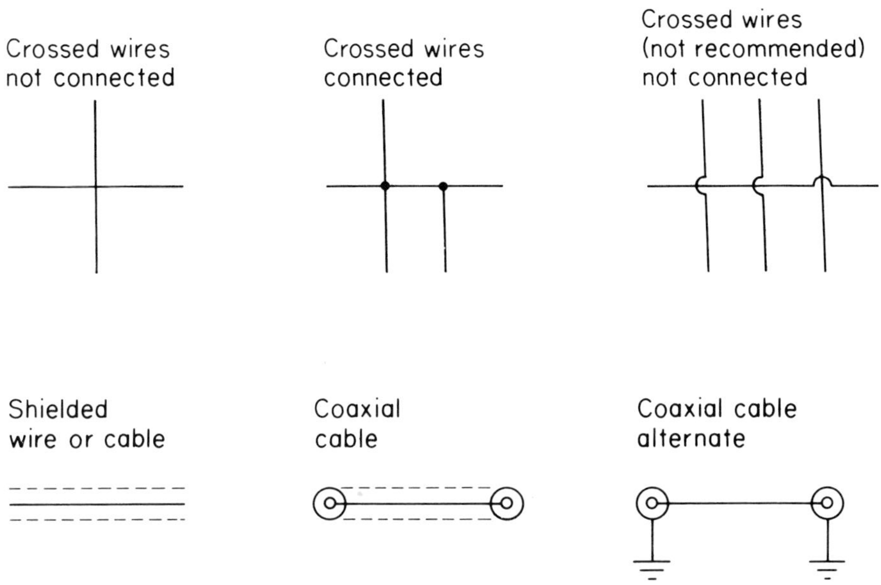


**FIGURE 2-1** Symbols for wires, cables and other conductors

is passed along the wire or conductor, or the atomic structure of wires. Simply accept the fact that current will pass from the negative terminal of a power source (battery, generator, etc.) through a conductor or series of conductors back to the positive terminal of the source. All that is needed is a complete circuit.

For example, as shown in Fig. 2-2, a very simple electrical circuit is formed when a battery is connected to a bulb or lamp by two wires (or conductors). When the circuit is *complete*, electrical current (generated by the battery due to chemical action) passes from the negative terminal through the outgoing conductor or wire to the bulb, and through the bulb back to the battery's positive terminal. Current passing through the bulb wire elements causes the bulb to light.

If either of the wires is broken or disconnected, the circuit is *incomplete* and no current flows. If another wire or conductor is placed across the circuit as shown in Fig. 2-2c, the circuit is *shorted.* In effect, the bulb is removed or bypassed. Current still flows, but not through the bulb. Thus, the bulb does not light.

Another problem with the short circuit is that more current will flow than when the bulb is in the circuit. This is because the bulb has more *resistance* than the wire or short. The greater the resistance, the smaller the current flow, and vice versa. Current is measured in *amperes*, resistance is measured in *ohms*, and the force producing

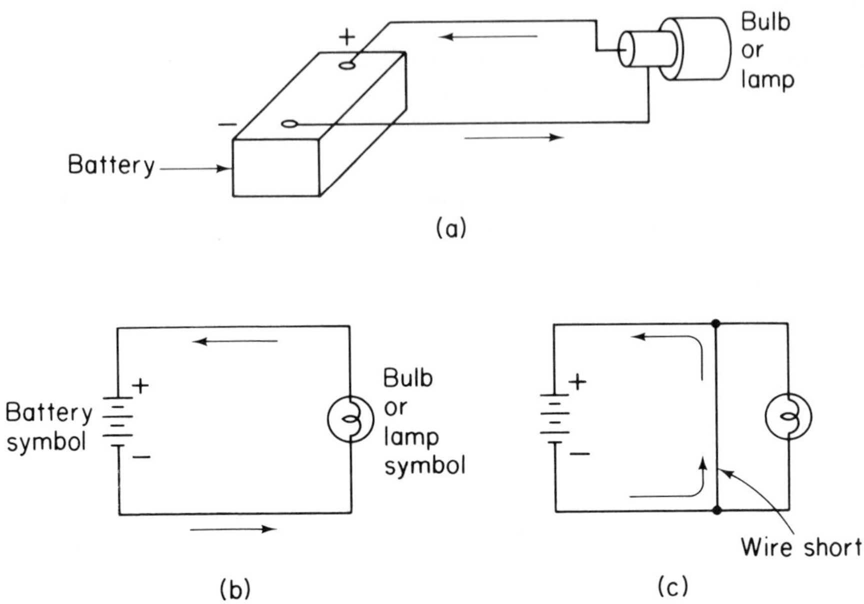


**FIGURE 2-2** Simple electrical circuit showing a battery connected to a lamp or bulb: (a) pictorial, (b) schematic, (c) short circuit.

the electrical current is measured in *volts*. There are definite relationships or laws governing current flow in circuits. *Ohm's laws* are the most commonly used rules to determine the relationships of voltage, current, and resistance. The basic Ohm's law formulas for direct current are given by

$$E = IR \qquad I = \frac{E}{R} \qquad R = \frac{E}{I} \qquad P = EI$$

where $I$ = current (amperes)
$R$ = resistance (ohms)
$E$ = force or potential across $R$ (volts)
$P$ = power (watts)

The calculations for Ohm's law are shown in Fig. 2-3, together with the commonly used multiples and submultiples of the basic units. The solutions for quantities printed in the center section are calculated by substituting known values in the formulas adjoining the respective center sections. For example, let us go back to our basic circuit of Fig. 2-2. Assume that a 3-V battery is used and that the bulb has a resistance of 1500 Ω. To find the amount of electrical current that flows in the circuit, divide the voltage ($E$) by resistance

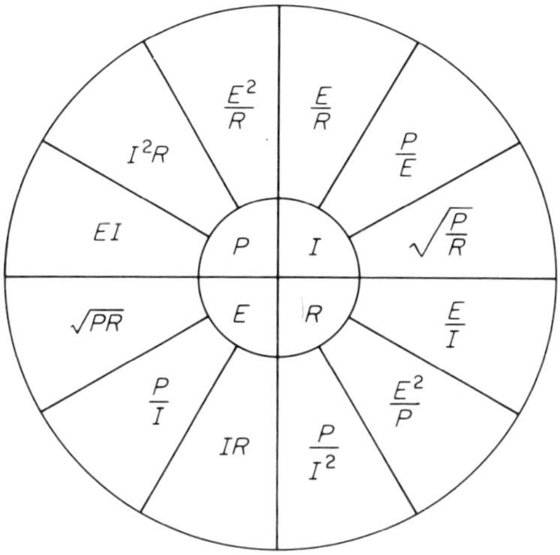


$E$ = voltage (volts)
$I$ = current (amperes)
$R$ = resistance (ohms)
$P$ = power (watts)

| Unit | Symbol | Multiple | Value |
|---|---|---|---|
| volt | $E$ | kilovolt (kV) | 1000 volts |
| volt | $E$ | millivolt (mV) | 1/1000 volt |
| volt | $E$ | microvolt ($\mu$V) | 1/1,000,000 volt |
| ohm | $R$ | kilohm (k$\Omega$) | 1000 ohms |
| ohm | $R$ | megohm (m$\Omega$) | 1,000,000 ohms |
| ampere | $I$ | milliampere (mA) | 1/1000 ampere |
| ampere | $I$ | microampere ($\mu$A) | 1/1,000,000 ampere |
| watt | $P$ | microwatt | 1/1,000,000 watt |
| watt | $P$ | milliwatt | 1/1000 watt |
| watt | $P$ | kilowatt | 1000 watts |
| watt | $P$ | megawatt | 1,000,000 watts |

**FIGURE 2-3** Ohm's law calculations with commonly-used multiples and submultiples of the basic units

($R$) to obtain a current ($I$) of 3/1500 = 0.002 ampere (A) or 2 milliamperes (mA). Ignore the fraction of an ohm of resistance in the wire conductors.

Now let us assume that a 6-V battery must be used with the bulb, which is rated at 3 V. One solution is to use wires or conductors that have 1500 $\Omega$ of resistance. Then 3 V will be dropped across the conductor, leaving 3 V for the bulb. This solution is somewhat impractical since the resistance of standard No. 30 copper wire is about 100 $\Omega$ per 1000 ft. This means that you would need about 15,000 ft of wire between the battery and bulb! The practical solution in this case is to use a *resistor* between the battery and bulb. Resistors are discussed in Sec. 2-2.

Before we get to resistors and their related symbols, however, make a note that *conductance* is the inverse of resistance, and vice versa. That is, conductance (abbreviated as $G$) is equal to 1 divided

$$\text{Conductance } (G) = \frac{1}{R}$$

$$\text{Resistance } (R) = \frac{1}{G}$$

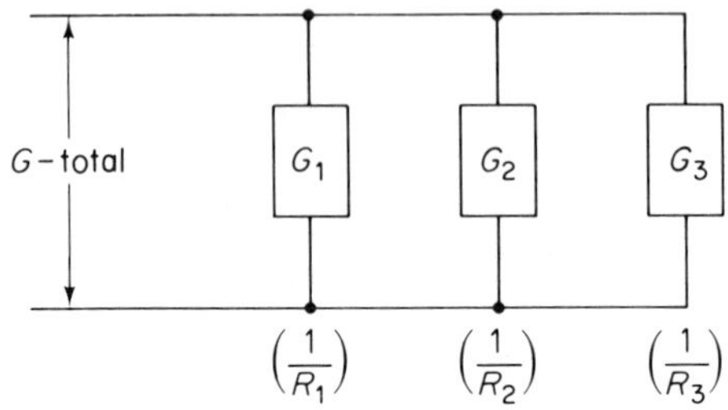


$$G\text{-total} = G_1 + G_2 + G_3$$

**FIGURE 2-4** Relationship of conductance to resistance

by resistance. Conductance is expressed in *mhos*, which is ohms spelled backward. Resistance is the inverse of conductance (1 divided by conductance). Figure 2-4 shows the relationship of conductance to resistance.

## 2-2 RESISTORS AND RESISTANCE ELEMENTS

Figure 2-5 shows the symbols of resistors and resistance elements. Figure 2-5 also shows the physical appearance of the two most common resistor elements, the fixed resistance and the variable resistance or potentiometer. Note that the adjustable contact (or potentiometer) symbol is preferred to both the variable resistance and the rheostat symbol. Actually, the rheostat is an old term not in common use today. In the strict sense, a true rheostat connection uses only two of the three potentiometer terminals (the adjustable contact or "wiper," and one end of the resistance element). Note that the heater element is not a true resistor but is a resistance element. That is, the heater element is designed to produce heat when current is passed through the element. Heat is also produced when current is passed through a resistor, but this is generally not a desirable condition.

Both fixed and variable resistors can be either *wire-wound* or *composition.* A wire-wound resistor is made up by winding wire on an insulating material. The resistance is determined by the composition of the wire material and the length of the wire. Composition resistors are formed when two leads are attached to a bar or cylinder of resistance material (usually, a composition of powdered carbon or a film material). The resistance bar is then encapsulated, with the leads protruding at the ends, as shown in Fig. 2-5.

Generally, composition resistors are used when the power to be handled by the resistor is less than 5 to 10 watts (W). Wire-wound resistors are used for higher powers. As shown in Fig. 2-3, power can be calculated when any two corresponding values of resistance, current, and voltage are known.

As shown in Fig. 2-5, the reference designation letter for resistors is *R*, and the values are given in ohms (indicated by the Greek letter omega). A well-drawn schematic will also give the resistor wattage, either by means of a common note or individually, and will give a tolerance for the resistance value (±5%, ±10%, etc.).

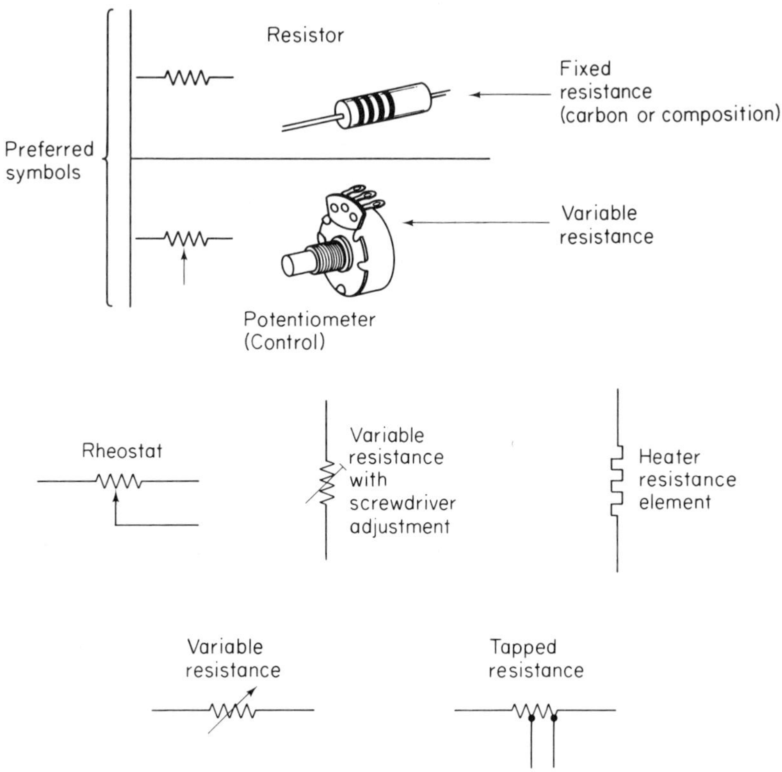


**FIGURE 2-5** Symbols for resistors and resistance elements

### 2-2.1 RESISTOR COLOR CODES

Most resistors of the molded-composition type are color coded. The coding is in the form of encircling bands grouped at one end, as shown in Fig. 2-6. The color coding for resistors is shown from left to right. Usually, four bands of color are present for the carbon-composition types, and five bands are used for the film types. In either case, the last color tells the tolerance that must be applied to the value obtained. Thus, if a rated 100-Ω resistor has a tolerance of 10%, the resistor can have an actual value between 90 and 110 Ω. Ths abbreviation GMV on resistor symbols stands for *guaranteed minimum value.* Where a value is marked *alternate* on the schematic, this indicates a coding that may have been used on schematics in the past but has generally been superseded in present-day schematics by the preferred coding designation.

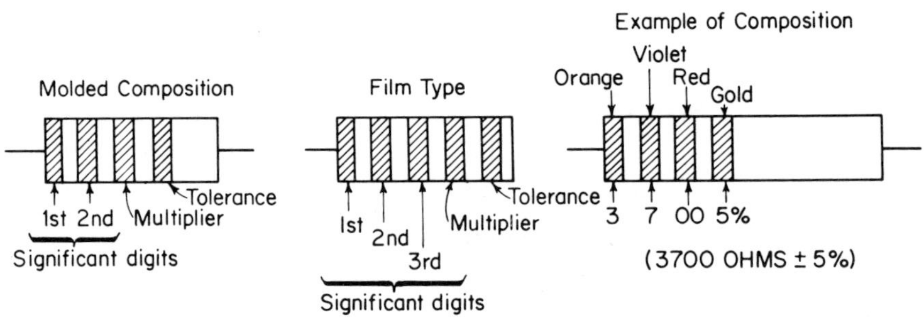


| Color | Digit | Multiplier | Composition tolerance | Film-type tolerance |
|---|---|---|---|---|
| Black | 0 | 1 | 20% | 0 |
| Brown | 1 | 10 | 1% | 1% |
| Red | 2 | 100 | 2% | 2% |
| Orange | 3 | 1000 | 3% | |
| Yellow | 4 | 10,000 | GMV | |
| Green | 5 | 100,000 | 5% (alt.) | 0.5% |
| Blue | 6 | 1,000,000 | 6% | 0.25% |
| Violet | 7 | 10,000,000 | 12.5% | 0.1 |
| Gray | 8 | 0.01 (alt.) | 30% | 0.05 |
| White | 9 | 0.1 (alt.) | 10% (alt.) | |
| Silver | | 0.01 (preferred) | 10% (pref.) | 10% |
| Gold | | 0.1 (preferred) | 5% (pref.) | 5% |
| No Color | | | 20% | |

**FIGURE 2-6** Resistor color coding

### 2-2.2 *FUNCTION OF RESISTORS IN CIRCUITS*

Resistors function to control the flow of electrical current, and to divide voltages. As discussed in other sections of this chapter, resistors can also be used to shape electrical waves. Although resistors are found on schematics in a variety of circuit combinations, these circuits are actually a version of the basic *series*, *parallel*, or *series-parallel* resistor networks shown in Fig. 2-7.

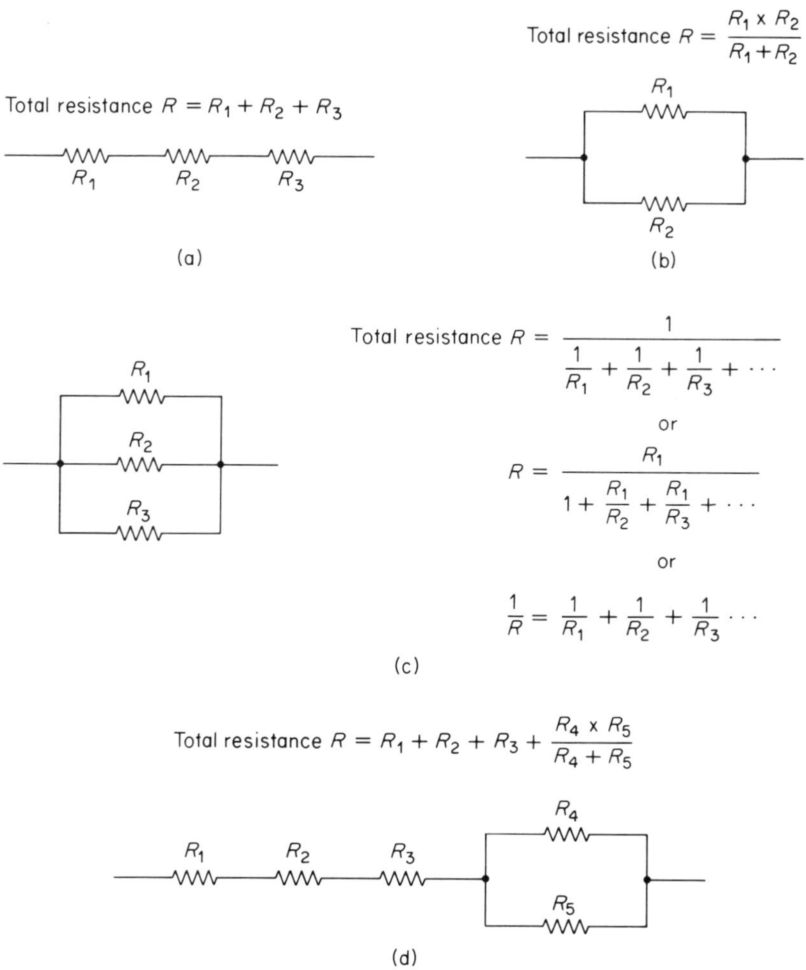


**FIGURE 2-7** Finding resistance in series, parallel, and series-parallel networks: (a) series, (b) two in parallel, (c) three or more in parallel, (d) series-parallel

**Resistor Networks.** When two or more resistors are connected in series (Fig. 2-7a), the total resistance is equal to the sum of the individual resistances.

When only two resistors are connected in parallel, use the equation of Fig. 2-7b to find the total resistance.

When three or more resistors are connected in parallel, use the most convenient of the three equations in Fig. 2-7c to find total resistance. For added convenience, make R1 the largest resistor in the parallel combination.

When series-parallel resistance values must be obtained, the series-resistance equation can be combined with the parallel-resistance equation, as shown in Fig. 2-7d.

In solving any parallel-resistance equation found on a schematic, where there is a 10:1 (or better) ratio between resistor values, the total resistance will be only slightly below the smallest resistor value. This can be used as a rough approximation of the total parallel-resistance value.

**Finding Resistance with Currents.** In practical schematic reading applications, it is often convenient to find resistance values by means of currents passing through a circuit. This is particularly true for parallel-resistance networks since the same voltage appears across each of the resistors. Figure 2-8 shows the equations necessary

$$I_1 = \frac{I_2 \times R_2}{R_1} \qquad I_2 = \frac{I_1 \times R_1}{R_2}$$

$$R_1 = \frac{I_2 \times R_2}{I_1} \qquad R_2 = \frac{I_1 \times R_1}{I_2}$$

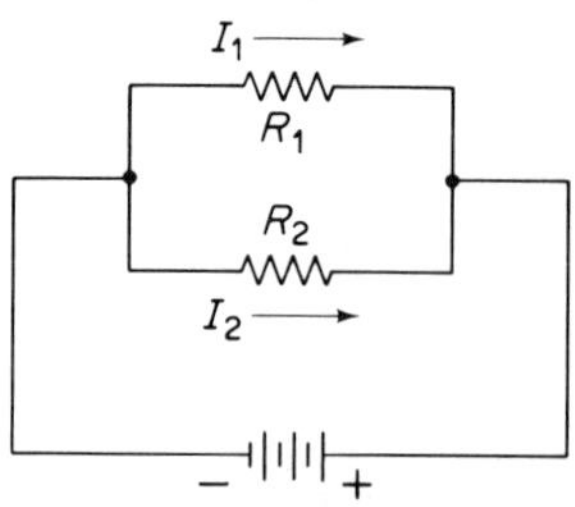


**FIGURE 2-8** Example of shunt law applications to find unknown current and/or resistance values in a parallel-resistance network

to find either current or resistance in a parallel-resistance network. This applies where there are two resistors and three of the four values are known. The four equations of Fig. 2-8 are based on what is sometimes referred to as the *shunt law* (the ratio of currents is inversely proportional to the ratio of resistances).

**Voltage Divider.** One of the most common resistance functions shown on schematics is that of a *voltage divider* such as shown in Fig. 2-9. The purpose of a voltage divider is to obtain a value of voltage less than that of the source. Fixed resistors are used where

$$\text{Output voltage} = E \times \frac{R_2 \parallel R_L}{R_1 + R_2 \parallel R_L}$$

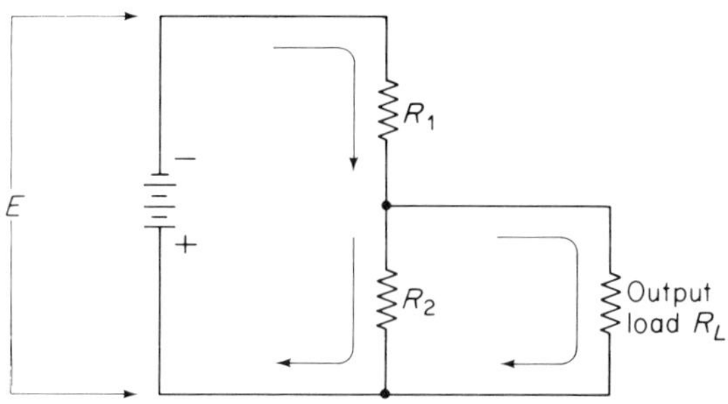


**FIGURE 2-9** Finding the output of a voltage divider network

a fixed voltage is desired. A potentiometer is used where a variable output is needed. Figure 2-9 also shows the equations necessary to find the output voltage for a typical voltage-divider network.

**Kirchhoff's Laws.** In many schematics, the arrangements of the resistors and applied voltages makes it almost impossible to solve such networks by simple application of Ohm's law. These problems can often be solved by applying *Kirchhoff's voltage law.* This law can be stated in many ways, two of which are:

> In any complete electrical circuit, the sum of all the voltages and all the voltage ($IR$) drops, taken with their proper signs, is zero; or *the sum of the voltages in a complete circuit is equal to the applied voltage.*

Figure 2-10 shows the relationship of the voltages and currents in a typical series circuit. In the series circuit, the same current ($I$)

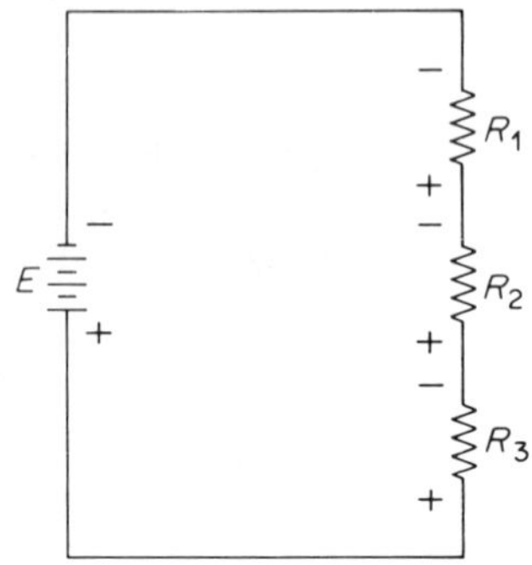


$$E = IR_1 + IR_2 + IR_3$$

$$E = -IR_1 - IR_2 - IR_3 = 0$$

$$-E = +IR_1 + IR_2 + IR_3 = 0$$

**FIGURE 2-10** Example of Kirchoff's voltage law

flows through each resistor. If the voltage (*IR*) drops across each resistor are added, their sum will be equal to the voltage of the battery. If this sum is subtracted from the battery voltage, the result will be zero. Also note that the polarity of the battery voltage and the polarity of the voltage drops across the resistors are opposite.

The following three pointers help in applying Kirchhoff's voltage law to a resistor network shown on a schematic:

In the more complicated schematics, it is not always easy to determine the direction of current. So simply *assume a direction* of current flow. If the assumption is incorrect, the answer for current strength will be numerically correct but will be a negative number (possibly backward from that which the schematic is trying to show).

Place *polarity markings* on all batteries and resistors in the circuit. The assumed current direction will not affect the battery polarities, but the voltage drop on resistors will be affected.

By working around the circuit, *set up each term of the equation.* Include all voltage sources and all voltage drops. In the equation, precede each term by the sign found on *leaving* each particular battery or resistor.

In some schematics, it is often more convenient to solve a network by means of *Kirchhoff's current law.* This law can also be stated in several ways:

> In any electrical network, the algebraic sum of the currents that meet (entering and leaving) at a point is zero; or *the*

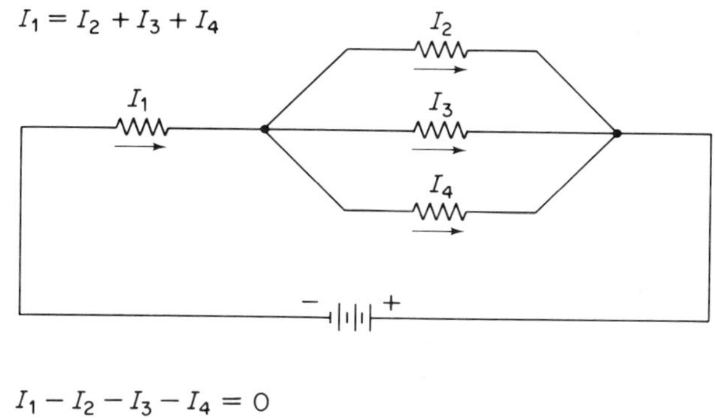


**FIGURE 2-11** Example of Kirchoff's current law

*algebraic sum of the currents flowing toward a junction is equal to zero.*

Figure 2-11 shows the relationship of the currents in a typical circuit.

**Aiding and Opposing Voltages.** When two or more voltage sources are connected in series, their effect is additive if the polarities are aiding. If the polarities are opposed, the smaller voltage is subtracted from the larger. If there are several voltages of various polarities, the total of one polarity is subtracted from the total of the other polarity. Figure 2-12 shows the relationship of the voltages in a typical circuit.

**Assumed Voltages.** It is sometimes convenient to solve resistance network problems with assumed voltages and then to apply Ohm's law. An example is shown in Fig. 2-13. Here, a 500-Ω resistor must be placed in parallel with an unknown resistor to produce an equivalent 100 Ω. If the two resistance values are known and it is desired to find the equivalent resistance, the equation of Fig. 2-7 can be used. Since it is desired to find one resistor value, the equation would have to be rearranged. Instead, assumed voltages can be used as follows.

Assume a voltage of 500 V across both resistors. This would produce 1 A in the known 500-Ω resistor $R = E/I$. The same 500 V would produce 5 A through the equivalent 100 Ω resistance. Since 5 A flows through the total network and 1 A flows through the known 500-Ω resistor, 4 A must be flowing through the unknown resistor. Since $R = E/I$, the unknown resistance must be 125 Ω: $125 = 500/4$.

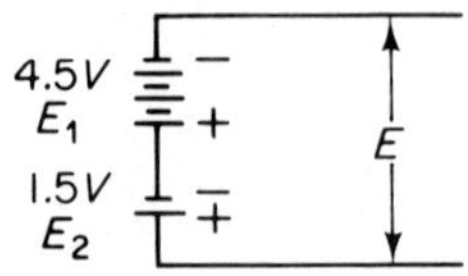

$E = E_1 + E_2$
(4.5 + 1.5 = 6)

Aiding voltages

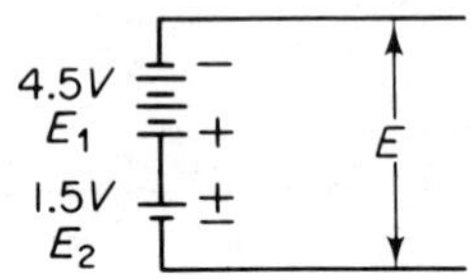

$E = E_1 - E_2$
(4.5 + 1.5 = 3)

Opposing voltages

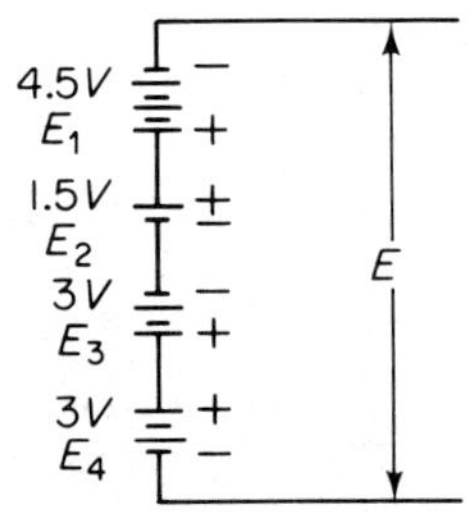

$E = E_1 + E_3 - E_2 - E_4$
(4.5 + 3 − 1.5 − 3 = 3)

Aiding and opposing

**FIGURE 2-12** Examples of aiding and opposing voltage in series

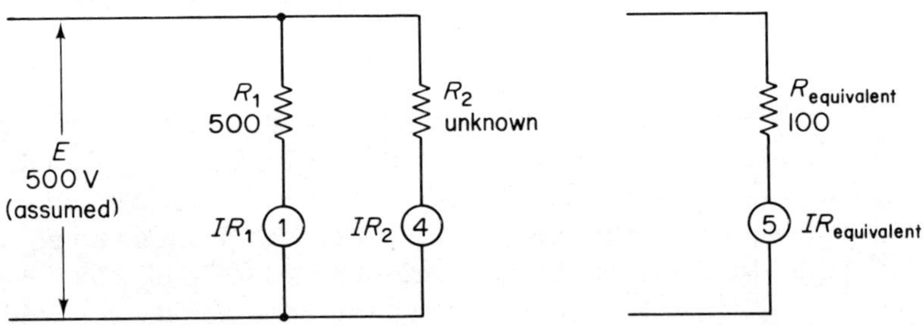


$$R_1 = \frac{500}{1} = 500$$

$$R_{\text{equivalent}} = \frac{500}{5} = 100$$

$$R_{2\ (\text{unknown})} = \frac{500}{4} = 125$$

**FIGURE 2-13** Solving resistance network problems with assumed voltages

### 2-2.3 *EXAMPLE OF RESISTORS IN CIRCUITS.*

Now let us see how resistors can be used in circuits using our example of the battery and bulb. (Sec. 2-1). Assume that a 3-V bulb must be used with a 9-V battery. This problem can be solved with a resistor and shown schematically as in Fig. 2-14.

The value of resistor R in Fig. 2-14a can be found as follows. Since the bulb is 3 V and has a resistance of 1500 Ω, there is a current of 2 mA flowing through the circuit: 3/1500 = 0.002 A = 2 mA. Resistor R must drop a total of 6 V, since the battery is 9 V and the bulb is 3 V. Thus, the resistance of R must be 3000 Ω: 6/0.002 = 3000. As a final check, find the total resistance of the circuit (ignoring the fraction of an ohm in the wires) by adding the individual resistances: 3000 + 1500 = 4500 Ω. Divide this into 9 V (battery) to find a current of 0.002 A (2 mA).

Now assume that you want to control the current in the circuit so that the amount of light produced by the bulb can be varied. In this case, the schematic would appear as shown in Fig. 2-14b, where

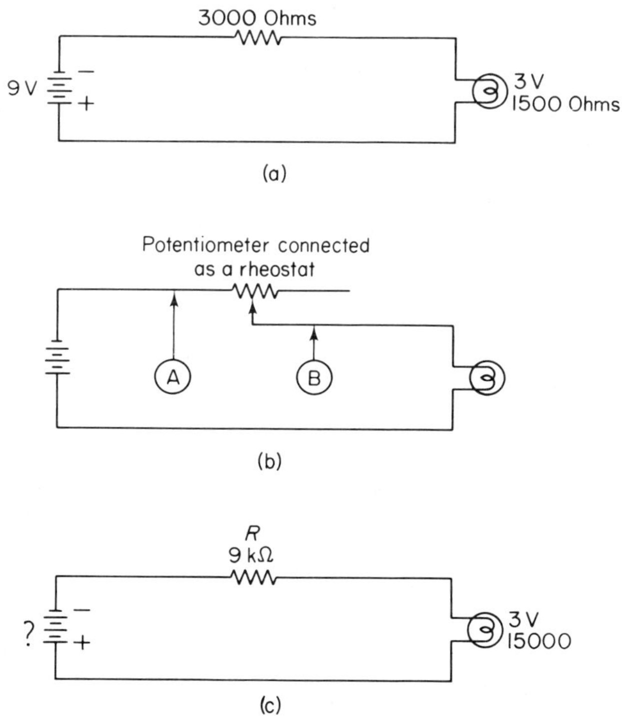


**FIGURE 2-14** Examples of resistors in schematic circuits

a potentiometer is connected as a rheostat. When the control shaft on the potentiometer is varied so that the total resistance of the potentiometer (between points A and B) is increased, less current will flow in the circuit since the voltage (battery) remains constant ($I = E/R$). This causes the bulb to dim (less light). The opposite occurs when the potentiometer resistance is decreased (more current, brighter glow).

Now one more example of using resistors in circuits. Assume that a schematic shows a resistor and bulb in series with an unknown voltage source, as in Fig. 2-14c. Can you find the voltage of the battery with the data given on the schematic? Since the bulb is 3 V and 1500 Ω, the current is still 2 mA. Since R is shown as 9 kΩ (9 kilohms or 9000 ohms) on the schematic, the voltage drop across R must be 18 V (9000 x 0.002 = 18). With 18 V across R and 3 V across the bulb, the total battery voltage is 21 V (18 + 3 = 21).

Whether you realize it or not, you are now reading and using electronic schematics. Admittedly, the circuits are elementary and are given to make a certain point. But the whole purpose of schematics is to make a point, usually to identify a circuit or to show how it works. Now let us go on to capacitors, the next most common symbol found on schematics.

## 2-3 CAPACITORS

Figure 2-15 shows the symbols and physical appearance of the three most common types of capacitors. These include the *electrolytic capacitor*, the *fixed* (paper, mica, or ceramic) *capacitor*, and the *variable capacitor*.

The electrolytic capacitor is used where a large value of capacitance is required (typically anything higher than about one microfarad, 1 μF). The two most common types of electrolytic capacitor are the *can type* and the *tubular type*, shown in Fig. 2-15. The can type often contains several capacitors in one package. Note that polarity is indicated on the electrolytic capacitor symbol (+ for the flat plate, — for the curved plate). This means that electrolytic capacitors must be connected into the circuit (with regard to polarity) as shown on the schematic. If polarity is reversed and the electrolytic is subjected to a large reverse voltage, the capacitor will be permanently damaged. Electrolytics are most often found in power supply circuits and in solid-state audio circuits, both of which require large-value capacitance.

The flat, rectangular fixed capacitor usually has a mica (or

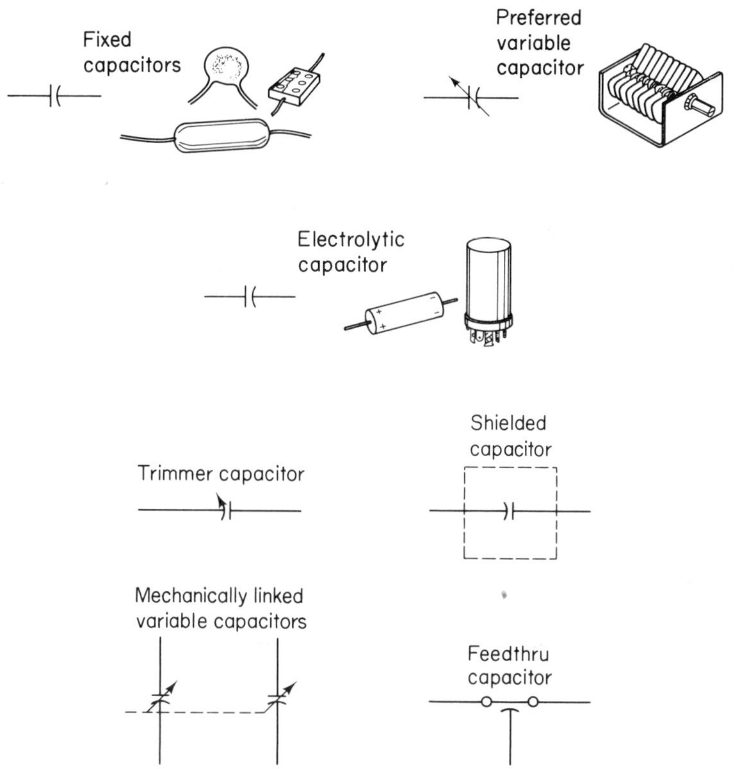


**FIGURE 2-15** Symbols for Capacitors

possibly a paper) *dielectric.* It should be noted that a capacitor is formed when two conductors are brought within close proximity of each other without touching. In such an instance, air is the medium of separation between two such conductors, and the air thus acts as an insulator. Other types of insulation, such as mica, paper, ceramic, and plastic, can be used. The insulation between the two conductors is called the *dielectric.*

Disk and tubular or radial fixed capacitors usually use ceramic or plastic as a dielectric. Note that the fixed-capacitor symbol does not show polarity. It is commonly accepted to show the curved plate connected to ground or the lowest voltage, but this does not hold true on all schematics.

As shown in Fig. 2-15, the preferred variable-capacitor symbol is with the arrow crossing both plates, whereas the trimmer capacitor has an arrow on the curved plate. However, these symbols are often mixed. A trimmer capacitor usually has only one fixed plate and one movable plate, which is adjustable by screwdriver. Although any number of variable capacitor sections can be ganged as shown, the maximum is usually three sections.

The reference-designation letter for capacitors is C, and the values are given in *microfarads* ($\mu$F) or *picofarads* (pF). The *farad* is the basic unit of capacitance, but since the farad is too large for practical applications, submultiples are used. The most common submultiples are the $\mu$F (one millionth of a farad) and the pF (one millionth of one millionth of a farad). The term pF has generally replaced the previously used term $\mu\mu$F on schematics. On very old schematics, the terms MF and MMF are sometimes used.

A well-drawn schematic will also give the capacitor *working voltage*, either by means of a common note or individually, and will give a tolerance for the capacitance values (±5%, ±10%, etc.) The working voltage (sometimes given as a maximum voltage) must not be exceeded or the capacitor may be damaged.

### *2-3.1 CAPACITOR COLOR CODES*

Both ceramic and mica capacitors are color-coded as to capacitance value, tolerance, temperature coefficient, and so on. The color-coding system for modern mica capacitors is shown in Fig. 2-16.

**FIGURE 2-16** Mica capacitor color codes

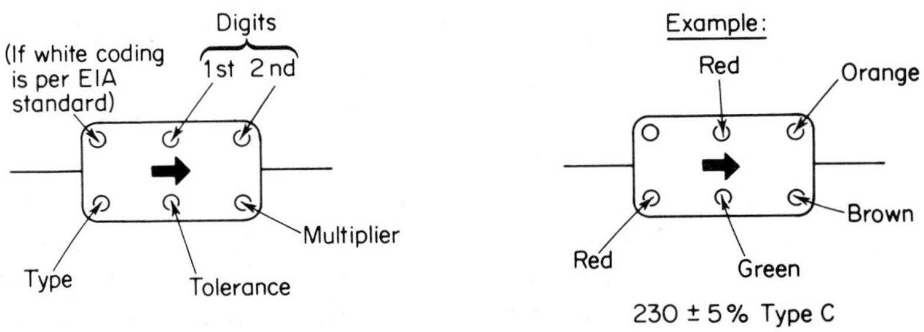


Capacitance Values in Picofarads (PF)

| Color | Digit | Multiplier | Tolerance | Type classification |
|---|---|---|---|---|
| Black | 0 | 1 | 20 % (±) | A |
| Brown | 1 | 10 | 1 % | B |
| Red | 2 | 100 | 2 % | C |
| Orange | 3 | 1000 | 3 % | D |
| Yellow | 4 | 10,000 | — | E |
| Green | 5 | — | 5 % | — |
| Blue | 6 | — | — | — |
| Violet | 7 | — | — | — |
| Gray | 8 | — | — | — |
| White | 9 | — | 10 % | — |
| Silver | — | 0.01 | — | — |
| Gold | — | 0.1 | — | — |

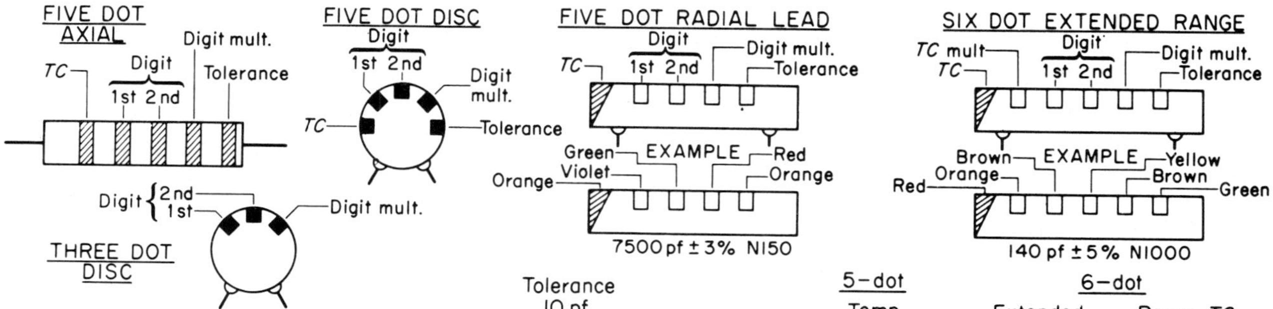


| Color | Digit | Multiplier | Tolerance 10 pf or less | Over 10 pf | 5-dot Temp. coeff. | 6-dot Extended sig. fig. | 6-dot Range TC mult. |
|---|---|---|---|---|---|---|---|
| Black | 0 | 1 | 2.0 pf | 20 % | NPO | 0.0 | −1 |
| Brown | 1 | 10 | 0.1 pf | 1% | N033 | — | −10 |
| Red | 2 | 100 | — | 2 % | N075 | 1.0 | −100 |
| Orange | 3 | 1000 | — | 3 % | N150 | 1.5 | −1000 |
| Yellow | 4 | 10,000 | — | — | N220 | 2.2 | −10,000 |
| Green | 5 | — | 0.5 pf | 5 % | N330 | 3.3 | +1 |
| Blue | 6 | — | — | — | N470 | 4.7 | +10 |
| Violet | 7 | — | — | — | N750 | 7.5 | +100 |
| Gray | 8 | 0.01 (alt.) | 0.25 pf | — | * | *** | +1000 |
| White | 9 | 0.1 (alt.) | 1.0 pf | 10 % | ** | — | +10,000 |
| Silver | — | 0.01 (pref.) | — | — | — | — | — |
| Gold | — | 0.1 (pref.) | — | — | — | — | — |

Capacitance values in picofarads (pf)

*General-purpose types with a temperature coefficient ranging from P150 to N1500.

**Coupling, decoupling and general bypass types with a temperature coefficient ranging from P100 to N750.

***If the first band (temperature coefficient) is black, the range is N1000 to N5000.

**FIGURE 2-17** Ceramic capacitor color codes

Note that the lower left-hand "type" color dot indicates the type or classification of the particular capacitor according to the manufacturer's specifications. Such specifications may include temperature coefficient, $Q$ factor, and related characteristics. The type designation of one manufacturer is not necessarily identical to that of another.

The color-coding system for ceramic capacitors is shown in Fig. 2-17. As shown, a temperature-coefficient color band or dot is included together with the capacitance and tolerance color coding. The temperature coefficient of ceramic capacitors is given in parts per million per degree Celsius (ppm/°C). A preceding letter N denotes negative-temperature coefficient (capacity decrease with an increase in operating temperature). The P designation indicates a positive temperature coefficient, and NPO indicates a negative-positive-zero type. As an example, an N220 designation indicates a capacity decrease with a temperature rise of 220 ppm/°C, and it shows by how much the value changes during the warm-up time of the device in which the capacitor is used. The NPO types are stable capacitors with negligible temperature effect on capacitance.

Where a value in Fig. 2-17 is marked *alternate*, it indicates a coding that may have been used in the past but for which the *preferred* coding designation is generally used on modern components.

### 2-3.2 FUNCTION OF CAPACITORS IN CIRCUITS

Capacitors function to store an electrical charge. *Direct current can be stored in a capacitor, but alternating current will pass around the capacitor.* As an example, if a capacitor is connected to a battery as shown in Fig. 2-18a, there is a surge of current in the direction of the arrows (from the battery's negative terminal to one capacitor plate, and from the other capacitor plate to the battery's positive terminal). The surge current will last until the capacitor is charged (as discussed in the following paragraphs).

If the battery is disconnected and a wire is connected across the capacitor plates, the stored current will flow through the conductor but in the opposite direction, as shown in Fig. 2-18b.

Now assume that alternating current is applied to the capacitor plates as shown in Fig. 2-18c and d. The capacitor will charge and discharge each time the alternating current changes polarity. Thus, alternating current will flow with a capacitor in the circuit. However, there is some opposition to alternating current by the capacitor (as discussed in the following paragraphs).

To sum up, when you find a capacitor in a schematic, keep in mind that the capacitor will block direct current (except for a momentary surge) but will pass alternating current.

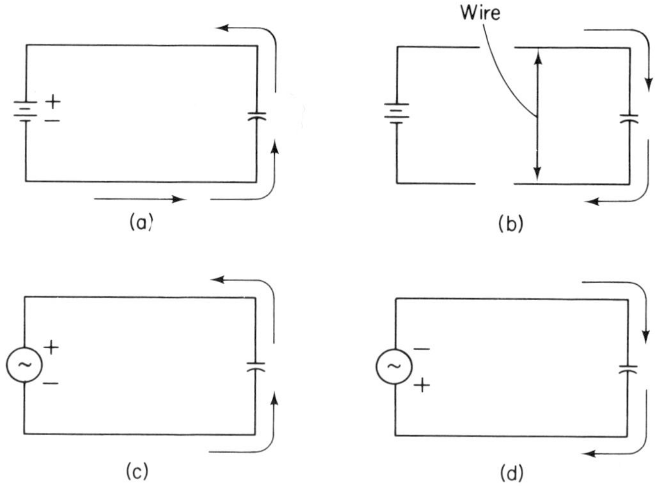


**FIGURE 2-18** Function of capacitors in circuits

**Time Constants for Capacitive-Resistive Circuits.** In a d-c circuit that is purely resistive (say, a resistor connected across the terminals of a battery), current should start to flow as soon as the voltage is applied. In a capacitive-resistive circuit (generally called an *RC* circuit) such as Fig. 2-19, the current starts at the maximum value permitted by the battery, and the voltage buildup across the capacitor lags the current as shown in Fig. 2-20. An *RC* time constant is the time required for voltage across the capacitor to reach 63% of the maximum value. In Fig. 2-20, curve *A* represents capacitor charge current, whereas curve *B* represents capacitor charge voltage. After five time constants, the capacitor is considered to have reached the full charge value (actually, 99.3% of full charge), the current flow in the circuit stops, and the voltage across the resistor becomes zero.

The calculations for capacitive time constants are shown in Fig. 2-19. For example, if a 100-Ω resistor is connected to a 10-μF capacitor and a battery as shown in Fig. 2-19, the capacitor will charge to 63% of the full value in 1000 microseconds (μs). You will find many *RC* combinations on schematics. It is not necessary for you to calculate the *RC* time constant to read the schematic. However, you must remember that the time constant will affect circuit operation, and is critical in many cases.

**Capacitor Networks.** Although capacitors are found on schematics in a variety of circuit combinations, these circuits are actually a version of the basic series, parallel, or series-parallel capacitor networks shown in Fig. 2-21.

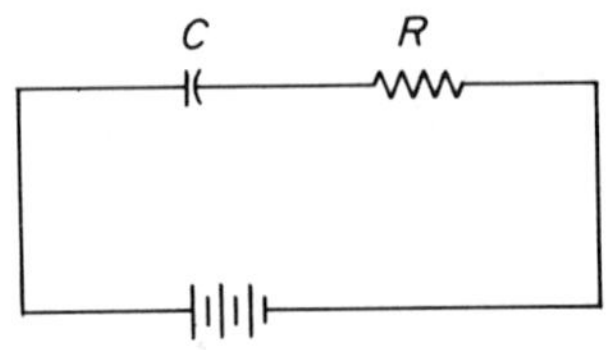


$TC$ (Time constant in seconds) $= R$ (Ohms) x $C$ (Farads)

$TC$ (Time constant in microseconds) $= R$ (Ohms) x $C$ (Microfarads)

$TC$ (Time constant in picoseconds) $= R$ (Ohms) x $C$ (Picofarads)

$$C = \frac{TC}{R}$$

$$R = \frac{TC}{C}$$

**FIGURE 2-19** Calculations for capacitive-resistive time constants

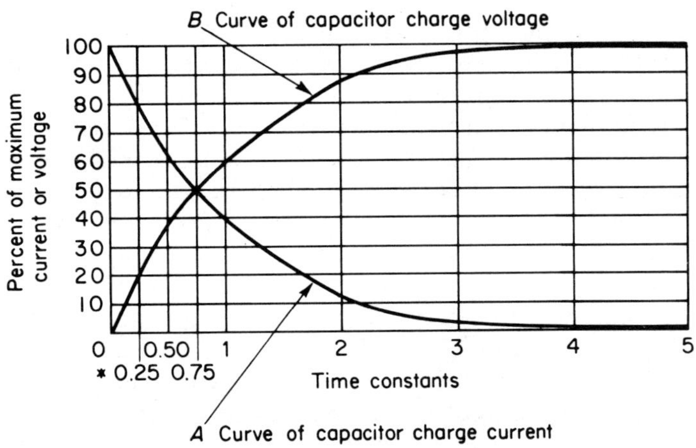


*The time indicated along the X axis is in relation to the instantaneous charge value of current or voltage, and does not indicate the time in seconds.

**FIGURE 2-20** Universal time constant chart for capacitance and resistance

The total capacitance of capacitors in parallel (Fig. 2-21c) is not affected by the order in which the capacitors are connected. Any number of capacitors can be added in this manner, provided that the same capacitance units (farad, microfarad, picofarad) are used.

$$\text{Total capacitance } C = \frac{1}{\frac{1}{C_1} + \frac{1}{C_2} + \frac{1}{C_3} + \cdots}$$

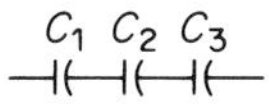


or

$$C = \frac{C_1}{1 + \frac{C_1}{C_2} + \frac{C_1}{C_3} + \frac{C_1}{C_4} + \cdots}$$

or

$$\frac{1}{C} = \frac{1}{C_1} + \frac{1}{C_2} + \frac{1}{C_3} + \frac{1}{C_4} \cdots$$

(a)

$$\text{Total capacitance } C = \frac{C_1 \times C_2}{C_1 + C_2}$$

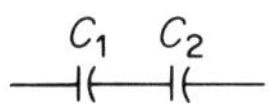


(b)

$$\text{Total capacitance } C = C_1 + C_2 + C_3$$

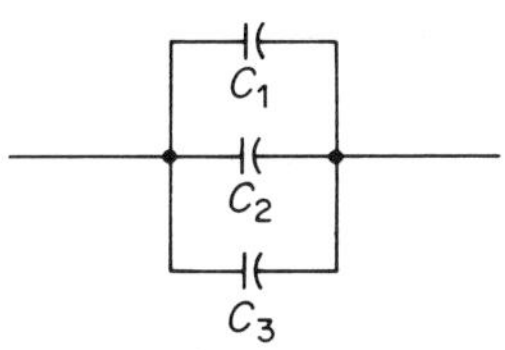


(c)

$$\text{Total capacitance } C = C_1 + C_2 + C_3 + \frac{C_4 \times C_5}{C_4 + C_5}$$

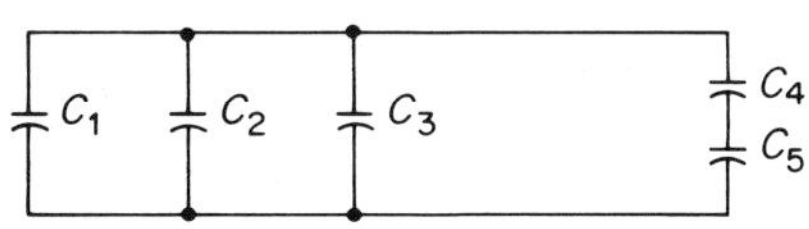


(d)

**FIGURE 2-21** Finding capacitance in series, parallel and series-parallel networks

When only two capacitors are connected in series, use the equation of Fig. 2-21b. The total capacitance is always less than the value of either capacitor.

When three or more capacitors are connected in series, use the most convenient of the three equations shown in Fig. 2-21a. For added convenience, make C1 the largest capacitor in the series.

When series-parallel capacitance values must be obtained, the series-capacitance equation can be combined with the parallel-capacitance equation, as shown in Fig. 2-21d.

In solving any series-capacitance equation found on a schematic, where there is a 10:1 (or better) ratio between capacitor values, the total capacitance will be only slightly below the smallest capacitor value. This can be used as a rough approximation of the total series-capacitance value.

**Capacitive Reactance and Impedance of Capacitors.** It is convenient to regard capacitive reactance as the opposition of a capacitor to alternating currents. Direct current will not pass around a capacitor. Therefore, the opposition is infinite. Pulsating current and alternating current will pass around a capacitor, but the process of alternately charging and discharging the capacitor produces some opposition. This opposition is termed *capacitive reactance*, is expressed in ohms and identified by the letters $X_C$ ($X$ for reactance, $C$ for capacitance).

If a capacitor were in a circuit where there is no d-c resistance, capacitive reactance could be substituted for resistance in the conventional Ohm's law equations. However, this is not practical. Any circuit has some d-c resistance (the capacitor wire leads, if nothing else). The resistance has the same effect as a resistor in series with the capacitor.

The combined effect of the capacitance and the effective resistance is known as the *impedance*, which is also expressed in ohms, and is represented by the letter $Z$. The impedance of a capacitor can be substituted directly in the basic Ohm's law equations:

$$E = I \times Z \qquad I = \frac{E}{Z} \qquad Z = \frac{E}{I}$$

where $Z$ is the impedance in ohms.

The relationship of capacitive reactance and impedance of capacitors, together with the calculations, is shown in Fig. 2-22. Again, it is not necessary to make these calculations to read the schematic. However, the calculations serve as a basis for understanding impedance circuits (discussed in Sec. 2-5).

**Alternating-Current Voltages in Capacitive Circuits.** Unlike voltages across resistors, it is not possible simply to add the a-c voltages across the capacitor and resistor in a series circuit to obtain the source voltage. Instead, the individual voltages must be added vectorially. Figure 2-23 shows the relationship of voltages, together with the calculations, for voltages in series $RC$ circuits.

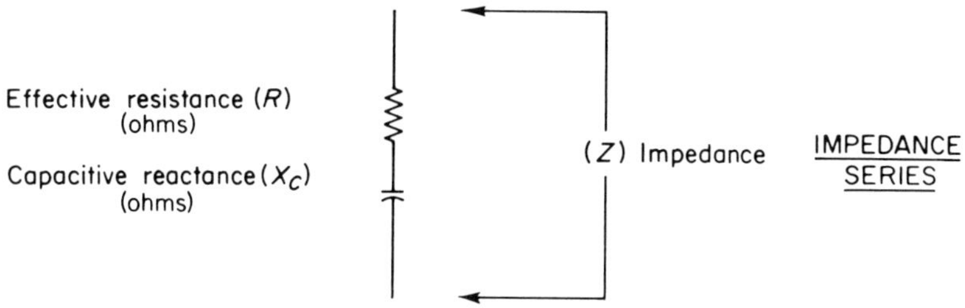


$$Z = \sqrt{R^2 + X_C^2}$$

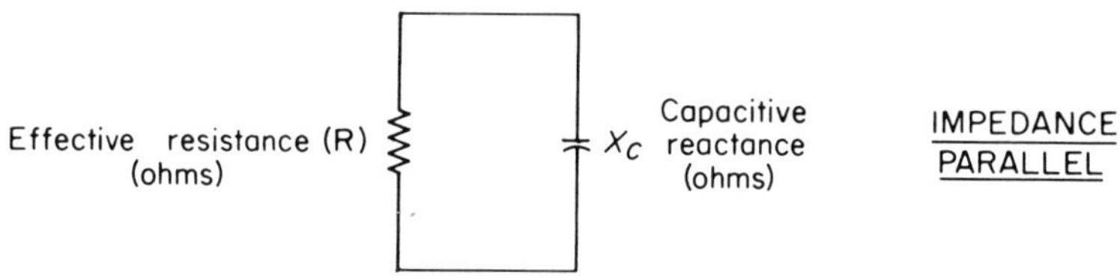


$$Z = \frac{R\ X_C}{\sqrt{R^2 + X_C^2}}$$

$$\frac{\text{Capacitive reactance}}{\text{Series or parallel}} \qquad X_C = \frac{1}{6.28\,FC}$$

$$X_C = \frac{1}{6.28 \times \text{frequency (in Hz)} \times \text{capacity (in farads)}}$$

or

$$X_C = \frac{159.2}{\text{frequency (in KHz)} \times \text{capacity (in microfarads)}}$$

$$C = \frac{1}{6.28FX_C} \qquad F = \frac{1}{6.28\,CX_C}$$

**FIGURE 2-22** Calculations for capacitive reactance and impedance in resistance-capacitance (RC) circuits

**Using Capacitors as Voltage Dividers.** Capacitors are often shown as voltage dividers on schematics, usually in a-c circuits. Capacitors can, however, also be used to divide direct currents capacitors will charge up to the d-c value. Either way, the source voltage is distributed among the capacitors in *inverse proportion*

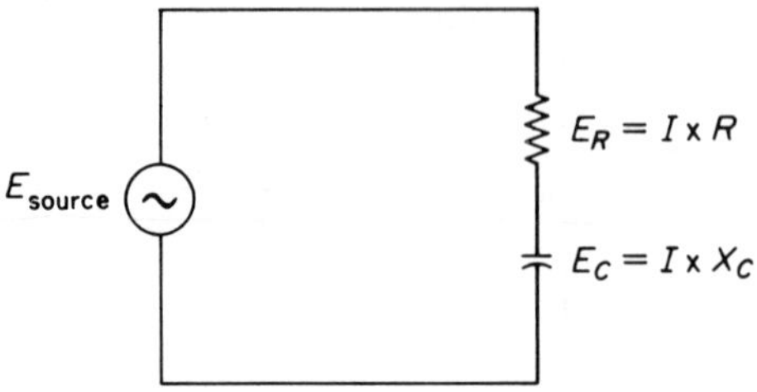


$$E_{source} = \sqrt{E_R^2 + E_C^2}$$

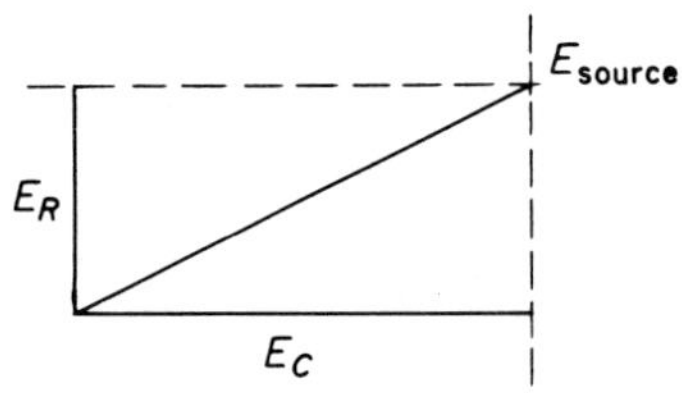


**FIGURE 2-23** Adding alternating current voltages in a series resistance-capacitance (RC) circuit

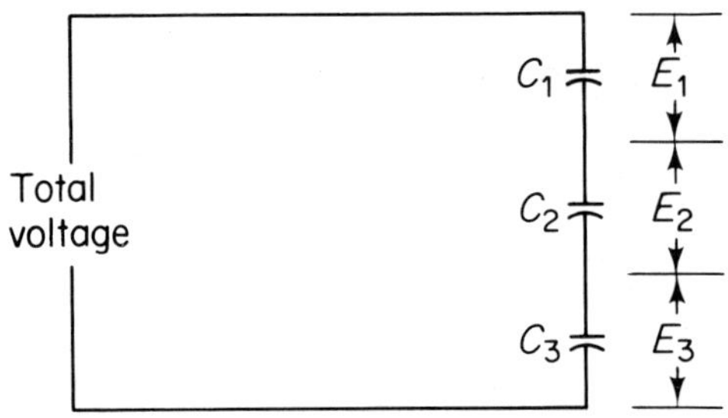


$$E_1 = \frac{\text{Total capacitance}}{C_1} \text{ X Total voltage}$$

$$E_2 = \frac{\text{Total capacitance}}{C_2} \text{ X Total voltage}$$

$$E_3 = \frac{\text{Total capacitance}}{C_3} \text{ X Total voltage}$$

$$\text{Total voltage} = E_1 + E_2 + E_3$$

**FIGURE 2-24** Calculating voltage across series capacitors

to the capacitance. That is, the capacitor having the largest capacitance has the smallest amount of voltage. This relationship is shown in Fig. 2-24.

## 2-4 INDUCTORS (COILS)

Figure 2-25 shows the symbols and physcial appearance of a typical inductor (more commonly known as a *coil*). The term "coil" is used because of the physical appearance. A basic coil is a few turns of wire, wound on a coil form, or possibily with an air core as shown. The term "inductor" is also used, since a coil has *inductance* (discussed later). There are two generally accepted styles for the basic inductor symbol, one with coil loops and one without.

When a magnetic core is added to a coil, lines are added to the symbol. When the magnetic core is adjustable (moved within the coil by means of a screwdriver adjustment), a dashed line with arrow is used. The dashed line is typically used to indicate a powdered iron core. The basic coil can be tapped and can be adjusted (either in steps or continuously), similar to a resistor.

The reference designation letter for inductors is *L* and the values are given in *henrys* (H). Commonly used submultiples of the henry are the *millihenry* (thousandth of a henry, mH), the *microhenry* (millionth of a henry, μH, and the *nanohenry* (billionth of a henry, nH). A schematic may also indicate the d-c resistance of a coil, expressed in ohms.

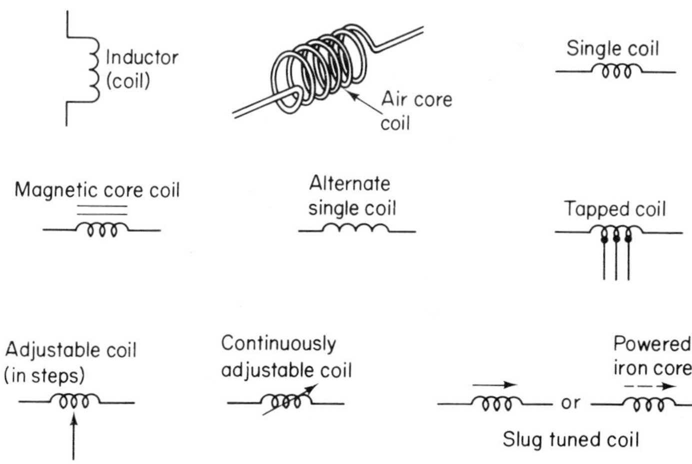


**FIGURE 2-25** Symbols for inductors (coils)

### *2-4.1 FUNCTION OF INDUCTORS (COILS) IN CIRCUITS*

The two most common uses for inductors in circuits are *to oppose the flow of alternating current* and *to form a resonant tuning circuit in conjunction with a capacitor.* Such resonant circuits are used to tune radio receivers and transmitters to specific operating frequencies, as discussed in Sec. 2-5.

**Inductive Reactance and Impedance of Inductors.** When current flows through an inductor (or coil), a magnetic field is set up around the inductance. If the current is alternating, the magnetic field will vary (build up and collapse). The buildup and collapse of the magnetic field will, itself, produce another current in the inductor. This self-induced current is *in opposition to the initial or external current.* This opposition is termed *inductive reactance,* is expressed in ohms, and is identified by the letters $X_L$ ($X$ for reactance, $L$ for inductance).

If an inductor were in a circuit where there is no d-c resistance, inductance reactance could be substituted for resistance in the conventional Ohm's law equations. However, this is not practical. Any circuit has some d-c resistance (the coil wire and leads have some resistance). The resistance has the same effect as a resistor in series with the inductor.

The combined effect of the inductance and effective resistance is known as *impedance,* which is also expressed in ohms, and is represented by the letter $Z$. The impedance of an inductor can be substituted directly in the basic Ohm's law equations:

$$E = I \times Z \qquad I = \frac{E}{Z} \qquad Z = \frac{E}{I}$$

where $Z$ is the impedance in ohms.

The relationship of inductive reactance and impedance of inductors, together with the calculations, are shown in Fig. 2-26. Although it is not necessary to make these calculations to read the schematic, the calculations serve as a basis for understanding impedance and resonant circuits (discussed in Sec. 2-5).

**Alternating-Current Voltages in Inductive Circuits.** Unlike d-c voltages across resistors, it is not possible simply to add the a-c voltages across the inductor and resistor in a series circuit to obtain the source voltage. Instead, the individual voltages must be added vectorially. Figure 2-27 shows the relationship of voltages, together with the calculations, for voltages in series $RL$ circuits.

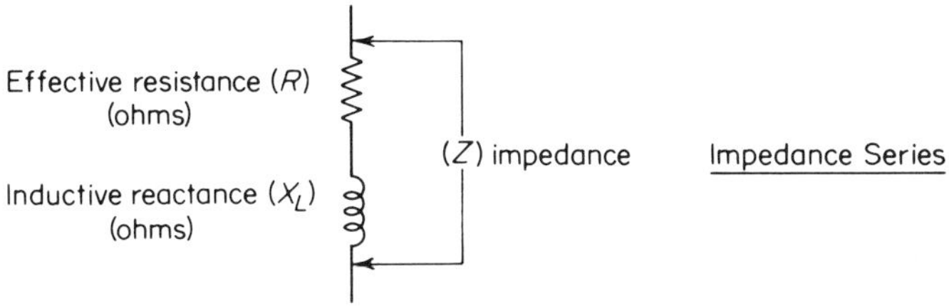


$$Z = \sqrt{R^2 + X_L^2} \qquad Q = \frac{X_L}{R}$$

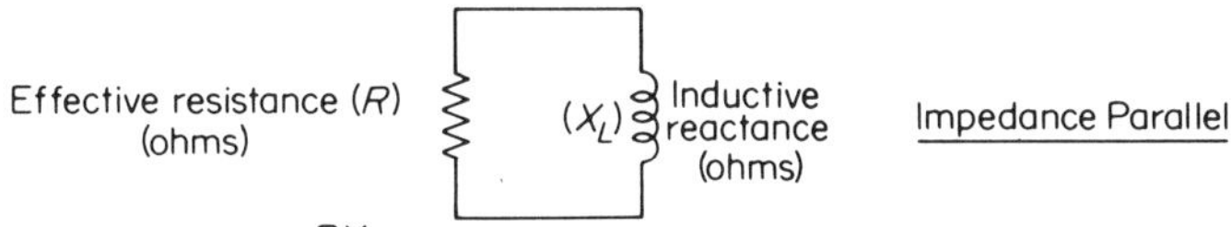


$$Z = \frac{RX_L}{\sqrt{R^2 + X_L^2}} \qquad Q = \frac{R}{X_L}$$

Inductive Reactance

series or parallel

$X_L = 6.28\ FL$

$X_L = 6.28$ x frequency (in Hz) x inductance (in henrys)
or 6.28 x (kilohertz, KHz) x (millihenrys, MH)
or 6.28 x (megahertz, MHz) x (microhenrys, $\mu$H)

$$L = \frac{X_L}{6.28\,F} \qquad F = \frac{X_L}{6.28\,L}$$

**FIGURE 2-26** Calculations for inductive reactance, *Q* and impedance in resistance-inductance (*RL*) circuits

**Calculating Inductance of an Air-Core Coil.** If you are interested in calculating the inductance of an air-core coil (the simple coil shown in Fig. 2-25), use the equations shown in Fig. 2-28. Note that the equations are based on the number of turns (as well as radius and length). However, the number of turns shown on the schematic usually has no relationship to the actual coil turns.

**Time Constants for Inductive-Resistive Circuits.** In a d-c circuit that is purely resistive (say, a resistor connected across the terminals of a battery), current should start to flow as soon as the voltage is applied. In an inductive-resistive circuit (generally called an *LC* circuit) such as Fig. 2-29, the current buildup lags the voltage, as shown in Fig. 2-30. An inductance-resistance time constant is the

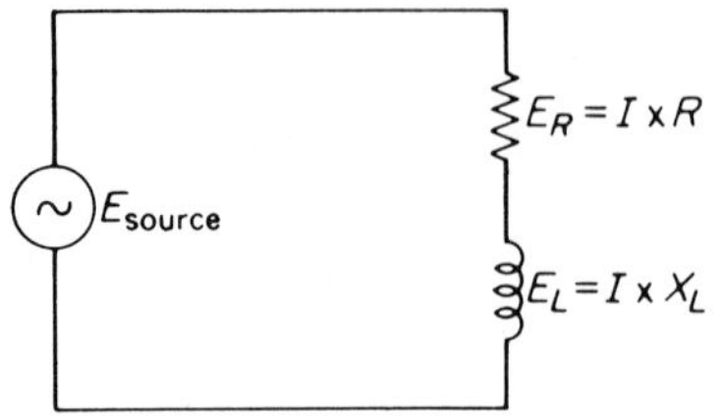


$$E_{source} = \sqrt{E_R^2 + E_L^2}$$

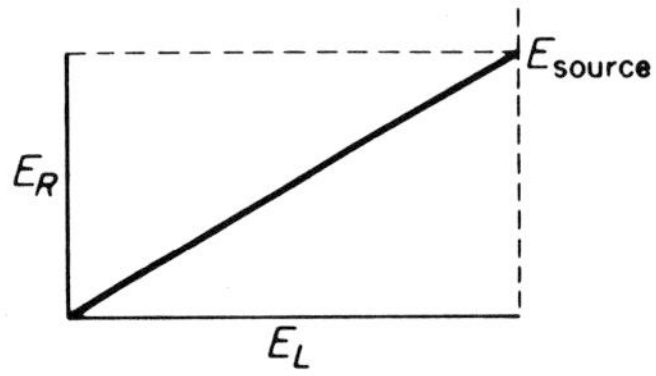


**FIGURE 2-27** Adding alternate current voltages in a series resistance-inductance (*RL*) circuit

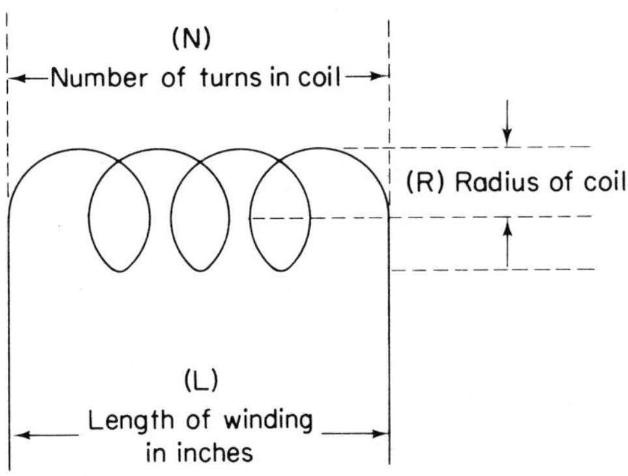


$$L \text{ (in } \mu\text{H)} \approx \frac{(\text{RN})^2}{9\text{R} + 10\text{ L (length)}}$$

When L (length) = 0.8 × R or R/L (length) = 1.25

$$\text{N} \approx \sqrt{\frac{17 \times \text{(inductance in } \mu\text{H)}}{\text{R}}}$$

$$L \text{ (in } \mu\text{H)} \approx \frac{(\text{RN})^2}{\text{R } 17}$$

**FIGURE 2-28** Calculations for inductance of single-layer air-core coil

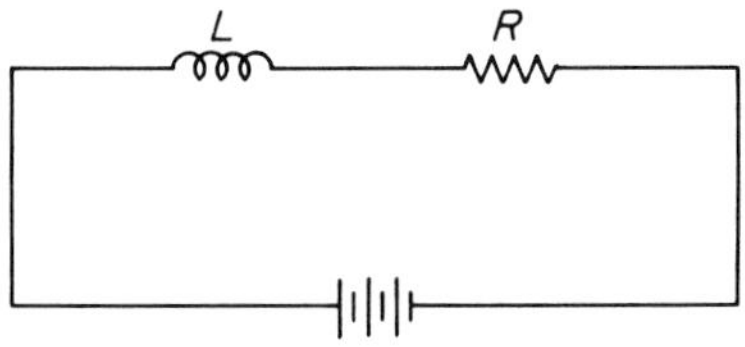


$$TC \text{ (Time constant in seconds)} = \frac{L \text{ (Henries)}}{R \text{ (Ohms)}}$$

$$TC \text{ (Time constant in milliseconds)} = \frac{L \text{ (Millihenries)}}{R \text{ (Ohms)}}$$

$$TC \text{ (Time constant in microseconds)} = \frac{L \text{ (Microhenries)}}{R \text{ (Ohms)}}$$

$$L = TC \times R$$

$$R = \frac{L}{TC}$$

**FIGURE 2-29** Calculations for inductive time constants

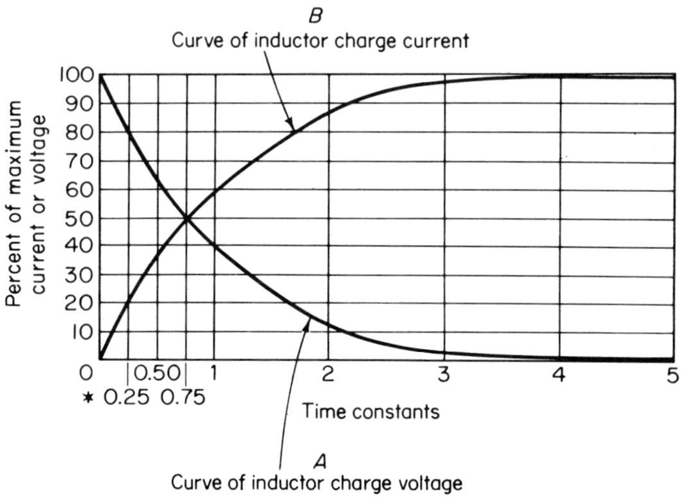


*The time indicated along the X axis is in relation to the instantaneous charge value of current or voltage, and does not indicate the time in seconds.

**FIGURE 2-30** Universal time constant chart for inductance and resistance

time required for current through the inductor to reach 63% of the maximum value. The time constant also represents the time required for the voltage across the resistor to build up to approximately 63% of the battery (or other source) voltage, as well as the time required for the voltage drop across the inductor to drop to 37% of the value of the source voltage. By use of the time-constant formulas, coupled with applications of Ohm's law, all the important factors relating to the time constants of an inductor-resistor charging circuit can be obtained.

The calculations for inductive time constants are shown in Fig. 2-29. For example, if a 100-Ω resistor is connected to a 1000-μH inductor and a battery as shown in Fig. 2-29, the current through the inductor will reach 63% of the full value in 100 μs. You will find many *LC* combinations on schematics. It is not necessary to make *LC* time-constant calculations to read schematics, but you must remember that the time constant will affect circuit operation, perhaps critically.

**Inductor Networks.** As in the case of resistors and capacitors, inductors can be connected in series, parallel, and in series-parallel networks. Where *no interaction of magnetic fields* is produced by the inductors, the equations are the same as for resistors, as shown in Fig. 2-31.

The total inductance of inductors in series (Fig. 2-31a) is not affected by the order in which the inductors are connected. Any number of inductors can be added in this manner, provided that the same inductance units (henry, millihenry, microhenry) are used.

When only two inductors are connected in parallel, use the equation of Fig. 2-31b. The total inductance is always *less* than the value of either inductance.

When three or more inductors are connected in parallel, use the most convenient of the three equations in Fig. 2-31c. For added convenience, make L1 the largest inductor in the parallel combination.

When series-parallel inductance values must be obtained, the series-inductance equation can be combined with the parallel-inductance equation, as shown in Fig. 2-31d.

In solving any series-inductance equation found on a schematic, where there is a 10:1 (or better) ratio between inductor values, the total inductance will be only slightly below the smallest inductor value. This can be used as a rough approximation of the total parallel-inductance value.

**Mutual Inductance and Coupling Coefficient.** Unless inductors are completely surrounded by shielding, are placed at right angles to

Total inductance $L = L_1 + L_2 + L_3$

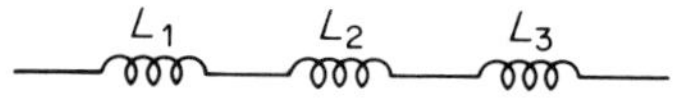


(a)

Total inductance $L = \dfrac{L_1 \times L_2}{L_1 + L_2}$

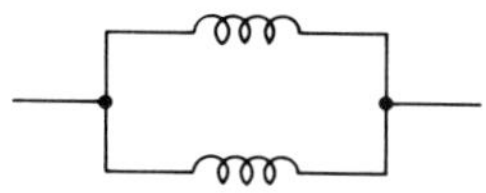

(b)

**FIGURE 2-31** Finding inductance in series, parallel and series-parallel networks: (a) series, (b) two in parallel, (c) three or more in parallel, (d) series-parallel

$$\text{Total inductance } L = \frac{1}{\dfrac{1}{L_1} + \dfrac{1}{L_2} + \dfrac{1}{L_3} + \cdots}$$

or

$$L = \frac{L_1}{1 + \dfrac{L_1}{L_2} + \dfrac{L_1}{L_3} + \cdots}$$

or

$$\frac{1}{L} = \frac{1}{L_1} + \frac{1}{L_2} + \frac{1}{L_3} \cdots$$

(c)

$$\text{Total inductance } L = L_1 + L_2 + L_3 + \frac{L_4 \times L_5}{L_4 + L_5}$$

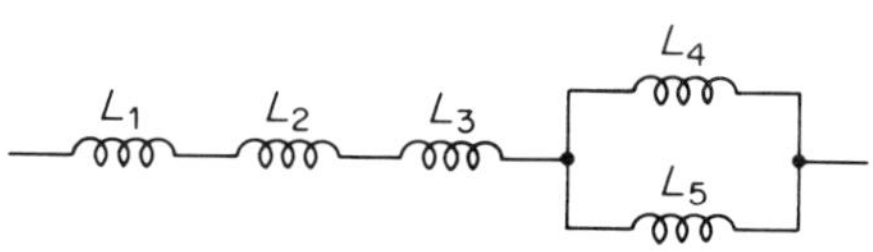


(d)

each other, or are sufficiently far apart, the magnetic field of one inductance will affect the other inductance. This effect is known as *mutual inductance*, and is measured in henrys (or mH and $\mu$H). The operation of transformers is based on mutual inductance, as discussed in Sec. 2-7.

Mutual inductance depends upon the self-inductance of the inductors, and upon the closeness (or efficiency) of the inductors. In transformers, the inductors are ideally as close as is practical. The effectiveness of the mutual coupling is known as the *coupling coefficient*, or *k* factor. A 100% efficient coupling (impossible in practical applications) is a *k* factor of 1; 90% efficiency is a *k* factor of 0.9. The calculations for mutual inductance and coupling coefficient are shown in Fig. 2-32.

**Effects of Mutual Inductance on Inductors.** Mutual inductance may add to the self-inductance of inductors or may subtract from the self-inductance, depending upon how the inductors are arranged. If the current flows in the same direction through each coil, the magnetic fields reinforce each other and the mutual coupling adds. If the current flows in opposite direction, the mutual coupling will subtract. The calculations for the effects of mutual inductance are shown in Fig. 2-32.

**Chokes and Radio-Frequency Chokes.** One of the primary functions of inductors is to oppose the flow of alternating current (to provide impedance to ac). This impedance can be increased if the inductor is wound on a magnetic core as is done when a *choke* is formed. One such choke (called a *filter choke*) consists of a coil wound on an iron core. The core acts to increase the natural impedance of the coil to ac. Filter chokes are most often used in power supply circuits (Chapter 4). Another choke (called a *radio-frequency choke* or RFC) consists of a series of coils or windings in a special formation (called a *toroid* winding), sometimes on a metal core. The special winding acts to increase impedance, especially to a-c volt-

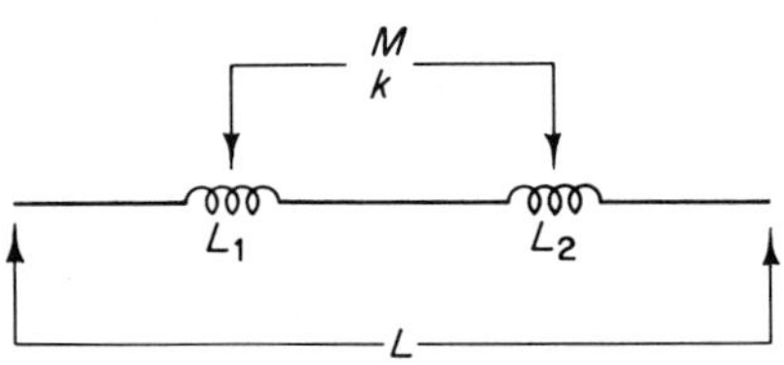


Mutual Inductance

$$M = k\sqrt{L_1 \times L_2}$$

Coupling Coefficient

$$k = \frac{M}{\sqrt{L_1 \times L_2}}$$

**FIGURE 2-32** Calculations for mutual inductance and coupling coefficient

ages at radio frequencies. RFCs are most often used in transmitter circuits (Chapter 5).

## 2-5 RESONANT-CIRCUIT SYMBOLS

The resonant circuit is one of the most common circuits found on schematics. In fact, it is impossible to find a CB or TV set schematic without resonant-circuit symbols. The basic resonant circuit consists of a capacitor and an inductance, connected either in series or parallel, as shown in Fig. 2-33. The main function of a resonant circuit is to tune the circuit or set to a particular frequency (either to pass or reject signals of certain frequencies). This is because the resonant circuit has the ability to accept or reject a-c signals and voltages of specific frequencies. (We discuss frequency in Sec. 2-6.)

If a capacitance and an inductance are connected in series or parallel and a signal of appropriate frequency is applied, the circuit is said to be *resonant when the inductive reactance is equal to the capacitive reactance.* The calculations for resonance are given in Fig. 2-33. As shown, resonance is dependent on three variables: capacitance, inductance, and frequency. Note that either (or both) the inductance or capacitance can be made variable (adjustable) in a practical resonant circuit. Also note that for any combination of inductance and capacitance, there is only one resonant frequency. However, for any frequency, there are many combinations of inductance and capacitance that will produce resonance.

It is not necessary to calculate the resonant frequency, capacitance, and inductance to read a schematic. However, it can be very helpful to understand the calculations if the frequency is omitted from the schematic. For example, assume that you are reading a receiver schematic and find a parallel resonant circuit at the input. The capacitance and inductance values are given but the receiver operating frequency is nowhere to be found. You can calculate the operating frequency using the equations of Fig. 2-33. Assume a capacitance of 0.002 $\mu$F and an inductance of 20 $\mu$H. The calculations of Fig. 2-33 show a resonant frequency of

$$\frac{159}{\sqrt{20 \times 0.002}} = 795 \text{ kilohertz (kHz)}$$

The term "kilohertz" is discussed in Sec. 2-6.

In a practical receiver resonant circuit, either the capacitance or inductance will be variable so that the reciever can be tuned over a given range. The calculations of Fig. 2-33 can be used to find this

### Resonance and impedance

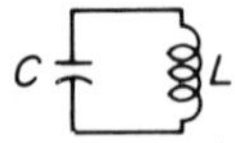

Parallel * (Infinite impedance) | Series (zero impedance)

$$F = \frac{1}{6.28\sqrt{LC}}$$

$$F\,(\text{kHz}) = \frac{10^6}{6.28\sqrt{L\,(\mu\text{H}) \times C\,(\text{pF})}}$$

$$F(\text{kHz}) = \frac{159}{\sqrt{L\,(\mu\text{H}) \times C\,(\mu\text{F})}}$$

$$L\,(\mu\text{H}) = \frac{2.54 \times 10^4}{F\,(\text{kHz})^2 \times C\,(\mu\text{F})}$$

$$F(\text{MHz}) = \frac{0.159}{\sqrt{L\,(\mu\text{H}) \times C\,(\mu\text{F})}}$$

$$C\,(\mu\text{F}) = \frac{2.54 \times 10^4}{F\,(\text{kHz})^2 \times L\,(\mu\text{H})}$$

*Approximate; accurate when circuit Q is 10 or higher

---

### Inductive reactance

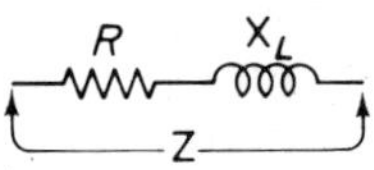


$$Z = \sqrt{R^2 + X_L^2} \quad Q = \frac{X_L}{R}$$

Series

$$L = \frac{X_L}{6.28\,F}$$

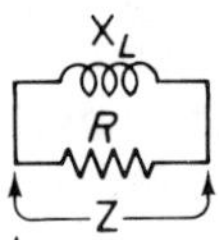


$$Z = \frac{RX_L}{\sqrt{R^2 + X_L^2}} \quad Q = \frac{R}{X_L}$$

Parallel

$$F = \frac{X_L}{6.28\,L}$$

$$X_L = 6.28 \times F(\text{Hz}) \times L(\text{H})$$

$$X_L = 6.28 \times F(\text{kHz}) \times L(\text{mH})$$

$$X_L = 6.28 \times F(\text{MHz}) \times L\,(\mu\text{H})$$

---

### Capacitive reactance

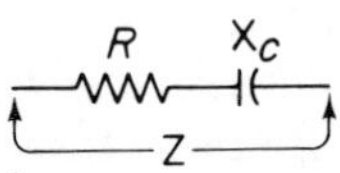


$$Z = \sqrt{R^2 + X_C^2} \quad Q = \frac{X_C}{R}$$

Series

$$F = \frac{1}{6.28\,CX_C}$$

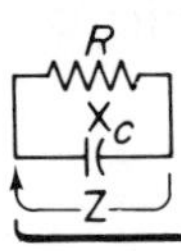


$$Z = \frac{RX_C}{\sqrt{R^2 + X_C^2}} \quad Q = \frac{R}{X_C}$$

Parallel

$$C = \frac{1}{6.28\,FX_C}$$

$$X_C = \frac{1}{6.28 \times F(\text{Hz}) \times C(\text{F})}$$

$$X_C = \frac{159}{F(\text{kHz}) \times C\,(\mu\text{F})}$$

**FIGURE 2-33** Resonant circuit equations

tuning range, even though it is not available on the schematic. For example, if the capacitance is variable, use the maximum and minimum capacitance values to find the ends of the tuning range.

### *2-5.1 SERIES VERSUS PARALLEL RESONANT CIRCUIT SYMBOLS*

In a series resonant, the impedance will be zero (theoretically) at the resonant frequency but high at all other frequencies. In a parallel resonant circuit, the impedance is (theoretically) infinite at resonance but low at all other frequencies.

A basic example of the resonant circuit as a tuning device is shown in Fig. 2-34, the schematic of a simple *field strength meter.* The function of such a meter is to measure the relative strength of radio signals at various locations and distances (in the field) from the antenna of a transmitting station.

In use, the meter is tuned to the frequency of the transmitter signal being radiated by the transmitting antenna. The signal is picked up by the meter antenna and applied to the resonant circuit of L1-C1. Signals at any frequency except the resonant frequency will pass through the parallel *LC* circuit to ground and do not register on the meter. Signals at the resonant frequency cannot pass through the parallel *LC* circuit and thus appear across the circuit. These signals are converted from alternating current to direct current by diode CR1 (as discussed) and are applied to the meter, which indicates the relative signal strength.

Another example of the resonant circuit as a tuning device is shown in Fig. 2-35, the schematic of a CB signal trap (or interference eliminator) connected at the antenna (input) of a TV set. The

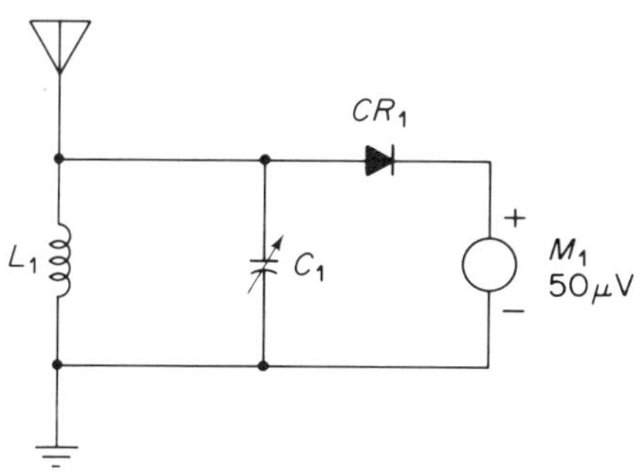


**FIGURE 2-34** Schematic of a simple field strength meter showing use of a parallel-resonant *LC* circuit

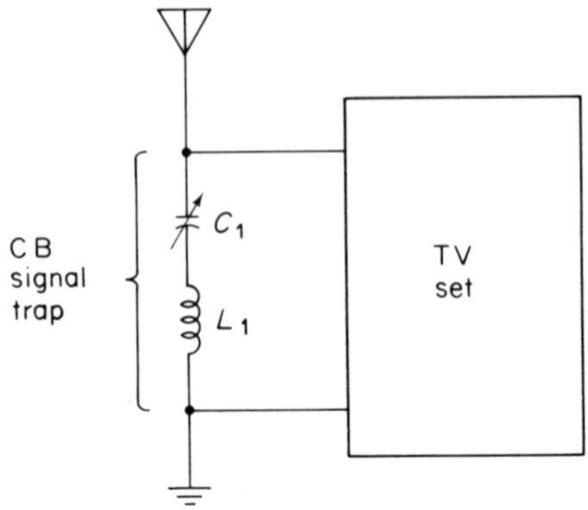


**FIGURE 2-35** Schematic of a CB signal trap (interference eliminator) showing use of a series-resonant *LC* circuit

function of such a tuning device is to prevent the signals of a CB set from interfering with normal TV reception.

In use, the trap is tuned to the frequency of the CB transmitter signal. When these signals appear at the TV antenna, the signals pass directly through the series *LC* circuit to ground and do not enter the TV circuits. Signals at all other frequencies cannot pass through the circuit and are thus applied to the TV set in the normal manner.

### *2-5.2 Q OF A RESONANT CIRCUIT*

A resonant circuit has a *Q*, or *quality factor*. The circuit *Q* is dependent upon the individual *Q* factors of the inductance and capacitance used in the circuit. The primary concern in a resonant circuit is that a high *Q* will produce a sharp resonant curve, whereas a low *Q* will produce a broad resonant curve. Usually, a resonant circuit is measured at points either side of the resonant frequency where the signal amplitude is down to 0.707 of the peak resonant value, but other reference points can be used. The calculations for resonant circuit *Q* are shown in Fig. 2-36.

In general, in a circuit where the d-c resistance is large in relation to the reactance (produced by the capacitor and inductor), the *Q* will be low (broad tuning). For this reason, you may find parallel resonant circuits on schematics with a resistance in parallel across the circuit. This broadens the circuit tuning. Such resistors are sometimes known as *swamping* resistors.

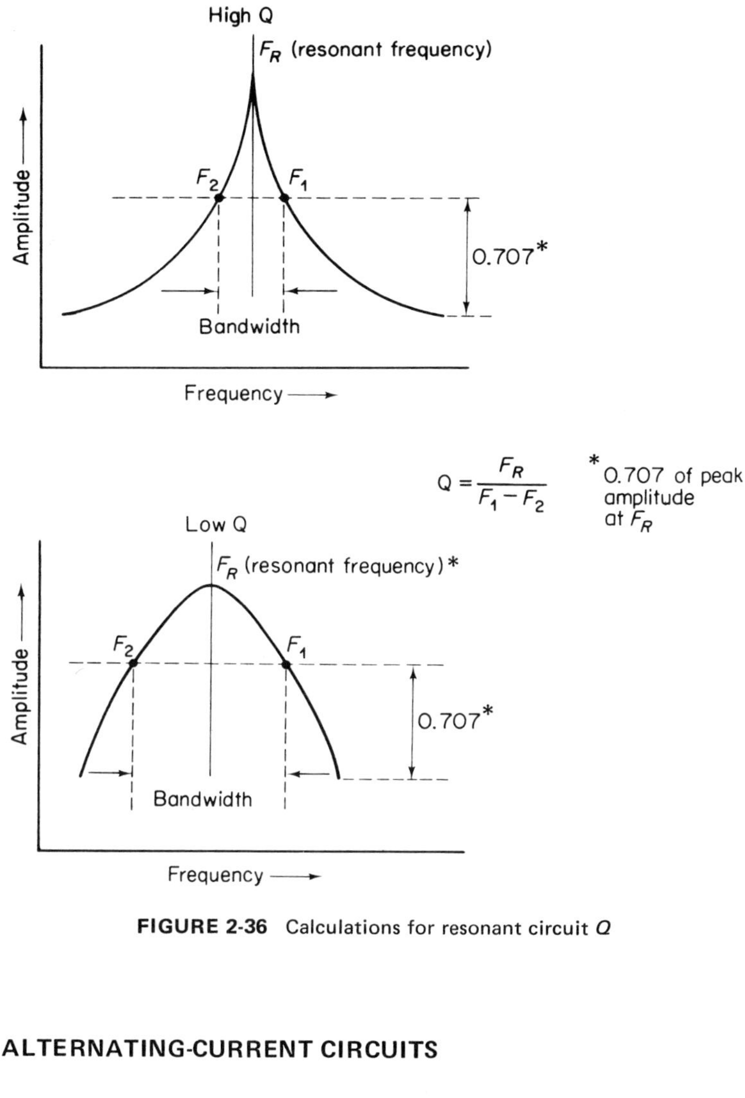


**FIGURE 2-36** Calculations for resonant circuit *Q*

## 2-6 ALTERNATING-CURRENT CIRCUITS

Before we go on with more symbols, let us review a-c circuits. As discussed, alternating current changes its polarity at specific intervals. That is, alternating current flows first in one direction and then reverses and moves in the opposite direction. The current (or voltage) assumes equal values on alternative directions. That is, at any given point in the circuit, the voltage appears to swing positive and negative by equal amounts. This condition is shown in Fig. 2-37.

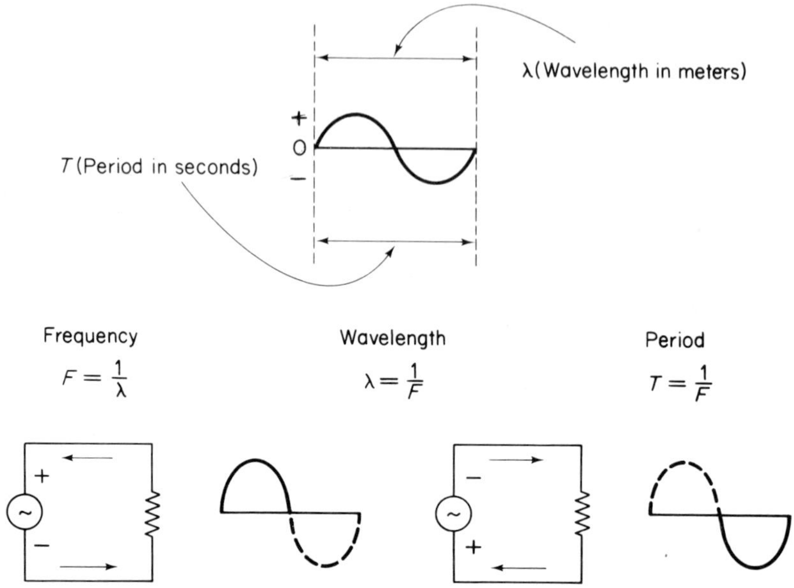


$$F(\text{in KHz}) = \frac{3 \times 10^5}{\lambda(\text{in meters})}$$

$$F(\text{in KHz}) = \frac{984}{\lambda(\text{in feet})}$$

**FIGURE 2-37** Relationship of wavelength, frequency and period in sinewaves

As illustrated, the distance from the point where the voltage or current crosses zero and starts in the positive direction, to the point where the voltage or current just returns to zero from the negative direction, is considered *one wavelength.* A wavelength can also be the distance between successive positive or negative peaks, or any corresponding points along the waveform. The usual symbol for wavelength is the Greek letter lambda (λ). Wavelength is usually expressed in meters (or submultiples of a meter).

The number of wavelengths per unit of time [usually 1 second(s)] is the *frequency* of the alternating current. Frequency is expressed in hertz. The term *hertz* has replaced the term *cycle.* When "hertz" is used, time units are implied. For example, kilohertz (1000 hertz or kHz) means kilohertz per second, and megahertz (1 million hertz, MHz) means megahertz per second.

The time duration of a single wavelength is the *period* of the alternating current. Period is usually expressed in seconds.

The calculations for wavelength, frequency, and period are given in Fig. 2-37. Note that these equations apply to voltages, currents, and signals of all frequencies and types.

The term *alternating current*, or *ac*, is usually applied when voltages, currents, and signals are in the 60-Hz range (which is the operating frequency of house power). The term *audio frequency* (AF) is applied to any voltage, signal, or current up to about 20 kHz. Above that point, the term *radio frequency* (RF) is applied. If RF signals are applied to an antenna, they will be radiated (they become radio waves). Below the RF range, signals will not radiate

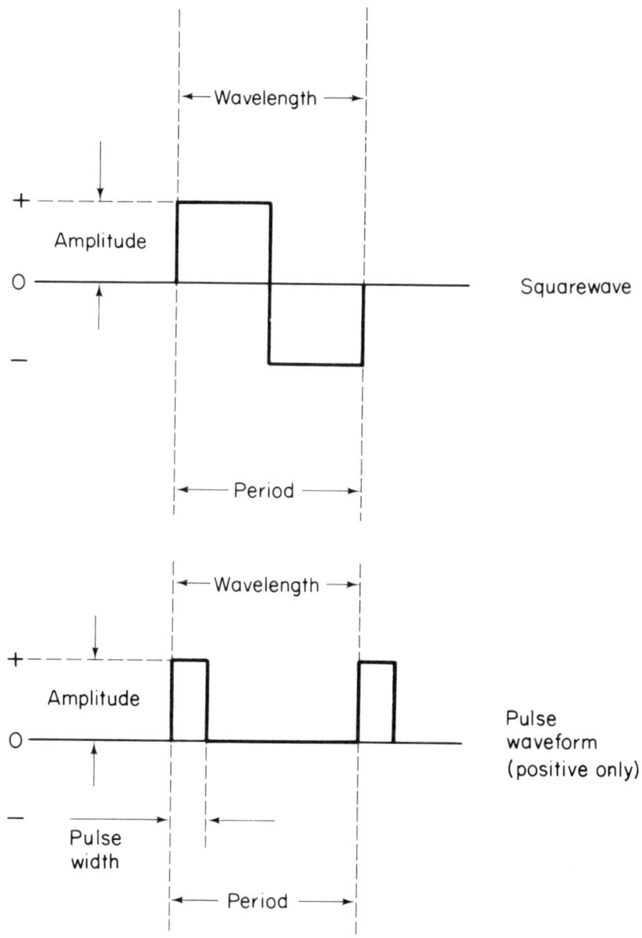


**FIGURE 2-38** Relationship of wavelength, frequency and period in squarewaves and pulse waveforms

from an antenna or any other conductor. However, a-c signals of any frequency will produce magnetic fields around conductors. The magnetic fields build up and collapse at the frequency of the signals passing through the conductor. Note that the terms "signals," "voltages," and "currents" are used interchangeably.

The waves shown in Fig. 2-37 are a form of *sine wave.* Voltages are also found in *square-wave* form and in *pulse form,* as shown in Fig. 2-38. In a true square wave, the positive and negative alternations are of the same value and length. This is not necessarily true of pulse waves. Also, pulse waves may be positive only or negative only, as shown. The calculations for frequency, wavelength, and period for sine waves shown in Fig. 2-37 can also be applied to true square waves.

### 2-6.1 ALTERNATING-CURRENT OHM'S LAW

Ohm's law equations for alternating current are essentially the same as for direct current (Sec. 2-1), except that impedance ($Z$) is substituted for resistance ($R$). Therefore,

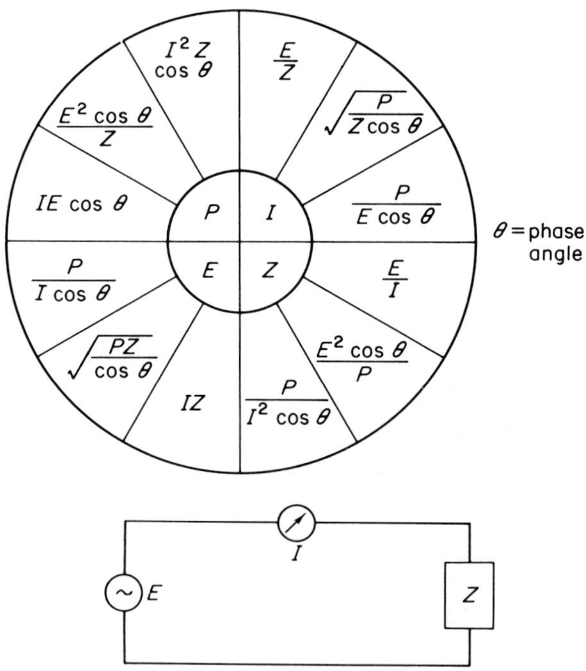


**FIGURE 2-39** Ohm's law calculations for alternating current

$$E = IZ \qquad I = \frac{E}{Z} \qquad Z = \frac{E}{I} \qquad P = EI$$

where $I$ = current (amperes)
$Z$ = impedance (ohms)
$E$ = voltage across $Z$ (volts)
$P$ = apparent power (watts)

The calculations for Ohm's law (ac) are shown in Fig. 2-39.

The term *apparent power* is applied when the reactive power factor of an a-c circuit is disregarded. To obtain the true power in an a-c circuit, it is necessary to multiply the apparent power by the cosine of the *phase angle.* (Phase angle is discussed later in this section.) Since the phase angle is always less than $90^{\circ}$ and the corresponding cosine is a fraction of 1, the true power will always be less than apparent power.

The ratio between the true power and the apparent power is known as the *power factor.* This can be expressed as

$$\text{power factor} = \frac{E \times I \times \text{cosine phase angle}}{E \times I}$$

or

$$\text{power factor} = \text{cosine phase angle} = \frac{R}{Z}$$

where $R$ = resistance in an a-c circuit (ohms)
$Z$ = impedance (ohms)

In those a-c circuits where there is no reactance (pure resistance), the phase angle is zero, so the apparent power is the true power.

### 2-6.2 ADMITTANCE, SUSCEPTANCE, AND CONDUCTANCE

As discussed in Sec. 2-1, *conductance* in a d-c circuit is the reciprocal of resistance and is expressed in mhos. A similar situation exists in a-c circuits. In most a-c circuits, however, the reactance must also be considered. For that reason, the terms *susceptance* and *admittance* are used more frequently.

Admittance is the reciprocal of impedance. Susceptance is the reciprocal of reactance. The calculations for admittance, susceptance, and conductance are shown in Fig. 2-40.

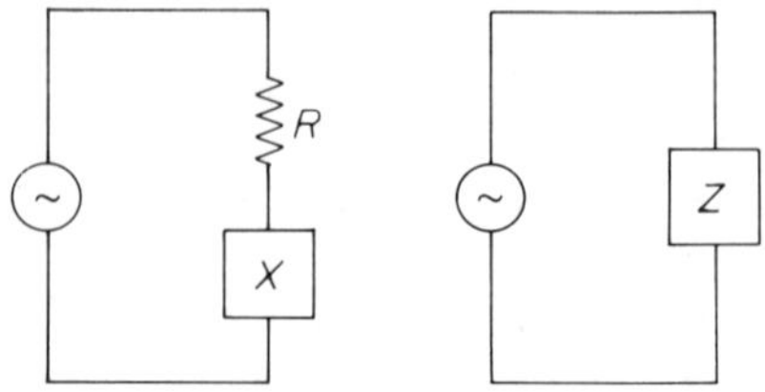


$$\text{Admittance } (Y) = \frac{1}{Z}$$

$$\text{Susceptance } (B) = \frac{X}{R^2 + X^2}$$

$$\text{Conductance } (G) = \frac{R}{R^2 + X^2}$$

$R$ = Resistance $X$ = Reactance $Z$ = Impedance

**FIGURE 2-40** Calculations for admittance, susceptance and conductance

### *2-6.3 PHASE-ANGLE RELATIONSHIPS*

In alternating currents, a pair of currents, a pair of voltages, or the voltage and current, need not be in step. One current may lead the other, one voltage may lag the other, or the voltage may lag the current. The amount of lead or lag is termed *phase angle* and is expressed by the Greek letter theta, $\theta$.

Since most alternating currents are sine waves, they have periodic waveforms where the lead or lag can be measured along the $X$ axis. Out-of-phase voltages (or currents) can be added vectorially to produce a resultant voltage (or current).

In a pure inductive circuit, the current will lag the voltage by $90^\circ$.

In a pure capacitive circuit, the current will lead the voltage by $90^\circ$.

These relationships are shown in Fig. 2-41.

Most a-c circuits are not pure resistive, inductive, or capacitive. Instead, the circuits are combinations of all three factors. The ratio of reactance to resistance determines the phase angle. This may be calculated vectorially or by means of trigonometry. We do not go into these calculations here since they do not apply to practical schematic reading.

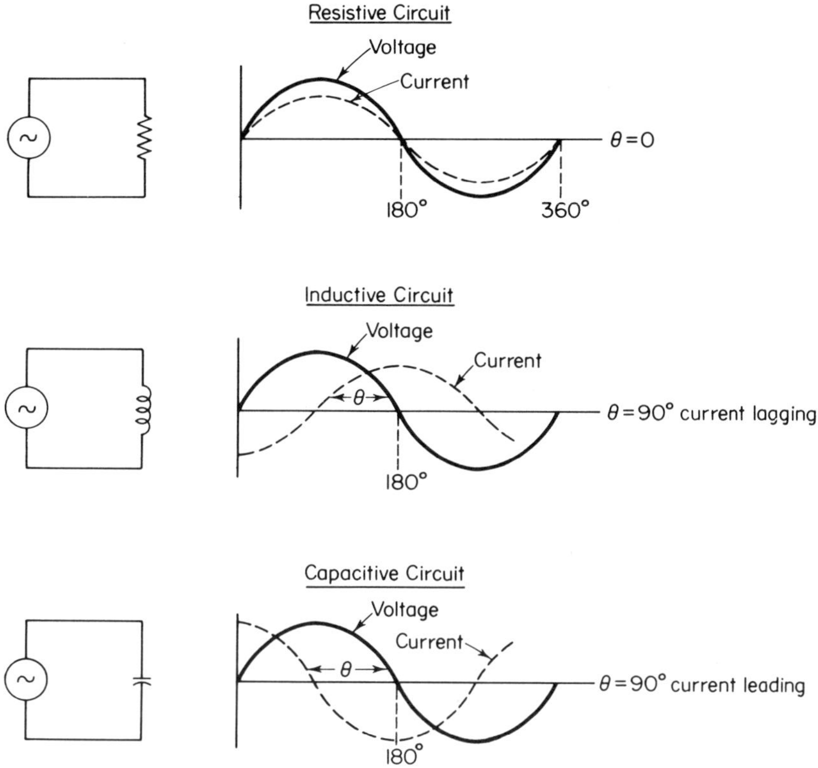


**FIGURE 2-41** Relationship of phase angle in pure resistive, inductive and capacitive circuits

**Three-Phase Alternating Current.** Most a-c circuits found on electronic schematics are *single-phase.* Two wires are required to complete the circuit, and the waveforms are as shown in Fig. 2-41. However, in power equipment (including some electronic power supplies), *three-phase* circuits are used. Three-phase systems (which usually involves generators, transformers, or motors) have three coils or sets of windings so arranged that the induced voltage for one set of coils differs from the others by a third of a cycle or $120^{\circ}$ (one full cycle or wave is considered as $360^{\circ}$). This is shown in Fig. 2-42.

Three-phase systems are classified according to connection into two groups: the *delta* (Δ) connection (so called because it looks like the Greek letter positioned sideways), and the Y (*wye*) or *star* connection. Sometimes the systems are mixed, as when a delta motor is powered by a wye generator through a wye-to-delta transformer.

Relationship of Phases (Wye or Delta)

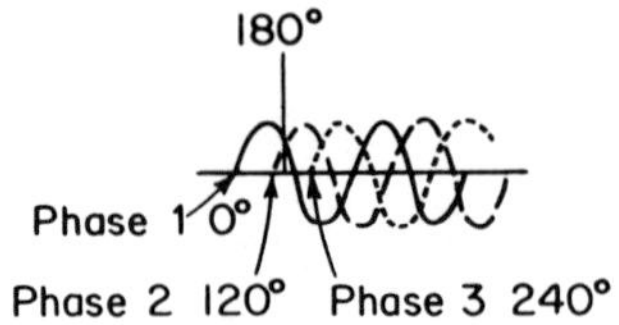


$\theta$ = Phase angle between voltage and current in any one phase

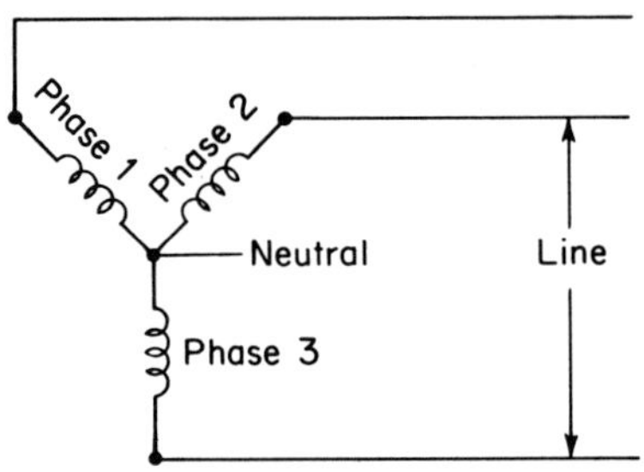


Star, or Wye, or Y Connections

$E_{phase}$ 1,2, or 3 = $\dfrac{E_{line}}{1.73}$

$E_{line} = E_{phase}$ 1,2 or 3 x 1.73

Power = $E_{phase}$ 1,2, or 3 x 1.73 x $I_{phase}$ x cosine $\theta$

$I_{line} = I_{phase}$

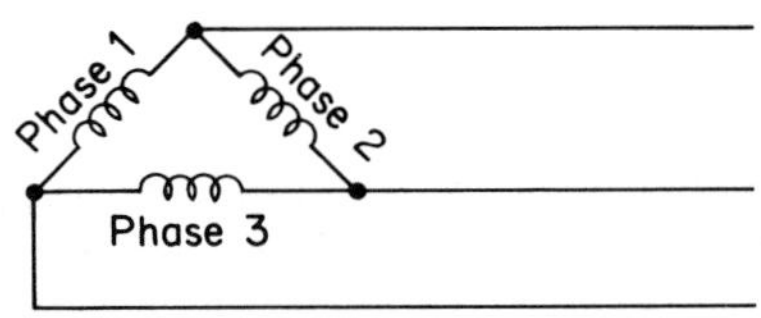


Delta Connections

$I_{phase}$ 1,2, or 3 = $\dfrac{I_{line}}{1.73}$

$I_{line} = I_{phase}$ 1,2 or 3 x 1.73

Power = $I_{phase}$ 1,2, or 3 x 1.73 x $E_{phase}$ x cosine $\theta$

$E_{line} = E_{phase}$

**FIGURE 2-42** Basic calculations for three-phase a-c systems

The main advantage of three-phase systems is that increased power can be obtained without a corresponding increase in individual voltages and currents. For example, in a wye system, this power increase is possible since the line voltage is 1.73 times the voltage across each phase. In a delta system, the power increase is possible because the line current is 1.73 times the current produced by each phase.

In heavy power systems, where both three-phase and single-phase are required, a neutral (ground) wire is often used as shown in Fig.

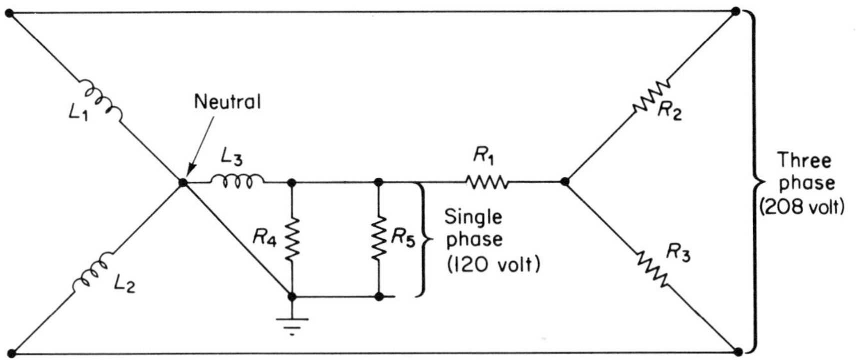


$$120 \times 1.73 = 208 \text{ V}$$

$$\frac{208}{1.73} = 120 \text{ V}$$

**FIGURE 2-43** Three-phase four-wire system connections

2-43. This is known as a *three-phase, four-wire* system. The neutral wire permits single-phase ac to be obtained from a three-phase system.

For example, each coil or phase of the generator (L1, L2, and L3) shown in Fig. 2-43 has a 120 V across it (this voltage is obtained with respect to the neutral wire and any one of the other wires). For the system shown in Fig. 2-43, the voltage drop across generator coil L3 is used to supply 120 V *single-phase* to the load circuits R4 and R5. These loads may be electric fans, heaters, or other devices requiring only 120 V single-phase. At the same time, a balanced load (represented by resistors R1, R2, and R3) is furnished to the three-phase source as shown.

## 2-7 TRANSFORMERS

Figure 2-44 shows the symbols and physical appearance of several common types of transformers. A transformer consists essentially of two (or more) coils placed close together in a common housing or on a common form. There are many coil arrangements in transformers, but the coils must be in parallel to produce maximum transformer action.

As discussed in Sec. 2-4.1 when two coils are placed close together, there is mutual inductance between the two coils. If an alternating current is passed through one coil (called the *primary winding*, or

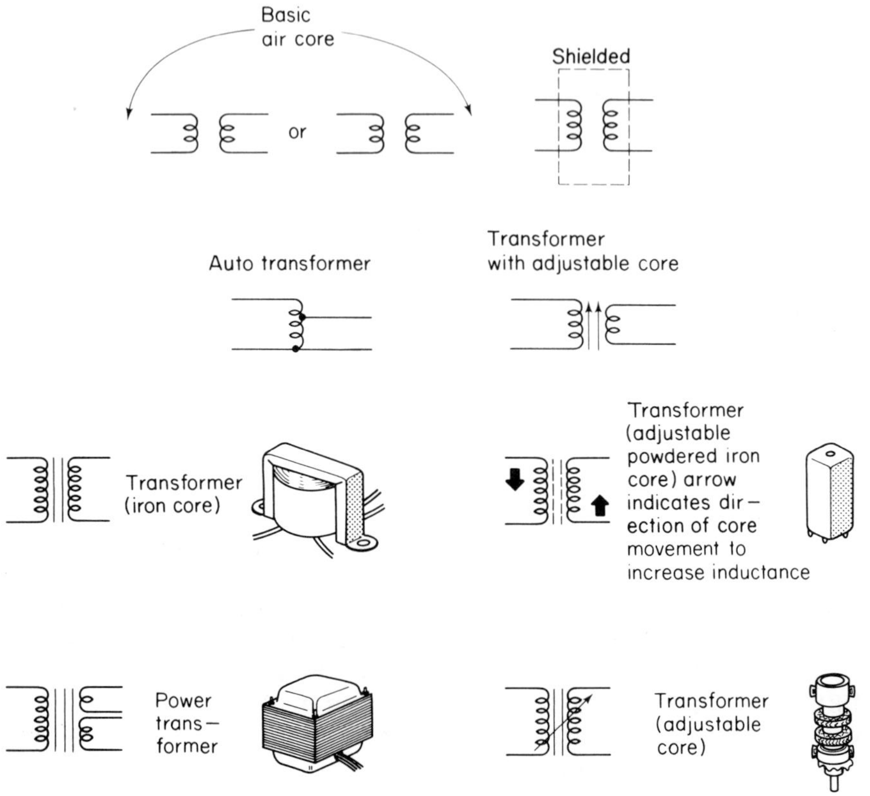


**FIGURE 2-44** Symbols for transformers

*primary*), a magnetic field is produced around the coil. The field builds up and collapses at the same frequency as the alternating current. (A direct current through a coil also produces a magnetic field, but the field remains fixed. For that reason, transformers operate only with alternating current.)

In a transformer, the other coil (called the *secondary winding*, or *secondary*), placed within the magnetic field, produces another electrical current (secondary current) at the same frequency as the primary alternating current. Note that in the *autotransformer* of Fig. 2-44, there is only one coil or winding. However, the winding is tapped so that one half of the winding functions as the primary, with the other half as the secondary.

As in the case of inductors, when a magnetic core is added to the transformer, lines are added to the symbol. Dashed lines usually indicate a powered iron core. A core formed of laminated iron plates is often used for audio transformers and power transformers. As shown in Fig. 2-44, there are several ways to show an adjustable core (sometimes called an *adjustment slug*). No matter what symbol is used, when the core (or slug) is moved between the coils (or within the coils), the inductance of each coil is changed (as is the mutual inductance between the coils).

The change in inductance with a change in core position makes it possible to tune the transformer and to use the transformer *tuned coupling device* between the two circuit stages. For example, the symbol of Fig. 2-45a is that of an intermediate-frequency (IF) transformer used in radio receivers, CB sets, TV sets, and so on. Note that there is a capacitor across each coil or winding, forming two resonant circuits (Sec. 2-6). Each circuit is set to a specific resonant frequency by an adjustable core. In this case, the capacitance remains fixed, so the resonant frequency changes with inductance. In some air-core tuned transformers (where there is no tuning core or slug), the inductances remain fixed, but the capacitors are adjustable. The symbol for such a transformer is given in Fig. 2-45b.

The reference-designation letter for transformers is T. The values shown on the schematic depend on the type of transformer. A power transformer will show the voltages produced on each winding. An audio transformer should show the impedances of each winding and (possibly) the d-c resistance of the windings. Tunable or adjustable transformers will usually show the center frequency for which the transformer is designed.

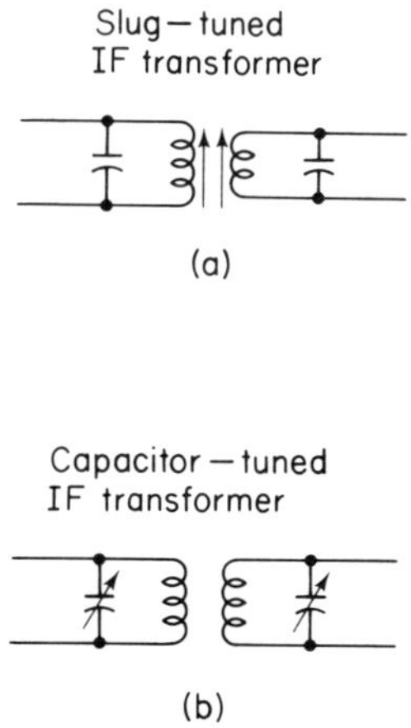


**FIGURE 2-45** Symbols for slug-tuned and capacitor-tuned IF transformers

### 2-7.1 FUNCTION OF TRANSFORMERS IN CIRCUITS

Transformers perform several functions in circuits. These functions are discussed as part of the overall circuit function in later chapters. However, the main function (other than coupling two circuits together) is to step up or step down voltages, currents, and impedances, which we discuss here.

The voltage, current, and impedance ratios of transformers are dependent upon the *turns ratio* of the primary and secondary windings,

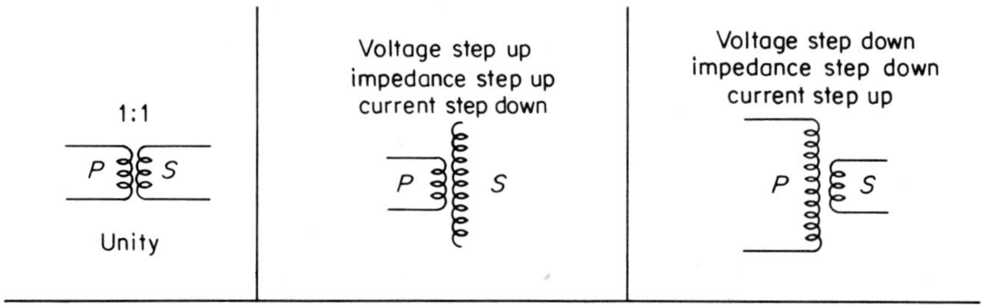


$E_P$ = Primary voltage  $N_P$ = Primary turns

$E_S$ = Secondary voltage  $N_S$ = Secondary turns

$I_P$ = Primary current  $Z_P$ = Primary impedance

$I_S$ = Secondary current  $Z_S$ = Secondary impedance

**Voltage Ratio**

$$\frac{E_S}{E_P} = \frac{N_S}{N_P} \qquad E_P \times N_S = N_P \times E_S$$

$$E_P = \frac{N_P \times E_S}{N_S} \qquad E_S = \frac{E_P \times N_S}{N_P}$$

$$N_S = \frac{N_P \times E_S}{E_P} \qquad N_P = \frac{E_P \times N_S}{E_S}$$

**Current Ratio**

$$\frac{I_P}{I_S} = \frac{N_S}{N_P} \qquad I_P \times N_P = N_S \times I_S$$

$$I_P = \frac{N_S \times I_S}{N_P} \qquad I_S = \frac{I_P \times N_P}{N_S}$$

$$N_S = \frac{N_P \times I_P}{I_S} \qquad N_P = \frac{I_S \times N_S}{I_P}$$

**Impedance Ratio**

$$\frac{Z_S}{Z_P} = \frac{N_S^2}{N_P^2} \qquad \text{Turns ratio} = \sqrt{\frac{Z_S}{Z_P}} \text{ or } \sqrt{\frac{Z_P}{Z_S}}$$

Where turns ratio is known:

$$E_S = \text{turns ratio} \times E_P$$

**FIGURE 2-46** Calculating voltage, current and impedance ratios of transformers

as shown in Fig. 2-46. Note that schematics rarely, if ever, show the turns ratio of transformers accurately.

The voltage across a transformer is stepped up when the secondary has more turns than the primary. The voltage is stepped down when the primary has more turns than the secondary.

The current across a transformer is stepped up when the primary has more turns than the secondary. The current is stepped down when the secondary has more turns than the primary. If you have been paying attention, you know that current is the opposite of voltage.

When the primary and secondary have the same number of turns, the transformer is said to have *unity coupling* or 1:1 coupling.

The impedance across a transformer is stepped up when the secondary has more turns than the primary and varies as the *square of the turns ratio*, not the turns ratio.

As an example, assume that you find a transformer on a schematic that gives the actual turns ratio of 1:4 (say 10 turns on the primary and 40 turns on the secondary). Whatever voltage is applied on the primary, the secondary voltage will be four times as much, and the secondary current will be one fourth of the primary current. However, the secondary impedance will be 16 times that of the primary impedance, as shown by the calculations of Fig. 2-46.

## 2-8 MISCELLANEOUS BASIC SCHEMATIC SYMBOLS

The following paragraphs serve as a guide to commonly used types of electronic parts. The symbols and related illustrations should prove helpful in reading schematics as well as in identifying the parts.

### *2-8.1 ANTENNAS AND GROUNDS*

Figure 2-47 shows the symbols for antennas and grounds. Note that the *general* antenna symbol is sometimes used even though the antenna is of a special type. A basic *dipole* antenna has two active elements connected to the lead-in, as shown. A *counterpoise* or directional antenna has reflective and directive elements in addition to the dipole. With such antennas, the dipole element is cut to a length that corresponds to the center operating frequency of the antenna (typically, a half wavelength). The reflective elements are made longer than the dipole, with the directive elements shorter. The loop antenna is often used with radio direction finders and with some portable radio receivers. Maximum signal strength

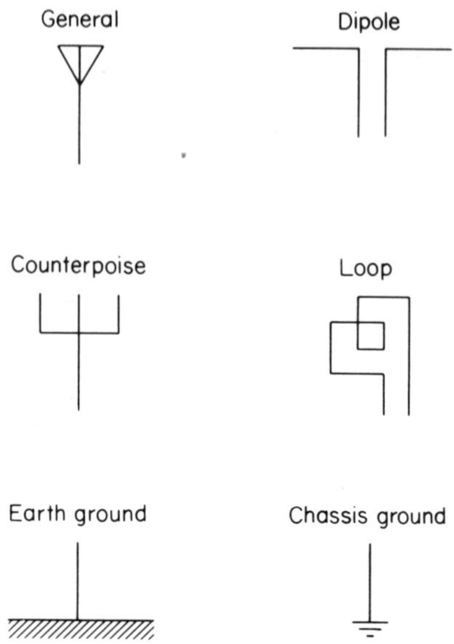


**FIGURE 2-47** Symbols for antennas and grounds

is obtained when the direction of the signal is at right angles to the loop. The letter A is sometimes used as the reference designation for antennas.

There are two basic ground symbols, *chassis* ground and *earth* ground. A chassis ground symbol can indicate connection to a metal chassis, a ground plane (metal strip around a circuit board), a common wire, or any common tie point (terminal post, binding post, etc.). A true earth ground symbol should be used to indicate connection to the earth (say to a water pipe or any conductor making physical contact with the earth).

### *2-8.2 ATTENUATORS AND PADS*

Figure 2-48 shows the symbols for attenuators and pads. Although there are many variations in attenuator circuits, there are only three basic types: *balanced*, *unbalanced*, and *unidirectional*. Attenuators and pads are resistive networks designed to perform either or both of two functions: to match dissimilar impedances, and/or to diminish signal strength. The terms "attenuator" and "pad" are used interchangeably.

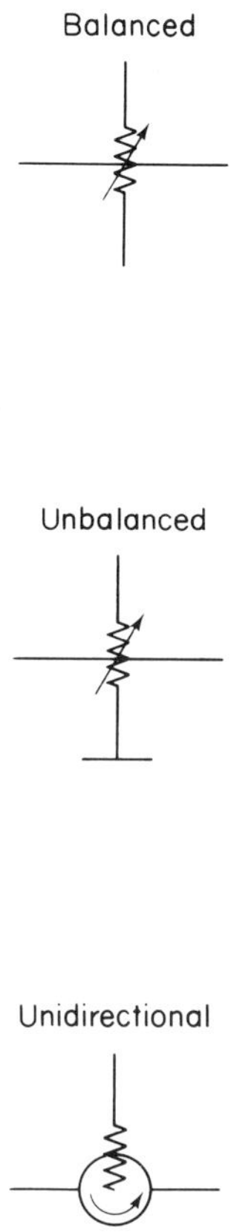


**FIGURE 2-48** Symbols for attenuators and pads

### *2-8.3 BATTERIES*

Figure 2-49 shows the symbols for batteries. Note that the polarity symbols (+ and -) may or may not be used. Also, the number of cells shown on the schematic does not necessarily indicate the true number of cells in the battery. The schematic symbols also do not indicate whether the battery is a dry cell, alkaline cell, lead-acid or nickel-cadmium type.

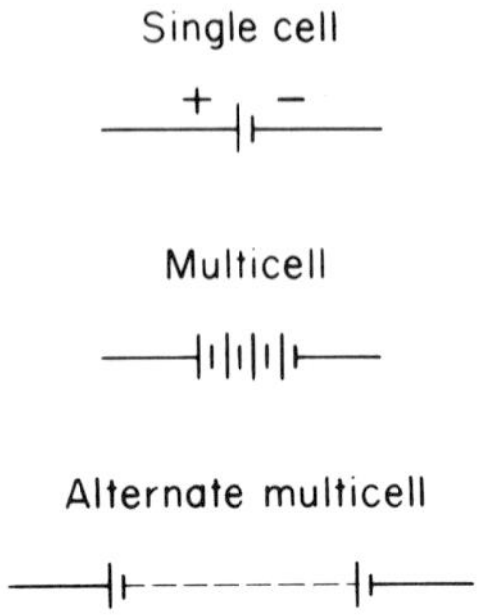


**FIGURE 2-49** Symbols for batteries

### *2-8.4 PIEZOELECTRIC CRYSTAL*

Figure 2-50 shows the symbols for a piezoelectric crystal. Such crystals function to control the frequency of radio transmitters. When such crystals are placed between two plates and an electrical current is applied, a mechanical strain is produced in the crystal (the crystal expands or contracts). In turn, this expansion or contraction produces an electrical current at the plates. The self-generated current causes further mechanical strain, and the two actions (strain and self-generated current) are mutually self-sustaining. The crystal produces an alternating current (usually at a radio frequency).

The frequency of the alternating current depends on the natural characteristics of the crystal (thickness, cut, etc.). Generally, *quartz* crystal is used because of its ability to withstand heat, vibration, and mechanical strain. The schematic should show the operating frequency of the crystal, but will probably not show the maximum voltage

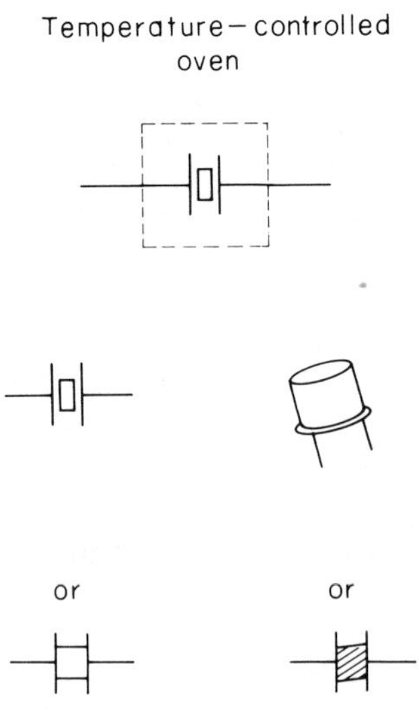


**FIGURE 2-50** Symbols for piezoelectric crystals

or any other characteristics. Sometimes, piezoelectric crystals are housed within thermostatically controlled "ovens" to maintain the crystal at a suitable temperature. Temperature changes produce frequency changes in the crystal. The letter Y is sometimes used as the reference designation for frequency-control crystals.

### *2-8.5 EARPHONES, HEADSETS, LOUDSPEAKERS, MICROPHONES, AND PICKUPS*

Earphones, headsets, and loudspeakers convert electrical signals into mechanical sound waves (audible to the human ear). Microphones and pickups convert mechanical sound waves into electrical signals. The symbols used for these devices are shown in Fig. 2-51, together with the basic operating principles for both devices.

In the case of earphones or headsets, an electrical current (corresponding in frequency and amplitude to the sound to be produced) is applied to an electromagnet (a coil wound over a magnetic core). The electrical current produces (or changes) a magnetic field around the core. The magnetic field, which varies in accordance with the electrical current in amplitude and frequency, causes the diaphragm to move back and forth (at a corresponding rate) and produce sound waves.

In a typical loudspeaker, the current to be reproduced is applied to a coil attached to the speaker cone. The coil is placed over a permanent magnet so that the coil and cone are attracted and repelled

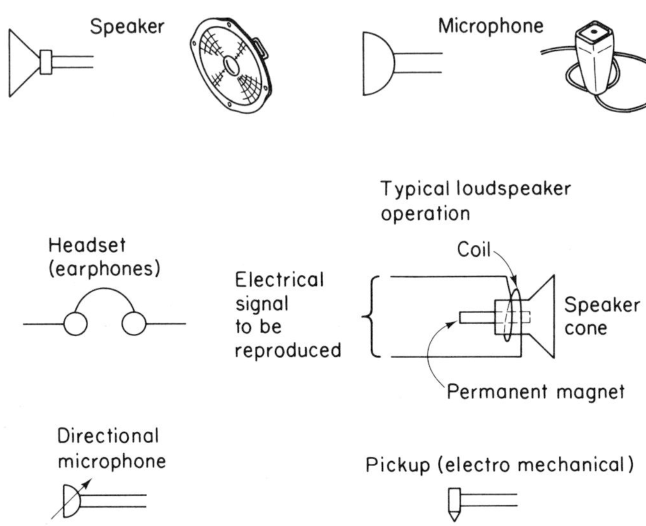


**FIGURE 2-51** Symbols for loudspeakers, headsets (earphones), microphones and pickups

by the magnet when the current is applied. This causes the cone to move back and forth, producing the sound waves.

There are many types of microphones and pickups (crystal, dynamic, ribbon, etc.), which we will not discuss in detail here. In all cases, the mechanical sound waves (or mechanical vibrations in in the case of a pickup) strike a diaphragm, causing it to move. This produces or changes an electrical current that varies in frequency and amplitude by a corresponding amount. The schematic symbols do not show the type of microphone or pickup.

The reference-designation letter for loudspeakers is LS. The letters HS are sometimes used as the reference designation for headsets and earphones. Special reference designations are used for pickups. On some schematics, the impedance of headsets and loudspeakers is given in ohms. This is not to be confused with the d-c resistance, which is usually not given on the schematic.

### *2-8.6 CIRCUIT BREAKERS AND FUSES*

Figure 2-52 shows the symbols for circuit breakers and fuses. As shown, circuit breakers can be *reset* by a pushbutton or handle. Circuit breakers can be *tripped* by thermal overload or magnetic overload. In either case, when the current through the circuit breaker exceeds the rating, the breaker is tripped and must be reset with the handle or pushbutton. The reference-designation letters for circuit breakers and fuses are CB and F, respectively.

### *2-8.7 LAMPS*

Figure 2-53 shows the symbols for lamps. Note that there is little standardization for lamp symbols, and the symbols are often interchanged. The reference-designation letter for lamps is I.

### *2-8.8 RELAYS*

Figure 2-54 shows the symbols for relays. As in the case of lamps, there is little standardization in relay symbols. However, the operating principles for all types are essentially the same. When current is applied to the coil portion of the relay (sometimes known as the relay *solenoid*), the coil becomes an *electromagnet*. That is, current passing through the coil sets up a very strong magnetic field around the coil. In the case of a relay, this field operates an armature, which, in turn, operates the relay contacts.

Although there are an infinite variety of relay contact arrangements, contacts are either normally open (NO) or normally closed

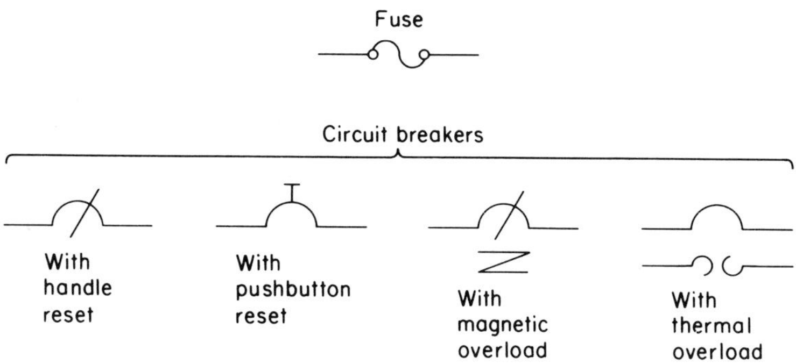


**FIGURE 2-52** Symbols for circuit breakers and fuses

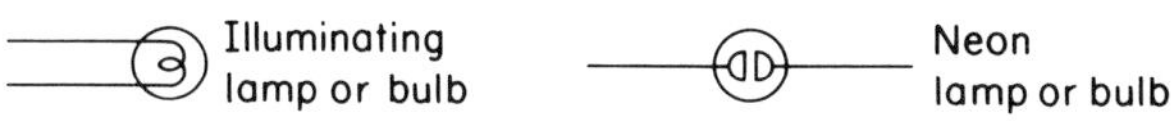


**FIGURE 2-53** Symbols for illuminating and neon lamps or bulbs

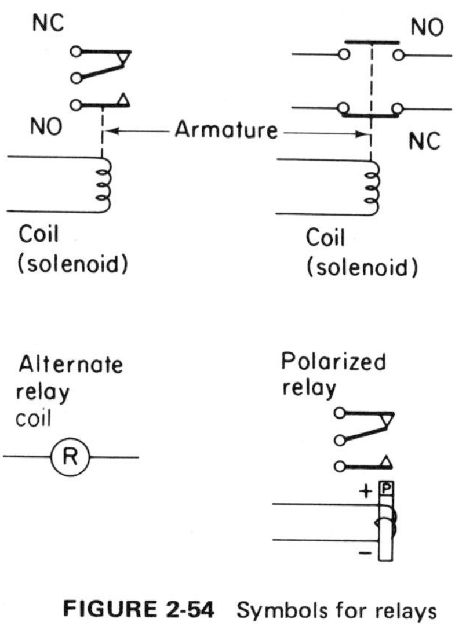


**FIGURE 2-54** Symbols for relays

(NC). An NO contact is closed when current is applied and opens when current is removed. An NC contact opens when current is applied and closes when the current is removed.

Relays are usually operated with direct current, although there are a-c relays. In most cases, the polarity of the direct current is of no consequence. However, there are some polarized type relays that can be operated only when the direct current is applied with a certain polarity, as shown.

The reference-designation letter for relays is K.

### *2-8.9 SWITCHES*

Figure 2-55 shows the symbols for switches. The most common types of switches found in electronics are the *toggle*, *slide*, *rotary*, and *pushbutton*. With toggle switches, each movable contact is called a *pole*. When only one movable contact is actuated, the switch is *single-pole*. A *double-pole* switch has two movable contacts that are actuated simultaneously. The positions in which the movable contacts may be placed to touch the fixed contacts are called *throws*. Figure 2-55 shows the symbol and physical construction of a *double-pole, double-throw* (DPDT) toggle switch, as well as the symbols for *single-pole, single-throw* (SPST); *double-pole, single-throw* (DPST); and *single-pole, double-throw* (SPDT) switches.

Figure 2-55 shows the two most common pushbutton switch symbols: NC (normally closed) push-to-close, and NO (normally open) push-to-open. Pushbutton switches are spring-loaded so that the contacts return to the position indicated by the symbol when released. For example, when you operate the pushbutton on a NO switch, the contacts close and complete the circuit. When you release the pushbutton, the spring opens the contacts, interrupting the circuit. The NC pushbutton switch operates in the opposite manner. Although only one set of contacts is shown in Fig. 2-55, a pushbuttton switch can have many sets of contacts, and the contacts can be mixed (one or more sets of NO and NC contacts on the same switch).

The contact combinations for slide switches are essentially the same as for toggle switches. Slide switches have become popular for use in electronic equipment because their contacts are self-cleaning. Each time the slide is moved, corrosion is wiped from the contact surfaces.

Figure 2-55 also shows a rotary switch where a movable contact is in the form of a rotating arm which successively makes contact with a set of fixed contacts arranged in a circle. The arm and fixed contacts generally are located on an insulated wafer called a *deck*. The decks may be *ganged* so that several arms may be rotated simul-

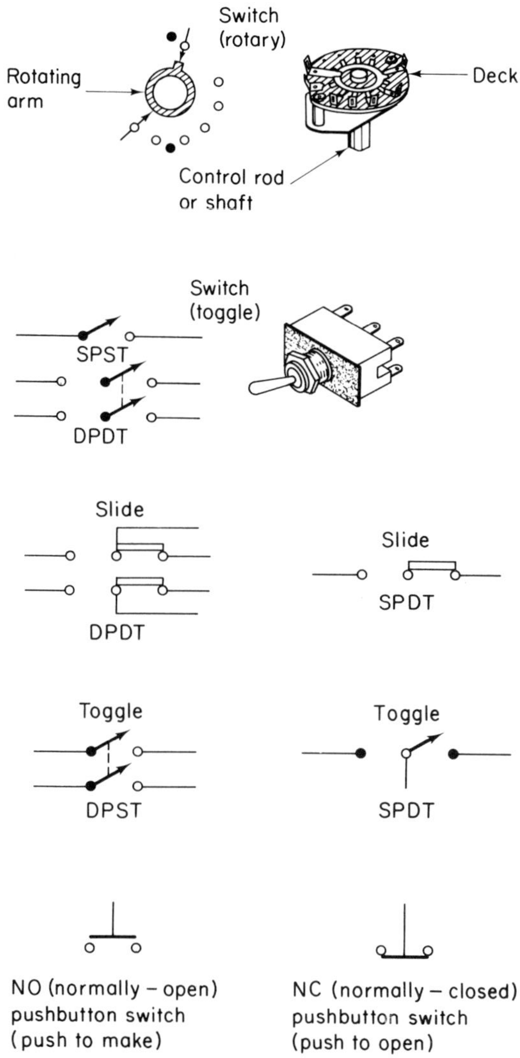


**FIGURE 2-55** Symbols for switches

taneously over their set of fixed contacts by means of a single control rod or shaft.

The reference-designation letter for switches is S.

### *2-8.10 CONNECTORS AND TERMINALS*

Figure 2-56 shows a variety of connector and terminal symbols used in schematics. Here again is an area where there is little standardization. For example, in a very strict sense, an *electrical*

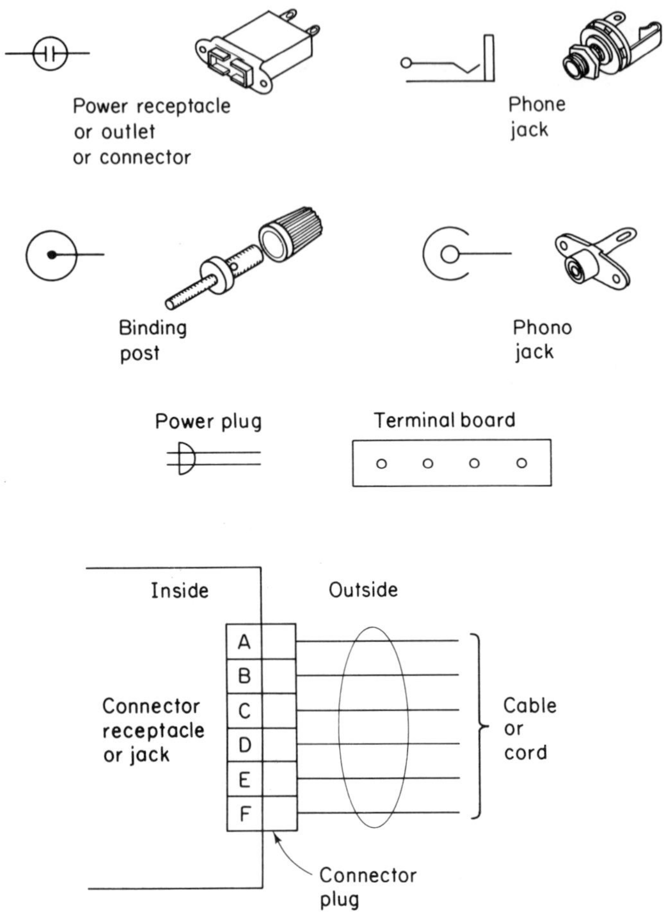


**FIGURE 2-56** Symbols for connectors and terminals

*connector* consists of a *plug* and a *jack.* The plug is movable (say at the end of a power cord or headset cord), and carries the reference-designation letter P. The matching jack is stationary (say on the front panel or chassis of a CB set), and carries the reference-designation letter J. However, the J and P designations are sometimes switched, or used for something else. (One CB manufacturer uses P to designate filters.) A jack is sometimes called a receptacle.

A *power receptacle* used for house power is sometimes called an *outlet* or a *connector.* However, the correct symbol is the two heavy bars within a circle, as shown. Such receptacles may carry the reference-designation letter J (or even P in a few cases).

Even the terminal board and binding post shown in Fig. 2-56 are sometimes called connectors. However, when there are several terminals on a common mounting, the entire part is called a terminal board or strip, and may possibly carry a reference-designation letter (most often TB).

Note that the phone jack may have more than one set of contacts. When a phone plug is inserted into a phone jack, the plug tip makes contact with the movable contact on the jack. The outer cylinder of the plug makes contact with the stationary contact of the jack. On phone jacks with multiple contacts, movement of the contact (caused by insertion of the plug) opens another set of contacts, breaking a circuit. In the example shown, an audio signal is normally applied through the jack contacts to a loudspeaker. When the plug is inserted, the loudspeaker contact is opened, and the audio signal is applied to the headset through the jack and plug.

### 2-8.11 METERS

Figure 2-57 shows some typical meter symbols. The basic meter symbol is a circle with wire leads or conductors as shown. On well-drawn schematics there will be some hint as to the meter's function and capabilities. For example, there may be the letters A (for ammeter) or V (for voltmeter) within the circle, or possibly a note such as "0-100 $\mu$A," indicating that the meter is capable of reading currents in the range 0 to 100 $\mu$A. The usual reference-designation letter for meters is M.

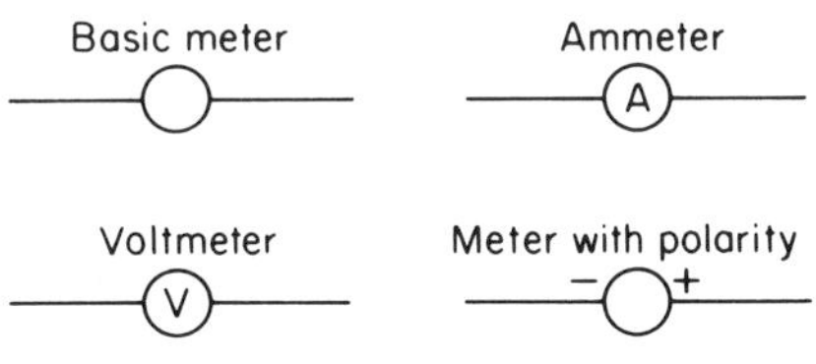


**FIGURE 2-57** Symbols for meters

### 2-8.12 MOTORS AND GENERATORS

Figure 2-58 shows the basic symbols for electrical motors and generators. Note that the symbol for motors is the same as for generators, so the symbols must be identified by the letters M and G.

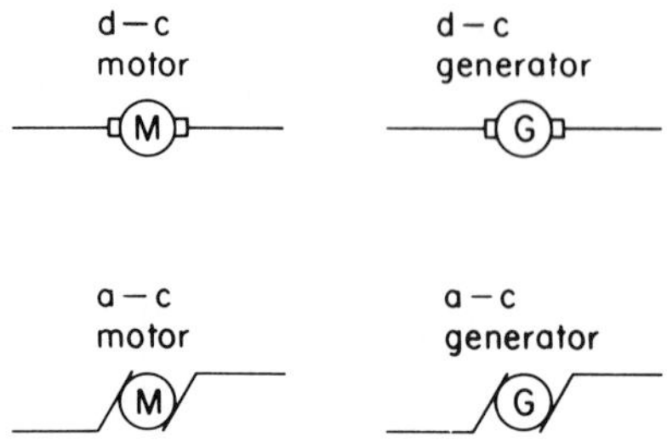


**FIGURE 2-58** Symbols for motors and generators

A basic d-c motor consists of an *armature* winding or coil and a *field* winding. The armature is the part that moves (rotates) when electrical current is applied. The field winding is usually stationary (usually a part of the motor housing. In an a-c motor, the rotating part is called the *rotor* and the stationary part is called the *stator.*

The symbols shown in Fig. 2-58 are of the armature (dc) and rotor (ac). It is common on electronic schematics to omit the field and stator windings. One exception is where motor control circuits are involved. This is because motor control is often obtained by controlling the field currents.

The function of a motor is opposite that of a generator. When current is applied to a motor, the armature or rotor rotates. Thus, electrical current or energy is converted to mechanical energy. With a generator, the armature or rotor is driven by mechanical energy (from a turbine, gasoline engine, windmill, etc.). This produces electrical current.

The reference-designation letter for motors is (sometimes) B.

## EXERCISES

**2.1.** When there is a dot at the junction of crossed wires, the wires are ______________.

**2.2.** What is the main purpose of an electrical conductor or wire shown on a schematic?

**2.3.** Referring to the short circuit shown in Fig. 2-2, will there be any current flowing through the lamp?

**2.4.** Referring to the circuit of Fig. 2-2 and assuming that there is a 2-mA current flow from the battery during normal operation, how much will the current flow from the battery decrease when there is a short circuit (Fig. 2-2c)?

**2.5.** If 3mA of current is flowing in the circuit of Fig. 2-2b and the battery is 3 V, what is the resistance of the lamp (ignoring the resistance of the conductors or wires)?

2.6. What is the main difference between a rheostat and a potentiometer?

2.7. Resistance values are given in ______________.

2.8. What Greek letter is used to indicate ohms?

2.9. Besides resistance values, what might a well-drawn schematic indicate about resistors?

2.10. When you find the letters GMV on a resistor symbol in a schematic, what do you know about the resistor value?

2.11. Name two functions of resistors shown on schematics.

2.12. What are the three most common circuit arrangements for resistors on schematics?

2.13. If you find two identical resistors in parallel on a schematic, what is the total resistance of these resistors?

2.14. If you find a 10-Ω resistor in parallel with a 50,000-Ω resistor on a schematic, what is the approximate total resistance?

2.15. Referring to Fig. 2-9 and assuming a battery voltage of 3 V, what will the output voltage be if the values of R1 and R2 are identical?

2.16. If the current flow in Fig. 2-10 is 1 mA, the battery is 1.5 V, the value of R1 is 100 Ω, and the value of R2 is 200 Ω, what is the value of R3?

2.17. Referring to Fig. 2-14a, assume that the resistance *R* is increased to 7000 Ω. In what way will this affect the battery voltage?

2.18. If you find a capacitor with polarity symbols on a schematic, it is probably an ______________ capacitor.

2.19. If you find a capacitor with an arrow through the symbol, the dielectric used is probably ______________.

2.20. If you find a capacitor without an arrow and without polarity symbols, the dielectric is probably ______________.

2.21. If you find a capacitor with an arrow on the curved plate, the capacitor is probably a ______________.

2.22. If a schematic shows capacitor value as 100 pF + 5%, -10%, what is the capacitance range?

2.23. A capacitor will pass ______________ current but block ______________ current.

2.24. If you find a 200-Ω resistor connected to a 20-μF capacitor on a schematic, what is the approximate time for the capacitor to charge through the resistor?

2.25. When lines are added to the basic coil (inductance) symbol, what does this indicate?

2.26. If the lines added to a coil symbol are dashed, the core is probably __________.

2.27. If the lines added to a coil symbol have an arrow, the coil is __________.

2.28. What is formed when a coil is combined with a capacitor?

2.29. What is the basic effect of a coil added to an a-c circuit?

2.30. What is the basic effect of a coil added to a d-c circuit?

2.31. When a coil is connected in series with a capacitor, signals at the resonant frequency will be rejected by the circuit. True or false?

2.32. When a coil is connected in parallel with a capacitor, signals at the resonant frequency will be rejected by the circuit. True or false?

2.33. Assume that you find a coil on a schematic and that the following information is given: 6 turns, wound on a ½-inch air core (single layer), 0.2-inch long. Using the data of Fig. 2-28, find the approximate inductance.

2.34. What is the primary purpose of a choke?

2.35. What is the primary purpose of a radio-frequency choke?

2.36. A choke will pass __________ current but oppose __________ current.

2.37. What is the primary purpose of a resonant circuit?

2.38. Assume that you find a parallel-resonant circuit on a schematic and that the following information is given: the capacitor is 360 pF and the inductance is 10 $\mu$H. Using the data of Fig. 2-33, find the resonant frequency.

2.39. Assume that you find a series-resonant circuit on a schematic and that the circuit values are the same as in Exercise 2.38. What is the resonant frequency?

2.40. In the schematic of Fig. 2-34, what is the purpose of CR1? Of C1?

2.41. In the schematic of Fig. 2-35, what is the purpose of C1, what happens to signals at the resonant frequency, and what happens to signals not at the resonant frequency?

2.42. If you find a resistor in parallel with a parallel-resonant *LC* circuit, what is the probable function of that resistor?

2.43. What is the approximate frequency range of AF signals?

2.44. What is the approximate frequency range of RF signals?

2.45. In the schematic of Fig. 2-43, what voltage would you expect to measure across L1, and from the open end of L1 to the open end of L2?

2.46. What type of connections are shown by the schematic of Fig. 2-43 for L1-L2-L3 and R1-R2-R3?

2.47. What other type of three-phase connection, not shown on Fig. 2-43, is in common use?

2.48. Alternating currents or signals can be stepped up or down by any of the transformers shown in Fig. 2-44. True or false?

2.49. Which of the transformers shown in Fig. 2-44 is best suited to step up or step down direct currents?

2.50. In a slug-tuned transformer, when the slug is adjusted so that the slug moves further into the winding, is the inductance increased or decreased?

2.51. What is the main function of the IF transformers shown in Fig. 2-45?

2.52. If a transformer has 10 turns on the primary and 100 turns on the secondary, and 8 V is applied to the primary, what voltage would you expect to measure on the secondary? (Refer to Fig. 2-46.)

2.53. If a schematic shows a transformer with 7 turns on the primary and 2 turns on the secondary (such as the power transformer in Fig. 2-44), and 7 V is applied to the primary, what voltage would you expect to measure on the secondary? (Refer to Fig. 2-46.)

2.54. If the polarity symbols are not shown on the schematic of a battery, how can you tell which terminal is positive? (Refer to Fig. 2-49.)

2.55. What is the primary function of the crystals shown in Fig. 2-50?

2.56. Earphones, headsets, and loudspeakers convert ____________ into ____________.

2.57. Microphones and pickups convert ____________ into ____________.

2.58. If you find a resistance value given for a headset or loudspeaker on a schematic, the value is likely to be the d-c resistance. True or false?

2.59. When the coil or solenoid of a relay is energized, the NC contacts ____________ and the NO contacts ____________.

2.60. What kind of current would you expect to find passing through the coil of a relay?

2.61. What kind of current would you expect to find passing through the contacts of a relay?

2.62. When you push and release an NC pushbutton switch, the contacts first close (when you push) and then open (when you release). True or false?

2.63. If you find on a schematic, a meter symbol containing the letters mV, what would you expect the meter to measure or indicate?

# 3

# Some Active-Device Schematic Symbols

In this chapter, we concentrate on *active-device* symbols. The symbols (and their related parts) discussed in Chapter 2 are essentially passive devices. That is, they supplement or aid the functions performed by active devices in electronic circuits. Active devices are those that perform the major circuit functions (amplification, oscillation, signal mixing, waveform generation, etc.).

At one time, *vacuum tubes* were the only active devices found in electronic circuits. (Vacuum tubes are also called *electron tubes* or simply *tubes*.) No matter what they are called, vacuum tubes have generally been replaced by solid-state devices. For that reason, we concentrate on solid-state devices in this book. However, since there are still many schematic diagrams that show vacuum tubes, and since it will not hurt any of you to understand the basics of vacuum tubes, we cover tubes briefly, starting now.

## 3-1 VACUUM TUBES (ELECTRON TUBES)

Figure 3-1 shows the symbols used for a variety of tubes. Not every possible vacuum-tube symbol is given, but the majority are here. The reference designation letter for vacuum tubes is V.

### *3-1.1 DIODES (VACUUM TUBE)*

The simplest tube is the *diode* with a *filament* (also known as a *heater*) and a *plate*. Vacuum tubes are similar to incandescent lamps.

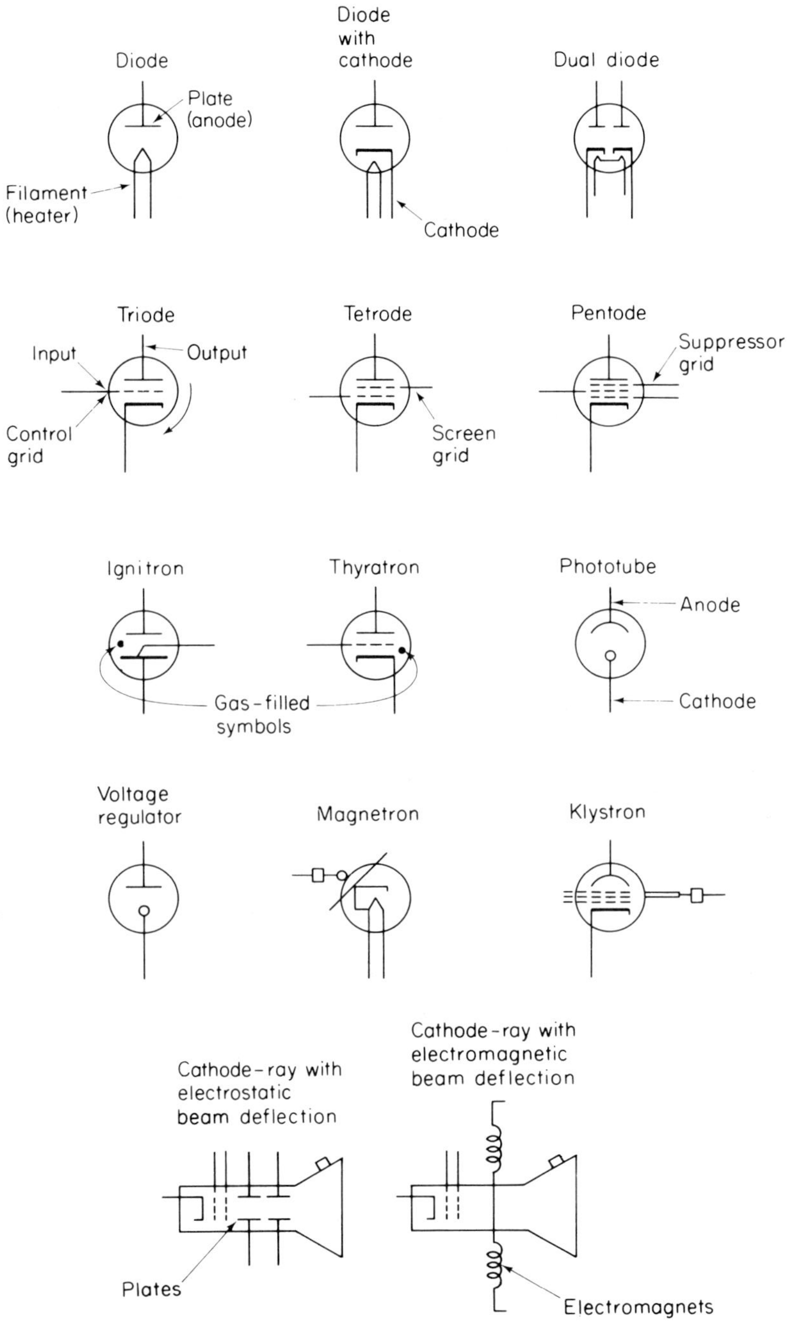


**FIGURE 3-1** Symbols for vacuum tubes

When current is passed through the filament, the filament heats and glows and gives off electrons. If the plate (sometimes called the *anode* when the tube is a diode) is made more positive than the filament, the electrons are attracted to the plate, and current flows from filament to plate. If a direct current is used, with the plate positive in respect to the filament, current continues to flow as long as the filament is heated and the plate voltage is applied.

If an alternating current is applied to a diode (between plate and filament), current flows only on the half-cycle (or half-wave) when the plate is positive in respect to the filament. Thus, the vacuum-tube diode acts as a rectifier and functions to convert ac into dc, in a manner similar to the solid-state diode discussed in Chapter 1.

Some vacuum-tube diodes have a *cathode*, in addition to the filament and plate (anode), as shown in Fig. 3-1. In such tubes, the filament (or heater) functions to heat the cathode, which, in turn, gives off the electrons. Current flows from cathode to plate when the plate is made positive in relation to the cathode. Note that the *dual diode* shown is essentially two diodes in one envelope but with only one filament.

### *3-1.2 TRIODES*

The *triode* tube is used when the flow of current (ac or dc) must be controlled. The triode is similar to the diode, except that a *control grid* is placed between the plate and cathode, as shown in Fig. 3-1. The control grid functions essentially as a gate to control electron flow. When the grid is made positive with respect to the cathode, more electrons flow, and the current flow between cathode and plate is increased. The opposite occurs when the grid is made negative with respect to the cathode. If the grid is made sufficiently negative, the flow of electrons (and current) is cut off completely. The voltages between grid and cathode are very small in relation to the voltage between cathode and plate. Thus, it is possible to control a large voltage with a small voltage and provide *amplification.*

In reading a schematic, the important point to remember about a vacuum-tube amplifier is that the *grid is the input* and the *plate is the output.* In some circuits, known as *cathode followers*, the output is taken from the cathode. Cathode followers do not provide any voltage amplification.

### *3-1.3 TETRODES AND PENTODES*

One problem with a triode is that the *interelectrode capacitance* (capacitance between the cathode, grid, and plate) is large. This can cause *feedback* between the elements (particularly plate to grid) and can restrict amplification (as well as cause undesired oscillation). The problem is solved by adding more grids between the grid and plate. These grids are made positive with respect to the cathode but not as positive as the plate. The additional grids cut down the interelectrode capacitance and also increase current flow from cathode to plate.

A *tetrode* is formed when a *screen grid* is added to the basic triode, as shown in Fig. 3-1. A *pentode* is formed when a *suppressor grid* and a screen grid is added to the triode.

### *3-1.4 BIAS VOLTAGE (VACUUM TUBES)*

The control grid of most triodes, tetrodes, and pentodes is provided with a *bias voltage*, which sets the amount of control available. The bias is a d-c voltage between grid and cathode, with the grid (usually) negative. The input signal applied to the grid must overcome the bias to produce an output at the plate (or cathode).

In some cases, the grid is made negative enough by the bias voltage so that the tube is never fully cut off, or fully turned on, by the incoming signal at the grid. This is known as *class A* operation of a tube. In other cases, the tube is biased just to cutoff, so that the positive swings of the incoming signal cause current flow, but never beyond the *saturation* point of the tube. This is *class B* operation. (Saturation is where the cathode-plate current flow cannot be increased, even with increased positive grid voltage.) In *class C* operation, the grid is biased well below cutoff so that only the positive peaks of the incoming grid signal cause current flow at the plate. *Class D* operation is where the incoming signal drives the tube into full saturation, and full cutoff, producing *square waves* or *pulsed waves* at the plate or cathode output. Class D is not commonly used in vacuum-tube circuit (or solid-state circuits, for that matter).

### *3-1.5 THYRATRONS AND IGNITRONS*

The thyratron and ignitron shown in Fig. 3-1 are *gas-filled* tubes used primarily to control motors and machinery where large currents are involved. These tubes have generally been replaced by solid-state *silicon-controlled rectifiers* (SCRs) except in very heavy duty power equipment.

### *3-1.6 PHOTOTUBES*

The phototube shown in Fig. 3-1 is essentially a diode with current flowing from cathode to anode. However, when light is shined on the cathode, the current flow is increased. Current flow decreases, or is cut off completely, when the light is removed. Phototubes have generally been replaced by solid-state photocells.

### 3-1.7 VOLTAGE REGULATORS

The voltage-regulator tube shown in Fig. 3-1 is a gas-filled diode. When the voltage across the cathode and plate (or anode) reaches a certain value, the gas is ionized and current flows. If the voltage exceeds this value, more current flows, reducing the voltage. Thus, the tube functions to regulate voltage and is usually used at the output of a power supply circuit. Voltage regulator tubes have generally been replaced by solid-state *zener* diodes and other regulator diodes.

### 3-1.8 MAGNETRONS AND KLYSTRONS

The magnetron and klystron shown in Fig. 3-1 are used only in *microwave equipment* (generally at frequencies above 1000 MHz). The magnetron tube is used where high power is required, such as the transmitter of a radar set or the oscillator of a microwave oven. The klystron is used for low-power applications, such as the oscillator of a radar receiver. Because of its highly specialized nature, microwave equipment is not discussed in this book.

### 3-1.9 CATHODE RAY TUBES (CRTs)

The cathode ray tubes shown in Fig. 3-1 are used in solid-state equipment, such as the picture tube or screen for TV sets, computer terminals, oscilloscopes, and radar sets. For that reason we will cover CRTs (also known as KRTs on some schematics) in more detail.

Figure 3-1 shows two CRT schematic symbols. Operation of the two types of CRTs are essentially the same. The major difference between the two is in beam deflection. With the *electrostatic beam deflection* CRT, the tube contains plates which deflect the beam. With the *electromagnetic beam deflection* CRT, the beam is deflected by *electromagnets* or *coils* placed around the tube. Electrostatic deflection is used in oscilloscopes and certain other applications. Electromagnetic deflection is used in TV sets, computer terminals, and radar screens.

Figure 3-2 shows the internal construction of a typical electrostatic deflection CRT. Electromagnetic CRT construction is essentially the same, except that the deflection plates are omitted and replaced by external coils. As with other vacuum tubes, the filament heats the cathode, which, in turn, emits electrons and causes current flow. The amount of current is set by the control grid. The two anodes are made positive in relation to the cathode, so the anodes accelerate the

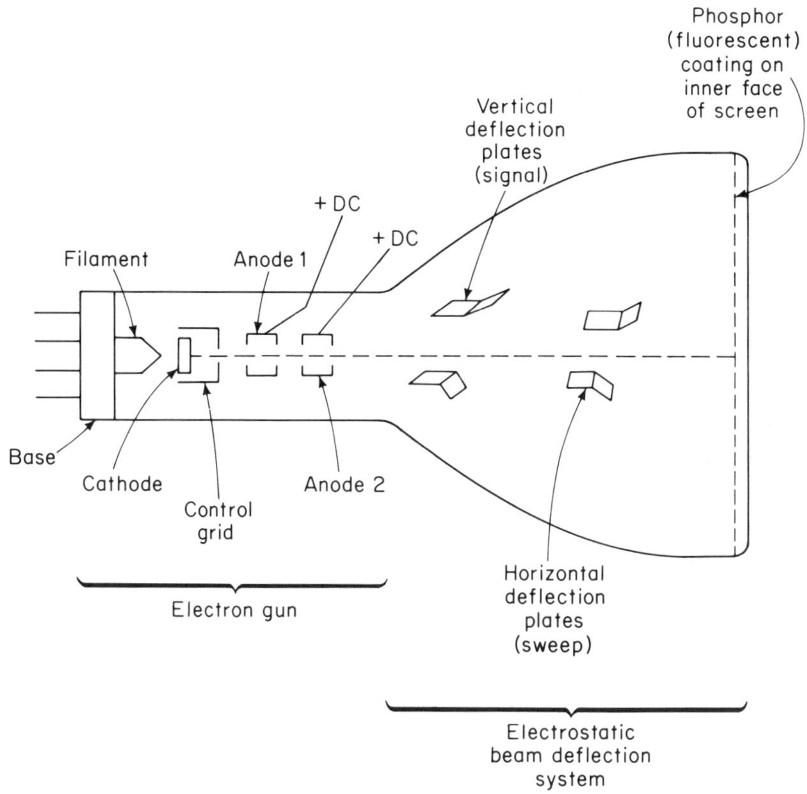


**FIGURE 3-2** Internal construction of cathode ray tube (CRT)

electrons and form them into a beam. The beam intensity is set by the control grid, which usually consists of a cylinder with a tiny circular opening to keep the electron beam small.

The beam is focused into a sharp pinpoint by controlling the voltage on the first anode (sometimes called the *focus* anode). The two anodes can be compared to a glass lens system, such as used in movie projectors, because the anodes focus the beam to a pinpoint at the face of the tube. A high voltage is applied to the second anode so that the electron beam speeds up for increased intensity and visibility when it strikes the tube face. The beam-forming section of the CRT is known as an *electron gun*.

The inside of the CRT face is coated with phosphor so that, when electrons strike this coating, the coating will *fluoresce* and emit light. When signals are applied to the deflection plates, the beam is swept across the CRT face at a rapid rate, but the light persists for just the right length of time to leave a complete trace of the signal applied to the deflection plates.

Electrostatic beam deflection is accomplished through the two pairs of parallel plates, placed between the second anode and the CRT face. (In electromagnetic beam deflection, the deflection coils are outside the tube but positioned between the second anode and CRT face.) When a voltage is applied to the horizontal plates, the negative potential on one of the horizontal plates repels the electron beam, whereas the positive voltage on the other plate attracts the beam. The horizontal voltage is generally of the *sweep* or *sawtooth type*, where the voltage gradually rises from zero, pulling the beam toward the positive horizontal plate. Thus, the electron beam is made to scan horizontally across the CRT face. Any voltage applied to the vertical deflection plates causes the beam to move vertically.

Figure 3-3 shows how an oscilloscope traces out an alternating current wave applied to the vertical input of the oscilloscope, when a horizontal sawtooth voltage (generated by a circuit within the oscilloscope) is applied to the horizontal plates. Figure 3-3a shows the alternating current (sine wave) applied to the vertical plates. Figure 3-3b shows the resultant trace on the CRT screen if the horizontal sawtooth sweep voltage is turned off. Under these conditions, the rising positive alternation of the input signal causes the electron beam to move upward. When the negative alternation of the input arrives at the vertical deflection plates, the electron beam is pulled

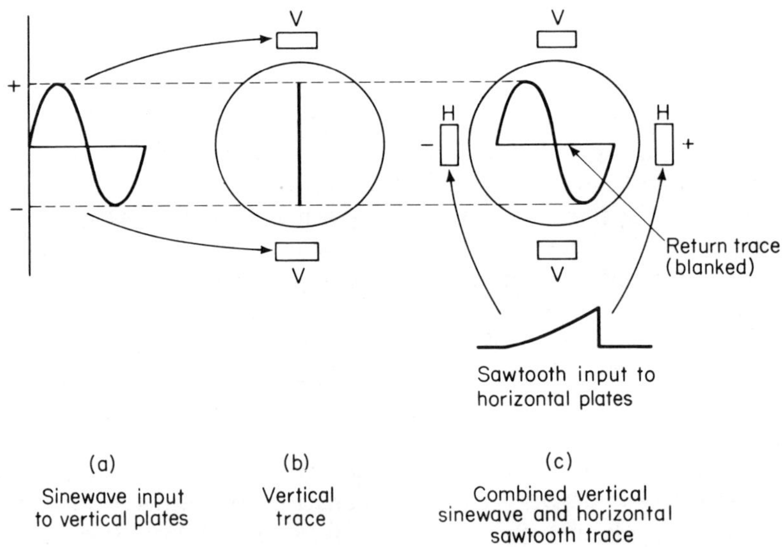


**FIGURE 3-3** How sinewave (vertical) and sawtooth (horizontal) signals are combined to produce an oscilloscope trace

downward from the center. The rapid rise and fall of the signal alternation causes the electron beam to move up and down the face of the CRT very rapidly, leaving a vertical line trace.

Figure 3-3c shows the results when the horizontal sawtooth sweep is on, together with the vertical input signal. The rising potential of the sawtooth voltage causes the right horizontal plate to become positive and the left plate to become negative. This, in turn, causes the beam to move from left to right (beam repelled by left negative plate and attracted by right positive plate). If one sawtooth occurs for each cycle of the alternating current applied to the vertical plates, the beam is pulled across the CRT face, once for each cycle. Thus, one complete alternation of the input is traced out as shown.

## 3-2 DIODES

Figure 3-4 shows the symbols for commonly used diodes. The reference-designation letter for diodes is CR, but may appear as D on some schematics. Note that the symbol for the basic diode (discussed in Chapter 1) is the same whether the diode is used as a *rectifier* in a power supply or a *detector* in a receiver circuit. The symbol remains the same no matter what the physical appearance or size. For example, a typical metallic rectifier diode (often called a *power diode*) is made of copper-oxide disks or cells. Selenium cells are also used for power diodes. The symbols for these power diodes are the same as for crystal junction *signal diodes* (also called *small-signal diodes*) used in a variety of circuits that do not require heavy current capability.

The symbols for *tunnel diodes* and *zener diodes* are different from the basic diode since they operate in a different manner. The tunnel diode is a *negative-resistance* device. That is, when the voltage

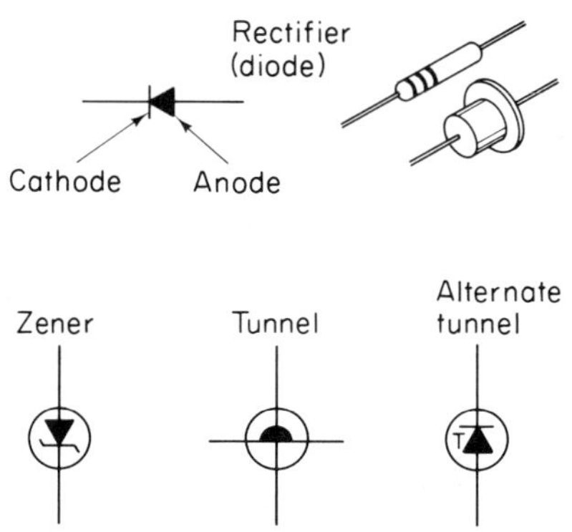


**FIGURE 3-4** Symbols for diodes

across the diode is increased, the current decreases (and vice versa), which is the reverse of a basic diode. Tunnel diodes can be used in oscillator circuits. However, tunnel diodes are not in general use at this time.

### *3-2.1 ZENER DIODES*

The zener diode shown in Fig. 3-4 functions primarily as a regulator, particularly in power supply circuits. A zener is similar to the basic diode, except for an "avalanche" condition. Like a basic diode, the zener passes current in the forward condition (anode positive) and prevents current flow in the reverse condition (anode negative). However, when sufficient reverse voltage is applied to a zener diode, the "avalanche" or "breakdown" condition occurs and the heavy reverse current flows. (Actually, any diode will operate similarly, but a zener is designed for this specific purpose.) With a zener, if the reverse voltage is increased above the breakdown point, additional reverse current flows and the voltage is dropped through a resistor to a fixed level (known as the *zener voltage* or *zener level*).

### *3-2.2 DIODE COLOR CODES*

In solid-state component designations, the number-letter combination 1N indicates a diode (2N indicates a two-junction or bipolar transistor, as discussed in Sec. 3-3). Some diodes are color-coded to identify the complete type. Figure 3-5 shows the various diode color-coding arrangements.

For the earlier type of diode which has two center digits (such as 1N34A), three color bands are used, as shown in Fig. 3-5a. The color codes are grouped toward the cathode end and, for the diode shown, are read from left to right. The first two color-code bands indicate the significant figures; the third color band identifies the letter.

For diodes having three central digits (such as the 1N100A), four color bands are used, as shown in Fig. 3-5b. The first three bands from left to right identify the first three significant digits, and the last band provides the letter designation.

When the diode has four central digits (such as the 1N3666A), five color bands are used, as shown in Fig. 3-5c. The first four bands from left to right identify the first four significant digits, and the last band provides the letter designation.

Figure 3-5d shows an example of diode color coding where there are three digits followed by a letter (1N445B).

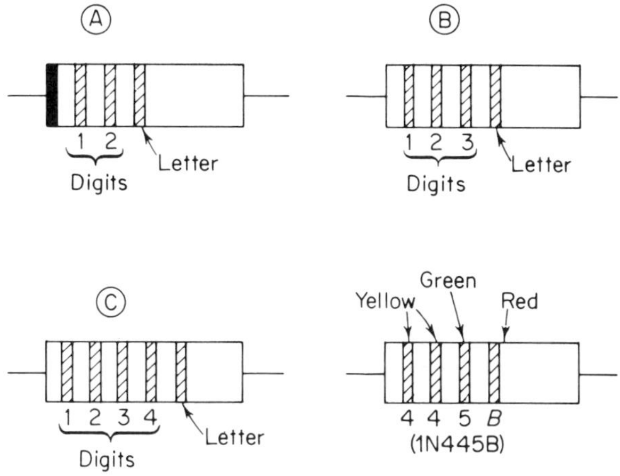


| Color | Digit | Suffix letter |
|---|---|---|
| Black | 0 | --- |
| Brown | 1 | *A* |
| Red | 2 | *B* |
| Orange | 3 | *C* |
| Yellow | 4 | *D* |
| Green | 5 | *E* |
| Blue | 6 | *F* |
| Violet | 7 | *G* |
| Gray | 8 | *H* |
| White | 9 | *J* |

**FIGURE 3-5** Diode color codes

## 3-3 TRANSISTORS (TWO-JUNCTION OR BIPOLAR)

Figure 3-6 shows the symbols and the block representations for two-junction or biploar transistors. Most present-day schematics use the symbols, but the block representation is found on some older schematics. The block is used here to show the basic construction of transistors. Like the basic diode (Sec. 3-2), with one junction, a transistor can be used to prevent (or limit) the flow of current in one direction. The prime use for a transistor as an active device, however, is to *control the amount* of current in a circuit, in a manner similar to that of a vacuum tube (Sec. 3-1). This is done by adding a second junction to the basic diode. For this reason, such transistors are called *two-junction* or *bipolar.*

As shown in Fig. 3-6, there are two possible arrangements for the two junctions: NPN, where a positive semiconductor material (usually silicon or germanium crystal) is placed between two negative

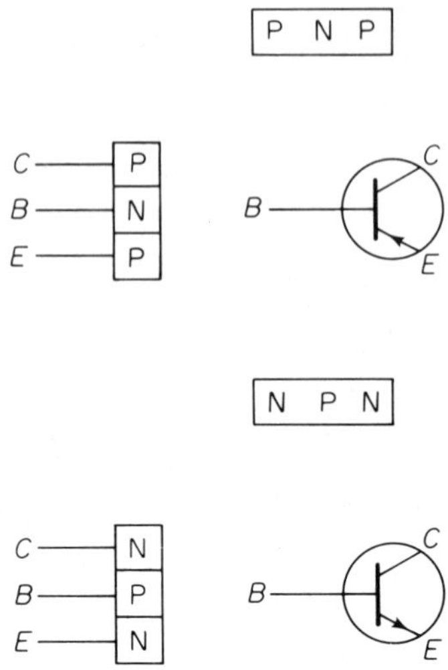


**FIGURE 3-6** Symbols and block representations for two-junction or bipolar transistors

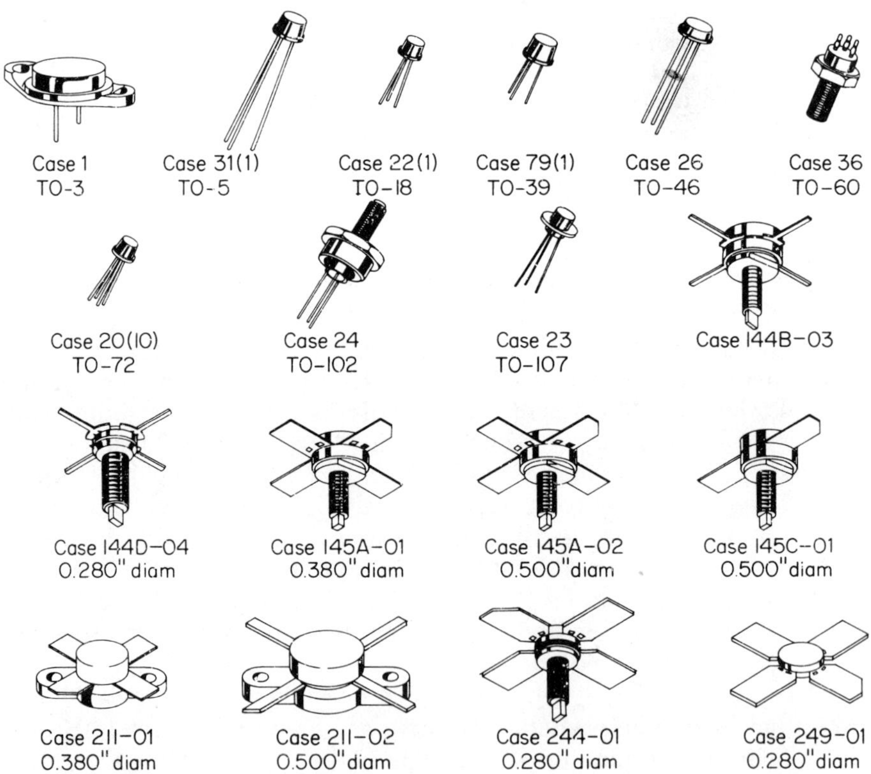


**FIGURE 3-7** Case styles for typical two-junction transistors (courtesy Motorola)

semiconductor materials (also silicon or germanium), and PNP, where a negative material is placed between two positive materials. With either arrangement, the transistor will have three elements: the *emitter*, which emits electrons; the *collector*, which collects electrons, and a *base*, which controls the flow of electrons. In simple terms, the collector is equivalent to a vacuum-tube plate, the emitter is equivalent to the cathode, and the base is equivalent to a grid.

The only difference between an NPN and a PNP transistor symbol is the arrowhead on the emitter. The arrowhead points away from the base for NPN and toward the base for PNP. Usually, the letters C (collector), E (emitter), and B (base) shown in Fig. 3-6 are omitted from schematics since the elements can be identified from the symbol. The appropriate letter portion of a transistor reference designation is Q. Transistors should be identified on schematics as Q1, Q2, Q103, and so on. However, you will find schematics where transistors are identified by the letters T, TR, or even V, although these letter designations are obsolete.

Figure 3-7 shows the physical appearance of some typical transistors. Note that the case styles shown represent only a small fraction of available transistors.

### *3-3.1 BASIC TRANSISTOR OPERATION*

Figure 3-8 shows how a two-junction or bipolar transistor operates in a basic circuit. The emitter-base junction will pass current easily because the junction is forward-biased (anode connected to positive, cathode connected to negative). The collector-base junction will not pass current (except for a small leakage current which you need not worry about in reading schematics) since the junction is reverse-biased (anode connected to negative, cathode connected to positive). The term *back bias* is sometimes used in place of reverse bias.

Note that the polarities of bias voltages for NPN differ from those of PNP, but the net results are the same. For example, in Fig. 3-8a (NPN), the emitter is negative with respect to base to produce forward bias. In Fig. 3-8b (PNP), the emitter is positive with respect to base to produce the same forward bias. Similarly, the collector has a reverse bias for both NPN and PNP, even though the polarities are reversed.

During normal operation of a transistor, using the bias arrangements shown in Fig. 3-8, current will flow between emitter and base, and between emitter and collector, but not between collector and base. Most of the current flows between emitter and collector because of the large voltage difference existing between these elements (the sum of the emitter-base voltage and the collector-base voltage).

Since very little voltage (approximately 0.2 V for germanium and 0.5 V for silicon) is required to produce a large current flow in the

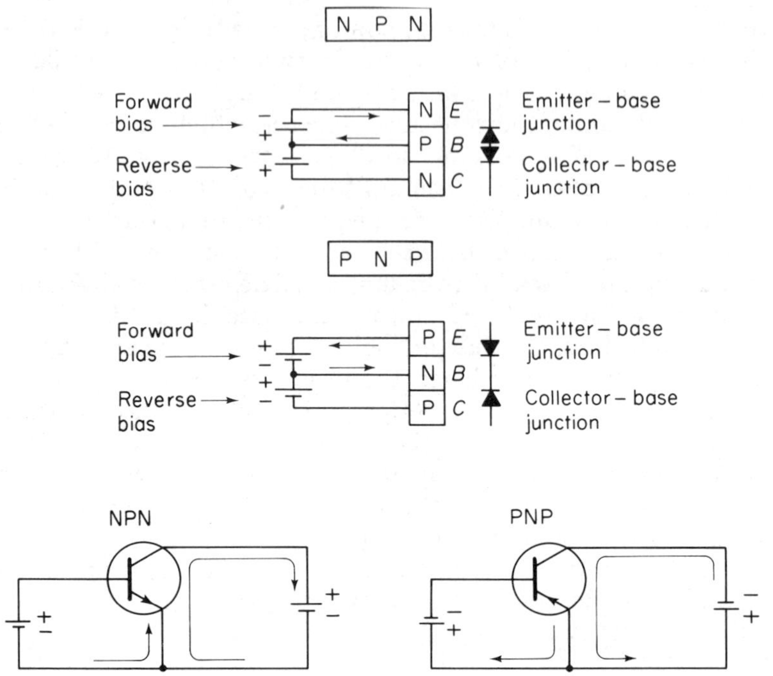


**FIGURE 3-8** Basic two-junction transistor bias circuits

emitter, input power to a transistor is low. Most of the emitter current flows in the collector circuit, where the voltage is made much larger. As a result, a relatively large amount of power can be controlled in an external load (connected in series with the collector circuit) by a small amount of power in the emitter circuit. The power gain of a transistor (the ratio of power output to power input) can be 40,000, or higher in some applications.

### *3-3.2 BASIC TRANSISTOR CIRCUIT CONNECTIONS*

There are three basic two-junction transistors circuits:

1. Common base (or grounded base).
2. Common emitter (or grounded emitter).
3. Common collector (or grounded collector).

Each of the three basic circuits is identified by the transistor element that is *common to both the input and output.* Although this element is frequently called the *grounded* element, the common element is not necessarily connected to ground. Each of the three circuits is discussed in the following paragraphs.

### *3-3.3 COMMON-EMITTER AMPLIFIER*

The common-emitter circuit of Fig. 3-9 is the most widely used two-junction transistor amplifier configuration. Note that Fig. 3-9 shows the power supply connections for both NPN and PNP transistors in *theoretical* circuits. In most practical circuits found on

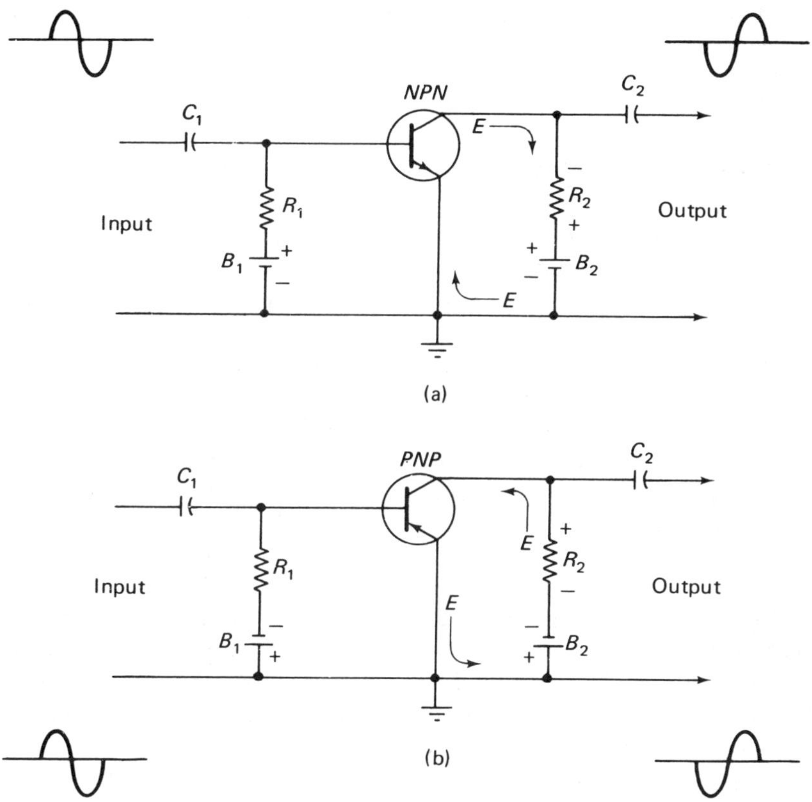


Typical Characteristics

Input resistance 500–1500 Ω
Output resistance 30–50 kΩ
Voltage gain 300–1000
Current gain 20–50
Power gain 25–40 dB

**FIGURE 3-9** Basic common-emitter amplifier stage

schematics, a single power supply is used. In the circuit of Fig. 3-9, the emitter is common to both the input and output circuits. Battery B1 forward-biases the emitter-base junction. Battery B2 reverse-biases the collector-base junction. In a practical circuit, the voltage of B2 is generally larger than that of B1, and both voltages are taken from the same source.

The input signal is applied between the base and emitter; the output signal appears between the emitter and collector. This provides a very low input impedance and a very high output impedance. The output signal is out of phase with the input. That is, a positive alternation at the input produces a negative alternation at the output, and vice versa.

Common-emitter circuits are the most often used since there is *current gain* and *voltage gain*, thus resulting in a large *power* gain. For example, if the output impedance is 10 times higher than the input impedance, there will be a power gain of 10, even if the base-emitter (input) and emitter-collector (output) currents are equal (no current gain). Since it is possible to control a large output current with a small input current in a practical common-emitter circuit (as it is possible to control a large voltage with a small voltage in vacuum tubes, Sec. 3-1.2), there is also current gain.

Capacitor C1 (called a *coupling capacitor* because it provides a signal path or coupling between the input and the transistor) is used to block any dc in the input signal. For example, the input might come from the collector of another transistor, which is connected directly to a voltage. The input signal voltage is developed across resistor R1, which also serves to provide a closed circuit for current flow in the emitter-base circuit. This is essential since a two-junction transistor is a *current-operated* device.

The flow of collector current produces a voltage drop, with the indicated polarity, across R2. The ac part of this voltage is the output signal of the amplifier. Capacitor C2 is another coupling capacitor, which blocks out the direct current produced by battery B2, but passes the signal (alternating current).

The phase relationship between the input and output signals is shown by the waves appearing in Fig. 3-9. If the input voltage increases the forward bias of the base-emitter junction (positive alternation), the total emitter current is increased, producing an increase in collector current in a like degree. This increase in collector current, in turn, produces a larger drop across R2. In the NPN circuit, the input voltage must be positive if it is to aid the forward bias of the emitter-base junction. In the PNP circuit, a similar action takes place but with reverse polarities. In all cases where the common emitter is used, as the input voltage rises, the output voltage falls, and vice versa. Thus, there is a *phase reversal* between input and output signals (the input and output signals are said to be *180° out of phase*).

**Practical Common-Emitter Circuit.** In a practical circuit, both junctions are biased from a single power source. A typical method of supplying bias to a common-emitter amplifier stage from a single supply is shown in Fig. 3-10, which is the schematic of an NPN transistor amplifier. Note that the power supply is labeled $+V_{CC}$ and $-V_{CC}$. The power supply or voltage source could be labeled B+ and B−, V+ or V−, or simply + and − on other schematics. On still other schematics the actual voltage of the supply might be given. Quite often, only one line of the supply is labeled, with the other line implied. For example, if the $-V_{CC}$ and its ground symbol were omitted from Fig. 3-10, all the ground symbols would be considered as at $-V_{CC}$, since the $+V_{CC}$ line is shown. Keep in mind that the ground symbols shown in Fig. 3-10 are common tie points (probably a metal chassis or common wire) and do not necessarily represent true earth ground (Sec. 2-8.1).

Since the emitter-base junction must be forward-biased and the collector-base junction reverse-biased, in the NPN transistor this means that the base must be negative *relative* to the collector and positive *relative* to the emitter. Thus, the collector is at the most positive potential (or voltage) of the circuit, and the emitter is at

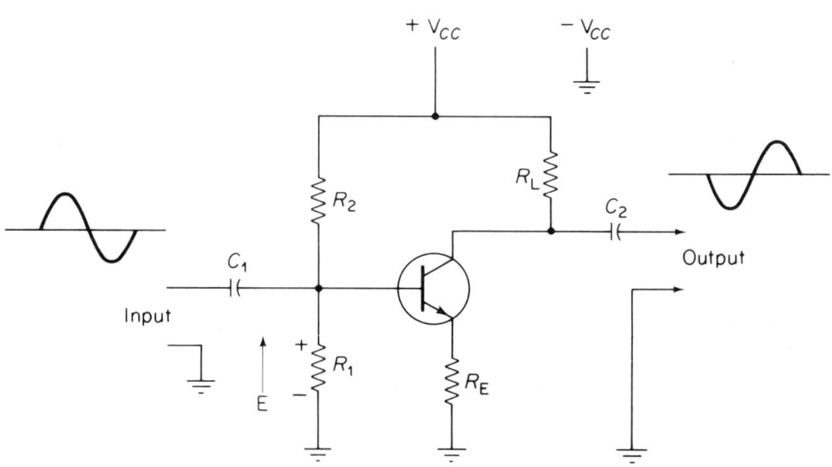


**FIGURE 3-10** Practical common-emitter amplifier operated from a single power supply

the most negative potential (or voltage). (As you have probably surmised, the terms *potential* and *voltage* are interchangeable in electronics.)

Note that when we say that the base must be less positive than the collector and more positive than the emitter, we are talking of *relative voltages.* For example, if the base voltage is +2 V and the emitter voltage is +1.5 V, the emitter is more negative than the base, even though both voltages are positive. Similarly, the emitter is more negative than the base if the emitter voltage is −2 V and the base voltage is −1.5 V.

Since the collector is connected to the positive side or line of $V_{CC}$, it is at the most positive potential. Similarly, since the emitter is connected to the negative side of $V_{CC}$, it is at the most negative potential. Resistors R1 and R2 form a voltage divider across $V_{CC}$. The flow of current through this divider produces a voltage drop across R1 with the indicated polarity. The remainder of the voltage is dropped across R2. For example, if $V_{CC}$ is 10 V and there is a 2 V drop across R1, there is an 8-V drop across R2. As a result, the voltage applied to the base is less positive than the collector voltage, although more positive than the emitter voltage.

If a PNP transistor is used in the circuit of Fig. 3-10, the action is the same except that the polarities of all voltages and the direction of current flow are reversed.

To summarize the common-emitter amplifier, the input signal is applied between the base and emitter and the output signal appears between the emitter and collector. This provides moderately low input impedance and a very high output impedance. The output signal is out of phase with the input. The common-emitter amplifier produces the highest power gain of all three configurations, with fairly high voltage and current gains.

**Gain in Two-Junction Transistor Amplifiers.** Before we go any further, we should resolve gain. Many terms are used to express the gain of common-emitter amplifiers as well as other amplifier configurations. For example, there are the terms *alpha* and *beta*, in addition to *current gain*, *voltage gain*, *resistance* (or *impedance*) *gain*, *power gain*, *voltage amplifier*, and *power amplifier*. The terms are interrelated and are often interchanged (properly and improperly). To minimize this confusion, the following is a summary of how these terms are used throughout this book.

*Alpha* and *beta* are applied to two-junction transistors connected in the common-base and common-emitter configurations, respectively. Both terms are a measure of the *current gain* of the transistor. Alpha is always less than 1 (typically 0.9 to 0.99). Beta is more than

1 and can be as high as several hundred (or more). The relationships between alpha and beta are

$$\text{alpha} = \frac{\text{beta}}{\text{beta} + 1} \qquad \text{beta} = \frac{\text{alpha}}{\text{alpha} - 1}$$

The terms "alpha" and "beta" do not necessarily represent the current gain of the circuit in which the transistors are used. Instead, the current gain of the circuit *cannot be greater* than the alpha and beta of the transistor. Schematics rarely, if ever, show either alpha or beta. At best, the schematics will show the transistor type (such as 2N6083). If you are interested, you can use the type number to find alpha and/or beta in the transistor data sheet. However, circuit gain is more often set by passive devices in the circuit (particularly the input and output resistances), as is discussed throughout this book.

*Current gain* can be applied to the transistor, or the complete circuit, and is a measure of change in current at the output for a given change in current at the input. Current gain is equal to the change in output current divided by change in input current. For example, when a 1-mA change in input current produces a 10-mA change in output current, the current gain is 10. Since alpha (common base) is always less than 1, there is no current gain in common-base amplifier circuits. (Instead, there is a slight loss.)

*Resistance gain* is simply the ratio of output resistance (or impedance) divided by input resistance (or impedance). For example, if the input resistance is 1000 Ω(1 kΩ) and the output resistance is 15,000 Ω(15 kΩ), the resistance gain is 15. The input and output resistances (or impedances) are dependent on circuit values as well as transistor characteristics. As discussed throughout this book, many amplifier characteristics are directly dependent on the relationship between transistor characteristics and values of circuit parts.

Resistance gain, by itself, produces no usable gain for the amplifier stage or circuit. However, the resistance gain has a direct effect on the circuit voltage and power gains. For example, using the previous values (current gain of 10, resistance gain of 15), it is possible to have a voltage gain of 150. Assume that the input resistance is 1 kΩ and a 1-mV signal is applied at the input. This results in an input current change of 1 μA. With a current gain of 10, the output current is 10 μA. This 10-μA current passes through a 15-kΩ output resistance to produce a voltage change of 150 mV. Thus, the 1-mV input signal produces a 150-mV output signal (a voltage gain of 150).

*Voltage gain* is equal to the difference in output voltage divided by a difference in input voltage. *Power gain* is equal to the difference in output power divided by the difference in input power. Except in a common-base amplifier, power gain is always higher than voltage gain since (as discussed in Chapter 1) power is based on the square of voltage (power = $E^2/R$) as well as the square of current ($I^2R$). Using the same values, the input power of the stage is 0.001 $\mu$W, the output power amplifier transistors have large current-carrying capacity but relatively low betas. Typically, there is at least one (and usually two) voltage-amplifier circuits or stages ahead of a power amplifier stage.

The function of a *voltage amplifier* is to receive an input signal consisting of a small voltage of definite waveform and to produce an output signal consisting of a voltage with the same waveform but much larger in amplitude. For example, as a radio wave cuts across the antenna of a radio receiver, the wave induces in that antenna a fluctuating voltage, usually on the order of microvolts. The voltage amplifier of the receiver amplifies this voltage to produce a similarly fluctuating voltage, which, in turn, operates the power-consuming loudspeaker.

Those transistors designed for voltage amplification usually have high betas with little current-carrying capability. On the other hand, power amplifier transistors have large current-carrying capacity but relatively low betas. Typically, there is at least one (and usually two) voltage-amplifier circuits or stages ahead of a power amplifier stage. This permits a low voltage input signal (say from an antenna, tape head, or industrial transducer) to operate a power-consuming device (radio or stereo loudspeaker or industrial servomotor).

### 3-3.4 COMMON-BASE AMPLIFIER

Figure 3-11 is the schematic of a basic NPN common-base amplifier, where both junctions are biased from a single source. In this configuration, the base is common to both input (emitter) and output (collector), and is connected through R1 to ground. Capacitors C1 and C2 are the input and output coupling capacitors, respectively. The input signal passing through C1 is developed across R3, which also serves to provide a closed circuit for current flow in the emitter-base circuit. The flow of collector current produces a voltage drop across $R_L$. The alternating current part of this voltage is the output signal, which passes through C2. The phase relationship between the input and output signals is shown by the wave.

The collector is connected to the positive side of source $V_{CC}$, placing the collector at the most positive potential. The emitter is at the most negative potential. Resistors R1 and R2 form a voltage divider across $V_{CC}$. Current flows through this divider, causing a voltage drop across R1 with the indicated polarity. This makes the

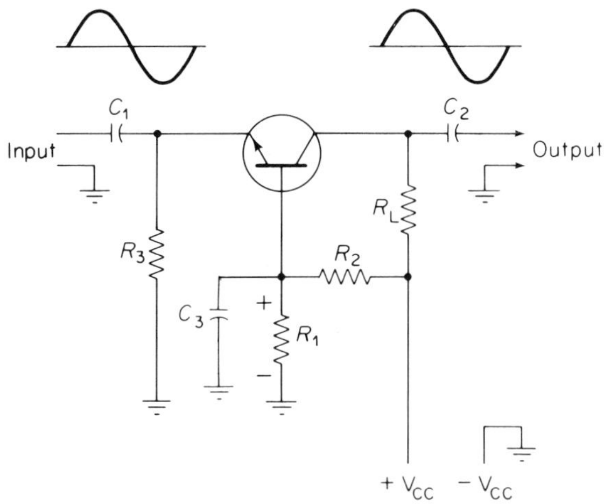


**FIGURE 3-11** Practical common-base amplifier operated from a single power supply

base less positive than the collector but more positive than the emitter.

To summarize the common-base amplifier, the input signal is applied between the base and emitter, and the output signal appears between the base and collector. This provides an extremely low input impedance and a very high output impedance. The output signal is in phase with the input. The common-base amplifier produces high voltage gains and modest power gains, even though there is no current gain. This is possible because of the *resistance gain*.

As an example, assume that the input resistance is 100 Ω, the output resistance is 15 kΩ, the current gain is 0.9 (less than 1), and the input signal is 1 mV. With 1 mV across 100 Ω, there is a 10-μA current change. With a current "gain" of 0.9, the output current is 9 μA. This 9-μA passes through a 15-kΩ output resistance to produce a voltage change of 135 mV. Thus, a 1-mV input signal produces a 135-mV output signal (a voltage gain of 135).

### *3-3.5 COMMON-COLLECTOR (OR EMITTER-FOLLOWER) AMPLIFIER*

Figure 3-12 is the schematic of a basic NPN common-collector amplifier, where both junctions are biased from a single source. The common-collector circuit is also known as an *emitter follower*, since the output is taken from the emitter resistance, and the output "follows" the input (in waveform and phase relationship).

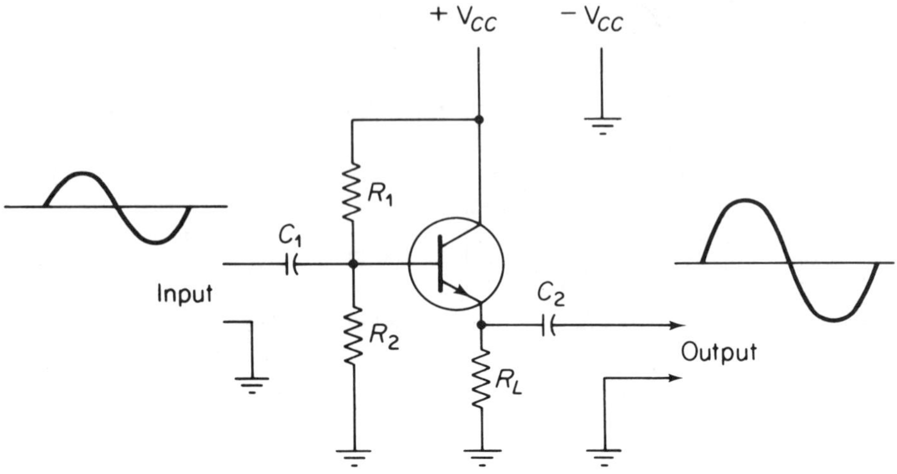


**FIGURE 3-12** Practical common-collector amplifier operated from a single power supply

Capacitors C1 and C2 are the input and output coupling capacitors, respectively. The input signal passing through C1 is developed across R2, which also serves to provide a closed circuit for current flow in the emitter-base circuit. The flow of emitter current produces a voltage drop across $R_L$. The a-c part of this voltage is the output signal, which passes through C2. The phase relationship between the input and output signals is shown by the waves.

Source $V_{CC}$ makes the collector positive relative to the base, thus reverse-biasing the collector-base junction. Resistors R1 and R2 form a voltage divider across $V_{CC}$, which forward-biases the emitter-base junction.

To summarize the common-collector (emitter-follower) circuit, the input signal is applied to the base and the output signal appears at the emitter. This provides extremely high input impedance and low output impedance. The output signal is in phase with the input. The common-collector produces modest current and power gains, even though there is no voltage gain. In general, the common-collector circuit gain (and hence the power gain) is limited by the current gain (beta) of the transistor.

### *3-3.6 TWO-JUNCTION TRANSISTOR BIAS NETWORKS*

Practically all two-junction transistor circuits require some form of bias (as discussed for vacuum tubes in Sec. 3-1.4). As a minimum, the collector-base junction must be reverse-biased. That is, current should not flow between collector and base. (Any collector-base current that does flow is a result of leakage or breakdown).

Under no-signal conditions, the emitter-base junction can be forward-biased, reverse-biased, or zero-biased (no bias). However, emitter-base current must flow under some conditions of operation. For example, some current flows all the time in class A. In class B and C amplifiers, current flows only in the presence of an operating signal.

The desired bias is accomplished by applying voltage to the corresponding transistor elements through bias networks, usually composed of resistors. The following paragraphs describe several basic bias networks. These circuits (or variations of them) represent most of the bias methods found on two-junction transistor schematics.

**Purpose of Bias Networks.** The bias networks serve more than one purpose. Typically, the bias network resistors (1) set the operating point, (2) stabilize the circuit at the operating point, and (3) set the approximate input/output impedance of the circuit.

The basic purpose of the bias network is to establish collector-base-emitter voltage and current relationships *at the operating point of the circuit.* (The operating point is also known as the *quiescent point, Q point, no-signal point, idle point,* or *static point.*) Since transistors rarely operate at this Q point, the basic bias networks are generally used as a reference or starting point for design. The actual circuit configurations and (especially) the bias network resistor values you find on schematics are generally selected on the basis of dynamic circuit conditions (desired output voltage swing, expected input signal level, etc.). Once the desired operating point is established, the next function of the bias network is to stabilize the circuit at this point.

### *3-3.7 TWO-JUNCTION TRANSISTOR BIAS STABILIZATION TECHNIQUES*

Bias stabilization networks are used to keep the circuit at (or near) the *Q* point, so that there will be no shift in operating characteristics of the circuit, and to prevent *thermal runaway*. Such thermal runaway can occur when a two-junction transistor is operated at or near its maximum power dissipation limits. When current passes through either transistor junction, heat is generated in the semiconductor material. Unless all of this heat is dissipated by the transistor case (an impossibility), the junction temperature increases. This, in turn, causes more current to flow through the junction, even though the voltage, circuit values, and so on, remain the same. In turn, this causes the junction temperature to increase even further, with a corresponding increase in current flow. If the heat is not dissipated by some means, the transistor will burn out and be destroyed.

There are several methods for providing bias stabilization of two-junction transistors. All of these methods involve a form of *negative feedback* or *inverse feedback*. That is, any change in transistor currents produces a corresponding voltage or current change that tends to offset the initial change. There are two basic methods for producing inverse or negative feedback: *inverse voltage feedback* and *inverse current feedback*. Current feedback is more commonly found on two-junction transistor schematics.

Figure 3-13 shows a typical inverse-current bias stabilization network using an NPN transistor. This is commonly called *emitter feedback*. Other specific bias networks using the same principle are discussed in Sec. 3-3.8. However, the use of an *emitter feedback resistor* in any bias circuit can be summed up as follows.

Base-emitter current (and, consequently, emitter-collector current) depends on the *differential in voltage* between base and emitter. If the differential voltage is lowered, less base current (and, consequently, less collector current) will flow. The opposite is true when the differential is increased. All current flowing through the collector (ignoring any leakage current) also flows through the emitter resistor. The voltage drop across the emitter resistor is thus dependent (in part) on the collector current.

Should the collector current increase (for any reason), emitter current and the voltage drop across the emitter resistor will also increase. This increase tends to decrease the differential between base

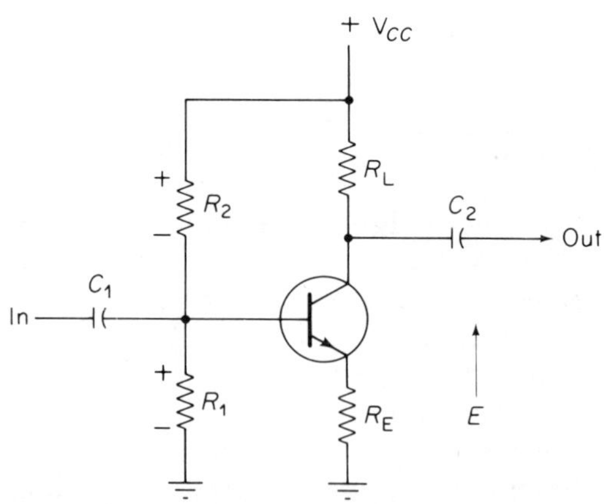


**FIGURE 3-13** Typical bias network with inverse-current feedback (emitter feedback)

and emitter, thus lowering the base current. In turn, the lower base current tends to decrease the collector current and offset the initial collector-current increase.

For example, assume that the drop across R1 is 1 V and that the drop across $R_E$ is 0.5 V. This results in a base-emitter differential of 0.5 V. Now assume that the collector current increases, raising the current through $R_E$ so that there is a drop of 0.8 V across $R_E$. This results in a differential of 0.2 V, which reduces both base and collector currents and tends to offset the initial increase in collector current.

Note that any form of negative or inverse feedback in an amplifier circuit tends to *oppose all changes*, even those produced by the signal being amplified. Thus, inverse or negative feedback tends to *reduce and stabilize gain.* This principle of stabilizing gain by means of feedback is found on many amplifier schematics, as discussed in Chapter 4.

### *3-3.8 SOME REPRESENTATIVE TWO-JUNCTION TRANSISTOR BIAS NETWORKS*

Figure 3-14 shows the schematics of typical two-junction transistor bias networks, all using inverse-current feedback (emitter feedback) of some form. Note that the bias network resistors, in addition to providing the required voltage relationships at emitter, base, and collector, also *set the approximate input and output impedances* (*Z*) of the circuit, as shown by the notes in Fig. 3-14. Similarly, when the circuits are used as amplifiers, *approximate voltage* (*V*) *and current* (*I*) *gains are set by resistance ratios.* The equations in the illustrations hold true for operation up to about 100 kHz (and higher in many instances). As operating frequencies increase beyond about 100 kHz, other factors enter into gain and impedance relationships, as discussed in Chapter 4.

The circuit of Fig. 3-14a provides maximum gain but least stability of all the emitter-feedback bias networks described here.

The circuit of Fig. 3-14b is essentially the same as the Fig. 3-14a circuit, but stability is increased by connecting base resistance $R_B$ to the collector rather than to the power supply. If the collector current increases, the drop across $R_L$ also increases, lowering the voltage at the collector. This lowers the base voltage and current, thus reducing the collector current. The feedback effect is combined with that produced by the emitter resistor $R_E$ to offset any variation in collector current. Because of the increased stability produced by the feedback, however, gain for the Fig. 3-14b circuit is slightly less than for the Fig. 3-14a circuit.

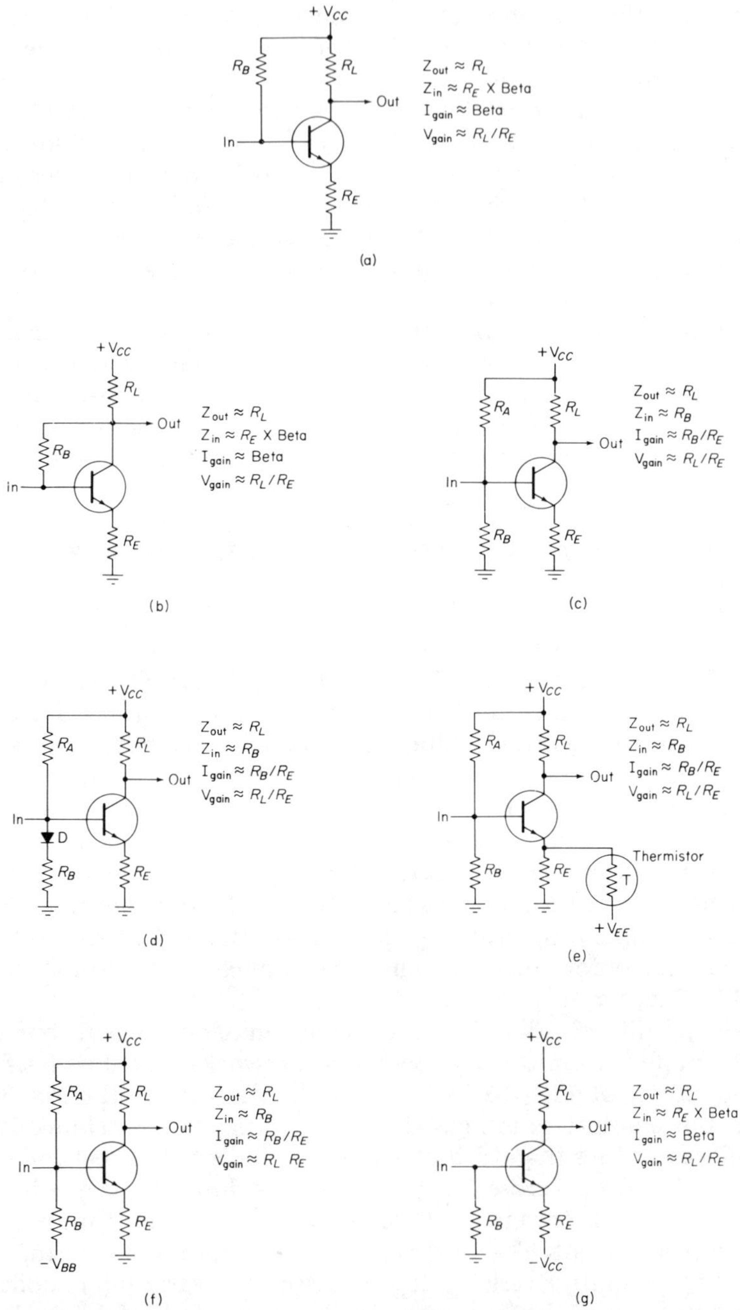


**FIGURE 3-14** Schematics of typical two-junction transistor bias networks

The circuit of Fig. 3-14c provides the maximum stability, but with a trade-off of lower gain. In addition to stability, the major advantage of the Fig. 3-14c circuit is that the input and output impedances, as well as voltage and current gain, are not dependent on transistor beta. Instead, circuit characteristics are dependent on circuit values.

The circuit of Fig. 3-14d has essentially the same characteristics as the Fig. 3-14c circuit, except that *temperature stability* is increased by connecting diode D between the base and $R_B$. Diode D (always forward-biased) is of the same material (silicon or germanium) as the base-emitter junction and is maintained at the same temperature. In practice, diode D is mounted near the transistor so that the base-emitter junction and diode D are at the same temperature. Thus, the voltage drops across diode D and the base-emitter junction are the same and remain the same with changes in temperature.

For example, should temperature rise, the emitter and collector currents of the transistor also tend to rise. The same temperature rise causes the forward resistance of diode D to decrease. As a result, the current flowing through $R_A$ increases, causing an increase in voltage drop across $R_A$. Since this voltage drop tends to reverse-bias the emitter-base junction, the net forward bias for that junction is reduced and the emitter-collector currents are reduced to normal (or tend to reduce to normal). Should the temperature fall, the action is in the reverse direction, and the emitter-collector currents are raised toward their normal values.

The circuit of Fig. 3-14e has essentially the same characteristics as the Fig. 3-14c circuit, except that the *negative temperature coefficient* (NTC) characteristics of a *thermistor* are used to provide temperature compensation. A thermistor is similar to a basic resistor (Chapter 2) except that the *thermistor resistance decreases with increases in temperature*, and vice versa. For best results, the thermistor is mounted near the transistor so that both devices are at the same temperature.

With the Fig. 3-14e circuit, the thermistor is used to vary the emitter voltage with temperature variations to minimize the effects of these variations upon the emitter current. Resistors $R_A$ and $R_B$ form a voltage divider to apply a portion of the supply voltage $V_{CC}$ in a direction to forward-bias the emitter-base junction.

Resistor $R_E$ and the thermistor form a second voltage divider across $V_{EE}$ (which is another voltage source). The direction of the voltage drop across $R_E$ is such as to reverse-bias the base-emitter junction. Since the forward bias applied to the base is larger than the reverse bias applied to the emitter, however, the net result is that the emitter-base junction is forward-biased.

Should the temperature rise, the emitter and collector currents tend to rise. The same rise in temperature reduces the resistance of the thermistor. This action permits more current to flow through the voltage divider. The increase in current flow increases the voltage drop across $R_E$, thus increasing the reverse bias being applied to the base-emitter junction. As a result, the net forward bias of the junction is reduced, thus reducing the emitter and collector currents toward their normal values. Should the temperature decrease, the action is reversed, thus preventing a decrease in emitter and collector currents.

The circuit of Fig. 3-14f has essentially the same characteristics as the Fig. 3-14c circuit. However, the Fig. 3-14f circuit is used in those special applications that require a negative and positive voltage, each with respect to ground, to control base current.

The circuit of Fig. 3-14g has somewhat different characteristics than all the previous networks. The Fig. 3-14g circuit is used in those special applications where it is necessary to supply collector-emitter current from both a positive and negative source. Since the transistor is NPN, the collector is connected to the positive source through $R_L$, while the emitter is connected to the negative source through $R_E$. If both sources are approximately equal, there is generally no voltage gain. The collector and emitter currents are approximately equal (ignoring the base current). Thus, if $R_L$ drops the positive source to half (say from 20 to 10 V), $R_E$ will be approximately twice the resistance of $R_L$. This will produce a loss, all other factors being equal.

The transistor bias networks described here represent those most commonly found on schematics. If you are interested in a thorough treatment of bias networks, as well as an in-depth look at transistors, your attention is invited to the author's *Handbook for Transistors* (Englewood Cliffs, N.J.: Prentice-Hall, Inc., 1976).

## 3-4 FIELD-EFFECT TRANSISTORS

Figure 3-15 shows the symbols for field-effect transistors (FETs). The basic FET consists of P and N junctions similar to that of two-junction transistors. However, the operation of FETs is quite different. The FET is sometimes known as the solid-state device that acts like a vacuum tube. This is because the characteristics of FETs are similar but definitely not identical to those of a pentode vacuum tube (Sec. 3-1.3).

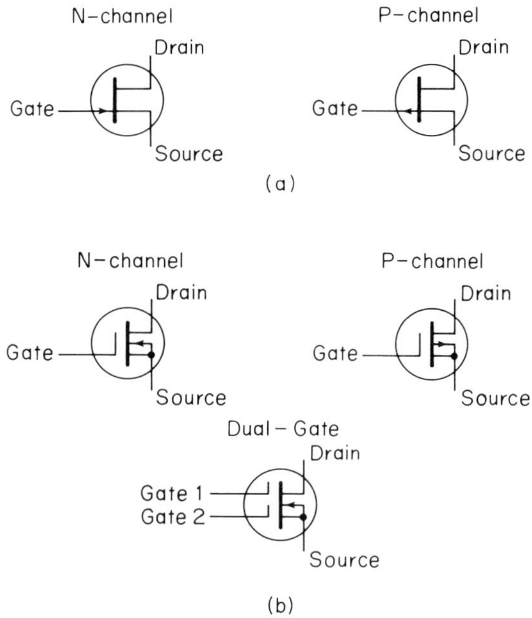


**FIGURE 3-15** Symbols for field effect transistors (FETs). (a) JFET, (b) MOSFET or IGFET

There are two basic types of FETs, known as the *junction field-effect transistor* (or JFET) and the *metal-oxide semiconductor field-effect transistor* (or MOSFET). The MOSFET is also known as the *insulated gate field-effect transistor* (or IGFET), because of its physical structure. We use the terms MOSFET and IGFET interchangeably throughout this book. The principles on which the JFET and MOSFET operate are very similar. The primary difference between the two is in the method by which the control element is made. This difference, however, results in a considerable difference in device characteristics.

### *3-4.1 JFET OPERATION*

The basic JFET is a bar of silicon (called the *substrate*) that acts like a resistor. To understand operation of the JFET, assume that the vertical bar in the JFET symbol is a variable resistance or channel. The amount of current that can pass through the resistance or channel is determined by the resistance value, since the voltage across the channel remains fixed. In turn, the channel resistance value is set by voltages applied to the resistance. The terminal into which current is injected is called the *source*, and is similar to the cathode of a vacuum tube or the emitter of two-junction transistor. The opposite terminal is called the *drain* and can be likened to a vacuum-tube plate or transistor collector. Unlike a vacuum tube, the polarity of the voltage applied to the drain and source of a FET can be interchanged. That is, the drain can be positive and the source negative, or vice versa.

The amount of drain-source current is set (primarily) by the *gate* voltage (or voltage between gate and source). The gate is similar to the vacuum-tube grid or transistor base, and is generally the input. The output can be either the drain or source, but is usually the drain. Whatever signal that appears at the gate is reproduced at the drain source. (If the gate voltage is ac, the drain-source current will also be ac.) Generally, the gate-source circuit is biased, as is the base-emitter circuit of a two-junction transistor.

Note that the arrow of the JFET symbol points away from the substrate in a P-channel device. This appears to be inconsistent with two-junction transistors, where the arrows point toward the substrate in PNP devices. This is because P-channel FETs are formed of N-type substrate material (and N-channel FETs use P-type material).

### *3-4.2 MOSFET (IGFET) OPERATION*

The MOSFET operates with a slightly different control mechanism than JFETs. The MOSFET gate, in effect, forms one plate of a "capacitor," the other plate being formed by the silicon substrate. However, voltages on the MOSFET gate still control drain-source current as in the case of a JFET. As shown in Fig. 3-15, MOSFETs are available in *single-gate* and *dual-gate* versions. That makes it possible to have two-channel control. Dual-channel MOSFETs are often found in circuits where two separate signals are involved, such as clippers, choppers, gated amplifiers, mixers, product detectors, color demodulators, and balanced modulators.

### *3-4.3 SPECIAL FEATURES FOUND IN FETs*

FETs have several special features that do not appear on the schematic, but do affect circuit operation. *Pinch-off voltage* ($V_P$) and *avalanche voltage* ($V_A$) are two important features. In general, when drain-source voltage is increased, the amount of drain-source current flow increases. However, when drain-source voltage is increased from zero, there is a point where further increases in drain-source voltage produce only slight increases in drain-source current.

The drain-source voltage that causes this current-limiting condition is called the pinch-off voltage, $V_P$. However, if the drain-source voltage is increased far beyond the $V_P$ point, the drain-source current flow becomes very heavy, and an avalanche condition occurs, somewhat like the zener avalanche condition described in Sec. 3-2.1. Both $V_P$ and $V_A$ are affected by the amount of bias applied between gate and source.

The *zero-temperature coefficient* (OTC) point is another feature found in FETs. If the gate source is biased at a specific voltage and is held constant at the voltage, the drain-source current will not vary with changes in temperature. Often, the values for FET bias circuits are chosen to provide OTC operation.

### *3-4.4 STATIC DISCHARGE PROBLEMS IN MOSFETs*

One problem with MOSFETs is that they are subject to damage by the discharge of static electricity across the gate "capacitor." If there is a large enough discharge, the capacitor breaks down permanently. This means that MOSFETs must be handled carefully. Usually, you must place one hand on the chassis or ground before touching a MOSFET. All leads of a MOSFET must be shorted together when the MOSFET is out of the circuit. With some MOSFETs, the static discharge problem is overcome by placing diodes between the gates and substrate so that the static electricity can discharge through the diode, rather than through the capacitor. Unfortunately, static-discharge diodes can also be a problem. Large input signals applied to the gate can be bypassed across the diodes. Keep in mind that the diodes may or may not show up on the schematic symbol.

### *3-4.5 GAIN IN FETs*

In two-junction transistors, gain is measured by the amount of change in output current (usually emitter-collector) for a given change in input current (usually base-emitter). In FETs, gain is measured by the amount of change in output current (drain source) for a given *change in input voltage* (gate source). This transfer or gain characteristic is usually given on FET data sheets as *transconductance* or *forward transconductance*, and identified by the letters $Y_{FS}$ or $g_{FS}$. The transconductance of a FET is usually not given on the schematic. As in the case of two-junction transistors, you must use the FET type number (3N153, etc.) and look up the gain on a data sheet.

### 3-4.6 FET OPERATING MODES

FETs can be operated in three modes: *depletion only*, *enhancement only*, and a combination of *enhancement and depletion.* The basic differences between these three modes can most easily be understood by examining the FET characteristics shown in Fig. 3-16. These characteristics, known as *transfer characteristics*, show the amount of drain-source current change for a change in gate-source voltage. Drain-source current, gate-source voltage, and drain-source voltage are identified by the letters $I_{DS}$ or $I_D$, $V_{GS}$, and $V_{DS}$, respectively.

To read a FET transfer characteristic, note the change in $I_{DS}$ for a given change in $V_{GS}$, at some particular value of $V_{DS}$. Use the transfer characteristic of the 2N3797 FET (shown in Fig. 3-16) as an example. The vertical center line represents the characteristics at a $V_{DS}$ of 10 V, since each division represents 2 V, and there are 5 divisions from zero volts. With a $V_{GS}$ of zero volts, the $I_{DS}$ is about 3 mA. If the $V_{GS}$ is increased to 2 V, $I_{DS}$ increases to 7mA. Thus, there is a 4-mA change in $I_{DS}$ for a 2-V change in $V_{GS}$, *at that particular operating point.* The amount of change is different for different values of $V_{DS}$ and $V_{GS}$. Keep in mind that transfer characteristics are not usually shown on schematics. However, transfer characteristics are used extensively on data sheets. This is true not only for FETs, but also for two-junction transistors, unijunction transistors, and most active devices.

Also as shown in Fig. 3-16, the *depletion-only* FET is classified as Type A and has considerable drain-current flow for zero gate voltage. No forward gate voltage is used. Maximum drain current flows when the gate-source voltage is zero. Drain current is reduced by applying a reverse voltage to the gate terminal. That is, $I_D$ decreases when $V_{GS}$ is increased.

The *depletion/enhancement* FET is classified as Type B and also has considerable drain-current flow for zero $V_{GS}$ (but not as much as Type A). Drain current is increased by application of a forward gate voltage and reduced by application of a reverse gate voltage. For a JFET used as Type B, drain current can be increased by gate voltage, only until the gate-source junction becomes forward-biased. At this point, a further increase in forward gate voltage will not produce an increase in drain current.

The *enhancement-only* FET is classified as Type C and has little or no current flow for zero gate voltage. Drain current does not occur until a forward gate voltage is applied. This voltage is known as the *threshold voltage* and is indicated in Fig. 3-16 as $V_{GS(TH)}$. Once

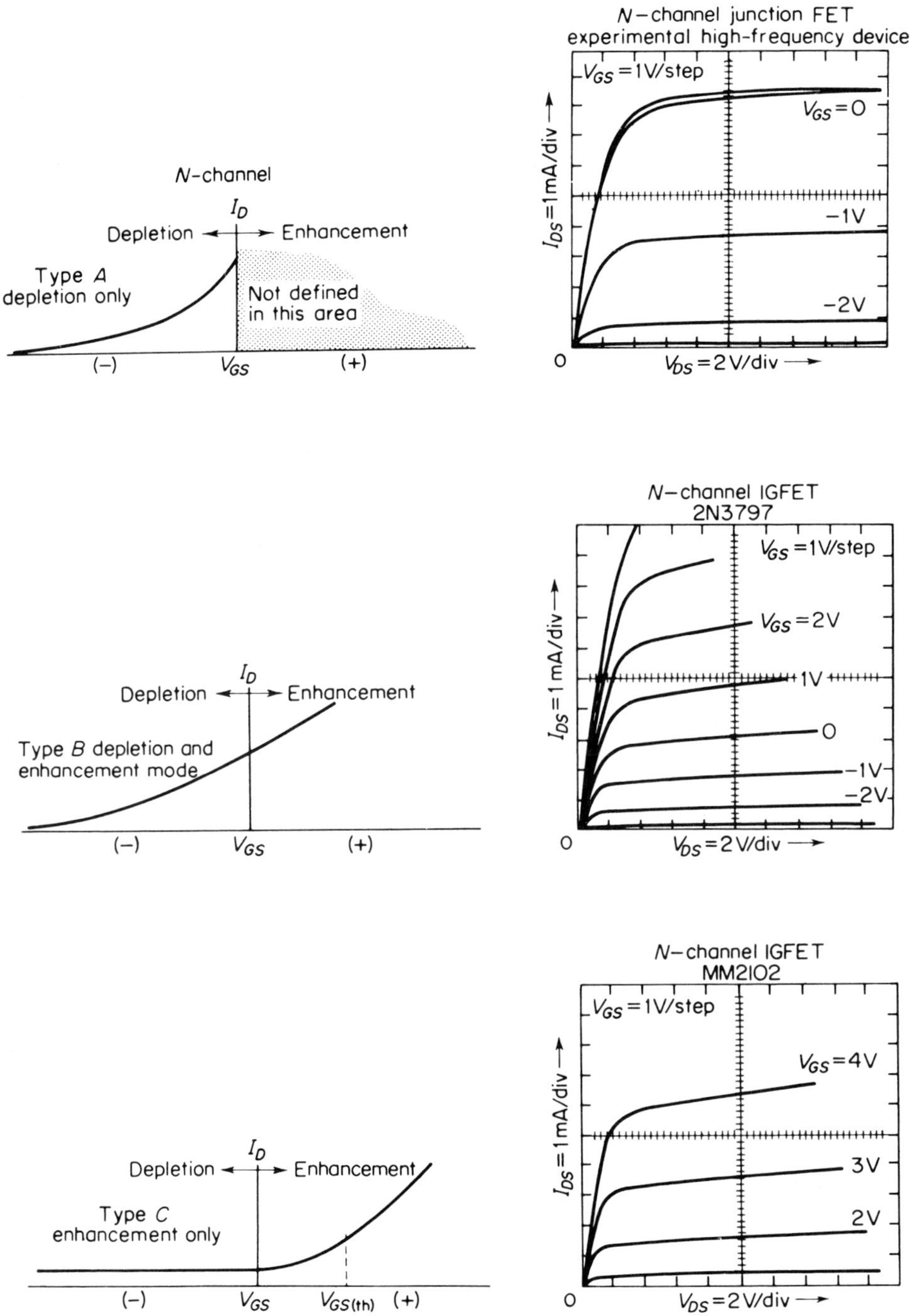


**FIGURE 3-16** Transfer characteristics and associated oscilloscope traces for Types A, B and C FETs. (courtesy Motorola)

the threshold voltage is reached, the transfer characteristics for a Type C FET are similar to those of a Type B.

Note that the schematic rarely, if ever, specifies the FET type of operation (Type A, B, or C). You must find the type information from the data sheet.

### *3-4.7 FET OPERATING REGIONS*

Both MOSFETs and JFETs have three distinct characteristic regions, only two of which are operational. Figure 3-17a, which is the output transfer characteristic of a depletion-type FET, illustrates the three different regions. Below the pinch-off voltage $V_P$, the FET operates in the *ohmic* or *resistance region*. Above $V_P$, the FET operates in the *constant-current region*, up to the *avalanche region* (where FETs are not operated). For this reason, the avalanche region is sometimes referred to as the *forbidden region*.

The curve of Fig. 3-17b illustrates how drain current varies with changes in gate-source voltage. For depletion FETs, the drain current decreases as the gate-source voltage is increased. For enhancement devices, the drain current is enhanced or increased as the gate-source voltage is increased.

If the FET is operated with a drain-source voltage below the pinch-off voltage, the FET characteristics vary considerably as the gate-source voltage is varied. This is shown in Fig. 3-17c, which is an enlarged view of the ohmic region for the FET shown in Fig. 3-17a.

In the ohmic region, the drain-source channel is actually a *voltage-variable* or *voltage-controlled resistor*. As shown in Fig. 3-17d, the drain-source resistance decreases with increasing gate-source voltage, for enhancement FETs. The converse is true for depletion FETs. MOSFETs are not generally operated in the ohmic region. However, MOSFETs are capable of operation below the pinch-off voltage.

### *3-4.8 BASIC FET CIRCUIT*

As in the case of two-junction transistors, FETs can be connected in one of three basic circuits: *common-source*, *common-drain*, and *common-gate*, which correspond to two-junction common-emitter, common-collector, and common-base, respectively. Figure 3-18 shows the three basic FET circuits, together with the approximate characteristics (input and output impedance, and voltage gain). Resistors R1 and R2 set the bias to operating point for the stage. Resistor $R_L$ is the load or output resistance. Resistor $R_S$ is the source resistance.

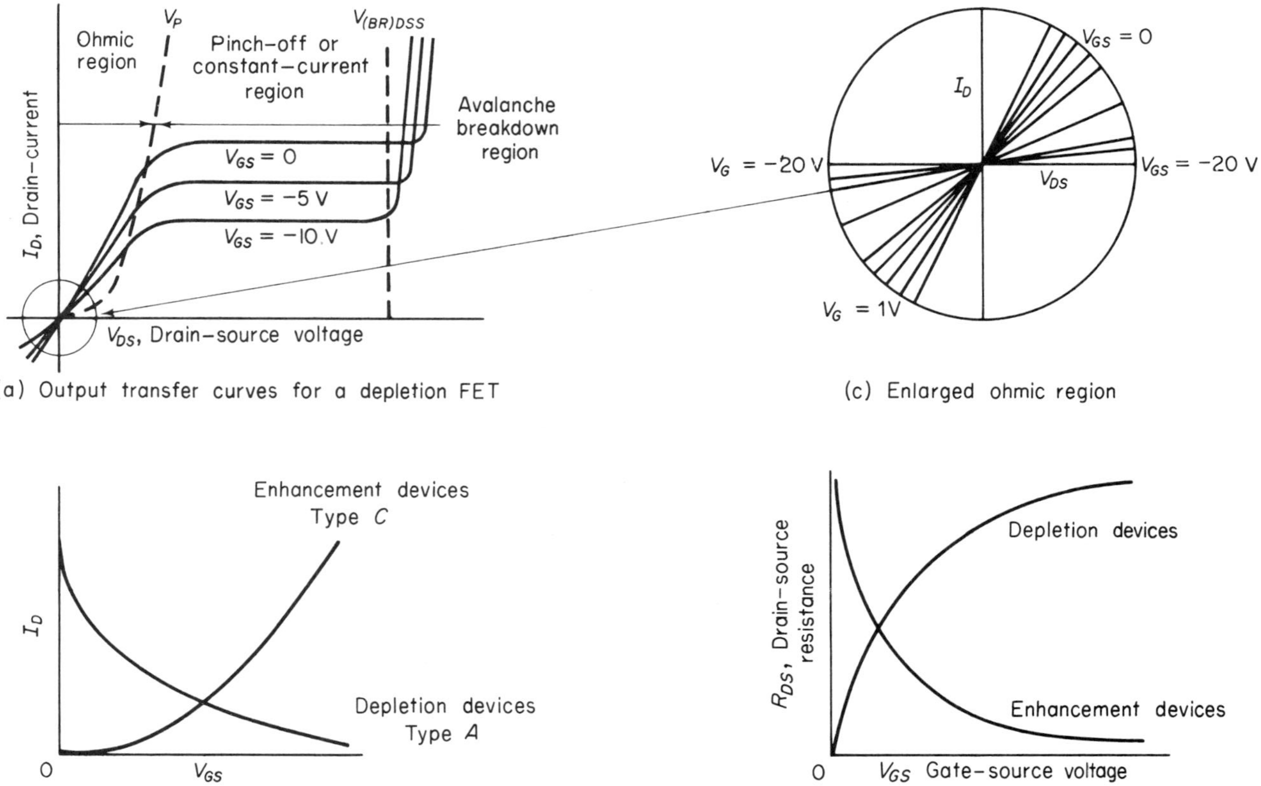


(a) Output transfer curves for a depletion FET

(c) Enlarged ohmic region

(b) Drain current versus gate voltage

(d) Channel resistance versus gate voltage

**FIGURE 3-17** Characteristics of FET operating regions (courtesy Motorola)

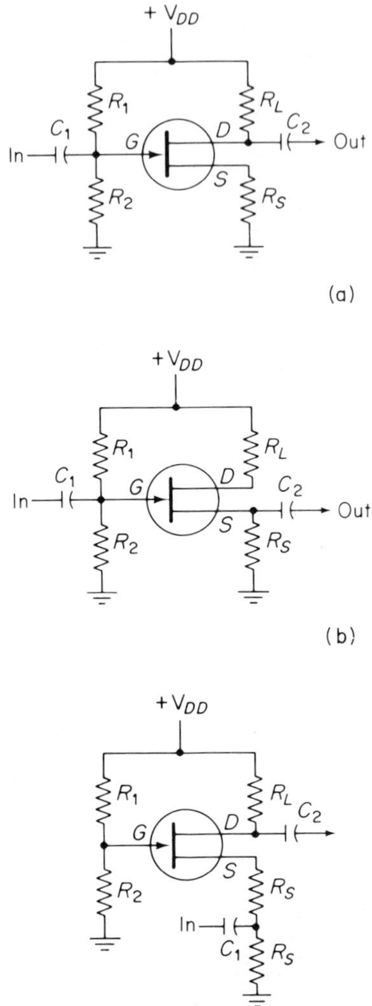


(a)

$$V_{gain} \approx \frac{R_S}{\frac{1}{y_{fs}} + R_S}$$

$$Z_{in} \approx R_1 \parallel R_2$$

$$Z_{out} \approx R_L$$

(b)

$$V_{gain} \approx \frac{R_S}{\frac{1}{y_{fs}} + R_S}$$

$$Z_{in} \approx R_1 \parallel R_2$$

$$Z_{out} \approx R_S \parallel \frac{1}{y_{fs}}$$

(c)

$$V_{gain} \approx \frac{R_S}{\frac{1}{y_{fs}} + R_S}$$

$$Z_{in} \approx R_S + \frac{1}{y_{fs}}$$

$$Z_{out} \approx R_L$$

**FIGURE 3-18** Schematics of basic FET circuits **(a) common-source, (b) common-drain, (c) common-gate**

Capacitors C1 and C2 are the coupling capacitors for input and output, respectively. The circuits of Fig. 3-18 are for N-channel JFETs. Simply reverse the polarity for P-channel JFETs. All MOSFETs may not conform exactly to the characteristics shown in Fig. 3-18, but are close.

## 3-5 UNIJUNCTION TRANSISTORS

Figure 3-19 shows the symbols for unijunction transistors (UJTs). The UJT operates on an entirely different principle from that of the two-junction transistor or the filed-effect transistor. The UJT is a *negative resistance* device. Under proper conditions, the input voltage or signal can be decreased, yet the output or load current will increase. Once the UJT is "turned on," it will not "turn off" until the circuit is broken or the input voltage is removed. For this reason, the UJT makes an excellent trigger source. The UJT can be biased just below the "firing" point. When a small trigger voltage (either intermittent or constant) is applied, the UJT will "fire" and produce a large output voltage pulse or signal that will remain on until the circuit is broken (by switching off the base voltage).

The *programmable unijunction transistor* (or PUT) is a device similar in structure to an SCR (silicon-controlled rectifier) described in Sec. 3-6. However, the PUT is used in UJT circuits, since the

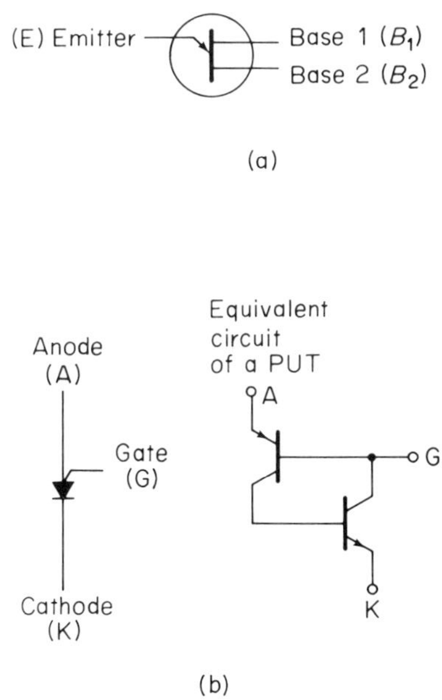


**FIGURE 3-19** Symbols for **(a)** unijunction transistors (UJTs) and **(b)** programmable unijunction transistors (PUTs)

characteristics of the two devices are similar. The trigger voltage of the PUT is programmable and can be set by an external resistive voltage divider network. The PUT is faster and more sensitive than the UJT and can replace the UJT in many applications (particularly as long duration timer, as discussed in Chapter 4).

### *3-5.1 UJT OPERATION*

As shown in Fig. 3-19, the UJT is a three-terminal device. The three terminals are emitter (E), base 1 (B1), and base 2 (B2). Note that the base 1 and base 2 leads are shown at right angles to the base because they are *nonrectifying* (not a junction as two-junction transistors). The emitter connection is represented by an arrow, however, because it is a rectifying PN junction. The arrow is slanted to indicate the emitting properties of the junction. The arrowhead is pointing toward the base, showing a P-type emitter and an N-type base.

### *3-5.2 PUT OPERATION*

As shown in Fig. 3-19, the PUT has three terminals: an anode (A), a gate (G), and a cathode (K). As seen from the equivalent circuit, the PUT is somewhat like two interconnected two-junction transistors. If the gate is made negative with respect to the anode, the PUT will switch from a blocking condition to the "on" state.

### *3-5.3 UJT AND PUT CHARACTERISTICS*

In reading schematics, the main point to remember about UJTs is that the emitter is the input, and the base 1, base 2 circuit forms the output. Once the UJT or PUT is triggered by a signal or voltage at the emitter, the device will turn on (current will flow between base 1 and base 2) and remain turned on until the base 1, base 2 voltage is interrupted. The main difference between the UJT and PUT functions is that the trigger point of the PUT is programmable, whereas the voltage required to trigger the UJT fixed, and is usually given as a percentage of the power supply voltage.

Although the basic function of the UJT and PUT is that of a switch, a *relaxation oscillator* is the primary building block in most UJT and PUT circuits (as discussed in Chapter 4). This basic relaxation oscillator provides *phase control of thyristors* such as SCRs and Triacs, discussed in the remaining sections of this chapter. For that reason, much UJT theory is based on the relationship to relaxation oscillator principles.

## 3-6 CONTROLLED RECTIFIERS AND THYRISTORS

The controlled rectifier is similar to the basic diode (Sec. 3-2), with one specific exception. The controlled rectifier must be "triggered" or "turned on" by an external voltage or signal. The controlled rectifier has a high forward and reverse resistance (no current flow in either direction), without the trigger. When the trigger is applied, the forward resistance drops to zero (or very low) and a high forward current flows just as with the basic diode. The reverse current remains high, and no reverse current flows, so the controlled rectifier will rectify a-c power in the normal manner. As long as the forward voltage is applied, the forward current will continue to flow. The forward current will stop, and the controlled rectifier will "turn off," if the forward voltage is removed.

There are a great number of controlled rectifiers in use. Many of these are actually the same type (or a slightly different version), but manufactured under a different trade name or designation. The term *thyristor* is applied to many controlled rectifiers. Technically, a thyristor defines any semiconductor switch whose action depends on PNPN regenerative feedback (Sec. 3-7). Thyristors can be two-, three-, or four-terminal devices.

### *3-6.1 SILICON- OR SEMICONDUCTOR-CONTROLLED RECTIFIER (SCR)*

With some manufacturers, the letters SCR refer to *semiconductor-controlled rectifier* and can mean any type of solid-state-controlled rectifier. However, SCR usually refers to *silicon-controlled rectifiers.*

If four semiconductor materials, two P-type and two N-type, are arranged as shown in Fig. 3-20a, the device can be considered as three diodes arranged alternately in series, as shown in Fig. 3-20b. Such a device acts as a conventional diode rectifier (Sec. 3-2) in the reverse direction, and as a combined electronic switch and series rectifier in the forward direction. Conduction in the forward direction can be controlled or "gated" by operation of the switch.

Figure 3-21 shows the circuit symbol, block diagram, and basic physical construction of a typical SCR. Note that there are two basic arrangements for the SCR: one with the gate terminal connected to the cathode and one to the anode. The cathode gate is the most common arrangement.

SCRs are most often used to control alternating current, but can be used to control a direct current (Chapter 4). Either a-c or d-c voltage can be used as the gate signal, provided that the gate voltage

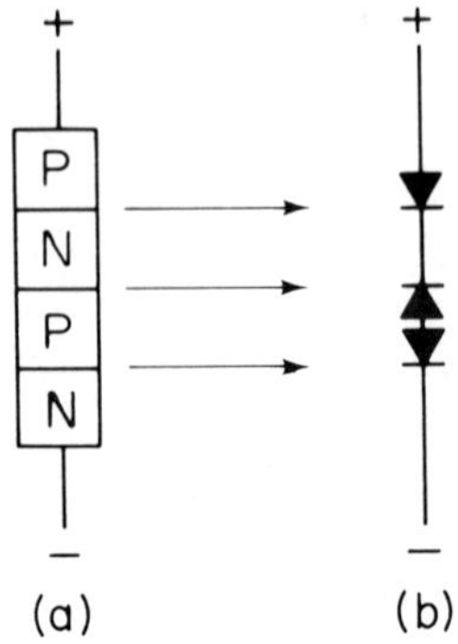


**FIGURE 3-20** Semiconductor material arrangement and equivalent diode relationship of an SCR

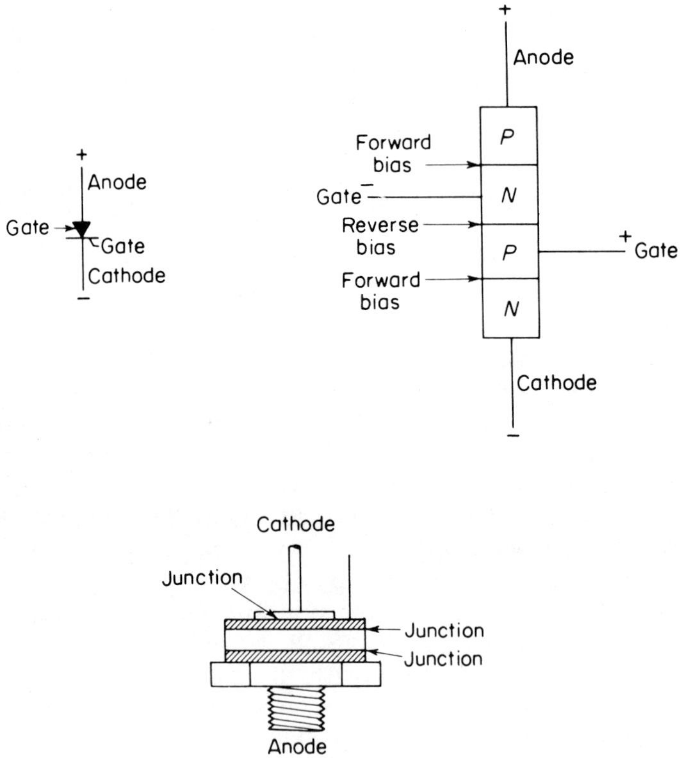


**FIGURE 3-21** Symbol, block diagram and basic construction of a typical SCR

is large enough to trigger the SCR into the "on" condition. In effect, the gate is the input, and the anode-cathode circuit is the output.

An SCR is used to best advantage when both the load and trigger are alternating current. With a-c power, control of the power applied to the load is determined by the *relative phase of the trigger signal versus voltage.* Because the trigger control is lost once the SCR is conducting, an a-c voltage at the load permits the trigger to regain control. Each alternation of alternating current through the load causes conduction to be interrupted (when the a-c voltage drops to zero between cycles), regardless of the polarity of the trigger signal.

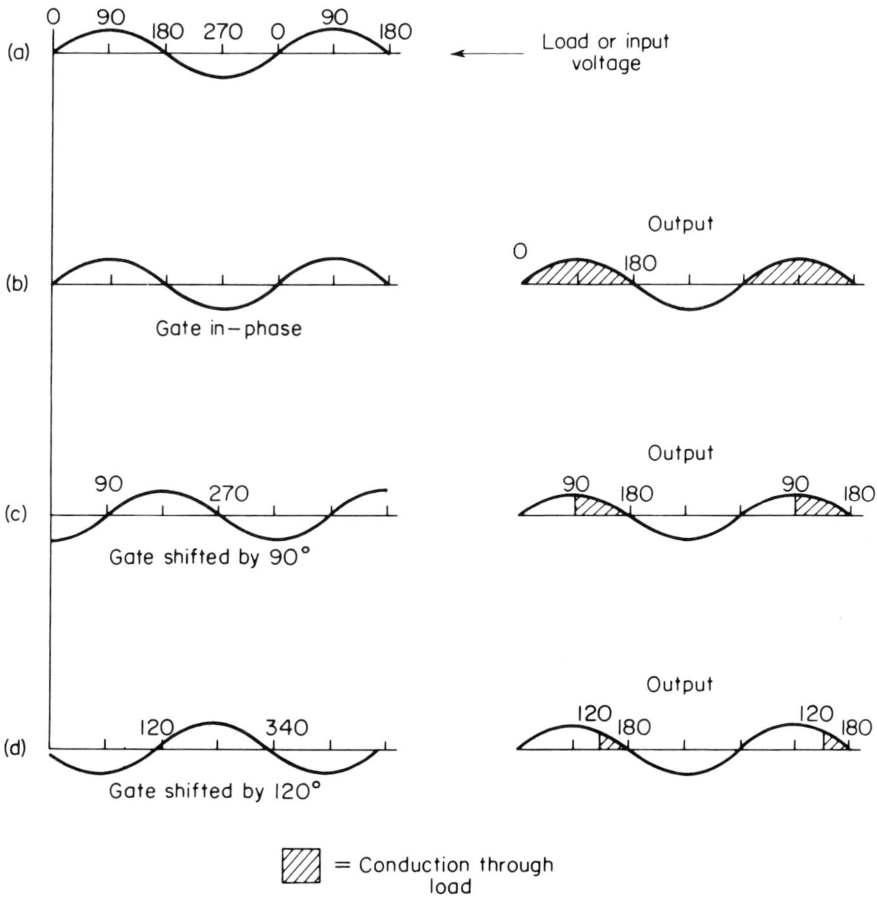


**FIGURE 3-22** Operation of an SCR with ac at both load and gate or trigger

**Phase Relationship between Gate and Load Voltages.** Figure 3-22 shows operation of an SCR with a-c voltages at the trigger circuit (gate) and load circuit (anode-cathode). If the trigger voltage is in phase with the a-c power input signal as shown in Fig. 3-22b, the SCR conducts for each successive positive alternation at the anode. When the trigger is positive-going at the same time as the load or anode voltage, load current starts to flow as soon as the load voltage reaches a value that will cause conduction. When the trigger voltage is negative-going, the load voltage is also negative-going, and conduction stops. The SCR thus acts as a half-wave rectifier in the normal manner.

If there is a 90° phase difference between the trigger voltage and load voltage (say load voltage lags trigger voltage by 90°) as shown in Fig. 3-22c, the SCR does not start conducting until the trigger voltage swings positive, even though the load voltage is initially positive. When the load voltage drops to zero, conduction stops, even though the trigger voltage is still positive.

If the phase shift is increased between trigger and load voltages as shown in Fig. 3-22d, conduction time is even shorter, and less power is applied to the load circuit. *By shifting the phase of the trigger voltage* in relation to the load voltage, it is possible to vary the output power, even though the voltages are not changed in strength (as discussed in Chapter 4).

### *3-6.2 PNPN-CONTROLLED SWITCH OR SCS*

The PNPN-controlled switch (often referred to as an SCS or silicon-controlled switch) is similar in operation to an SCR. However, the SCS is a PNPN device where all four semiconductor regions are made accessible by means of terminals. Figure 3-23 shows the circuit symbols, block diagram, and basic physical construction of a typical SCS. Note that two circuit symbols are used and that either may appear on schematics. Often, the SCS is used as an SCR, with the extra gate terminal not connected on the schematic.

In some control applications, the SCS can be considered as a transistor and diode in series. Figure 3-24 shows this arrangement. If a negative load voltage is applied to terminal 4, with a positive at terminal 1, the SCS does not turn on, no matter what trigger signals are applied. However, with a positive at terminal 4 and a negative at terminal 1, the SCS is turned on by either a positive voltage at terminal 2 or a negative voltage at terminal 3.

The SCS also has the ability to turn off by means of a gate signal. The arrangement is shown in Fig. 3-25. It should be noted that the gate turn-off method applies only when the conducting current is

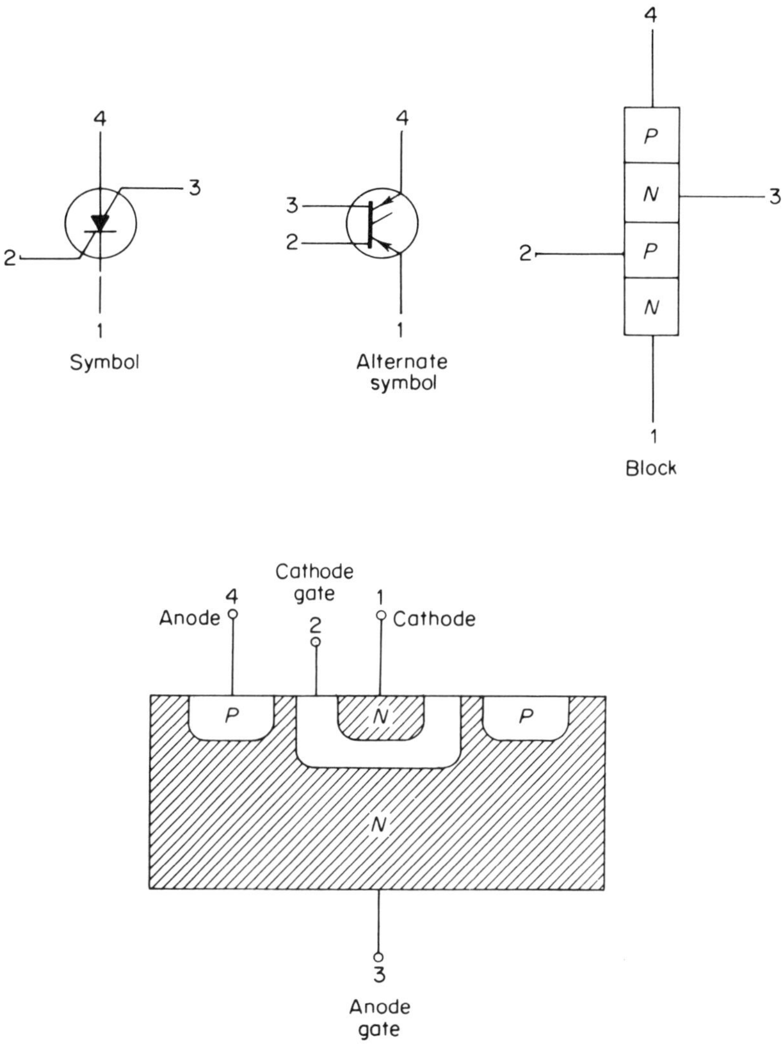


**FIGURE 3-23** Circuit symbols, block diagram and basic physical contruction of typical SCS (or PNPN controlled switch)

below a certain value. The SCS is often considered as two transistors (an NPN and a PNP) connected as shown in Fig. 3-26. Both transistors are connected so that the collector output of the NPN feeds into the base input of the PNP, and vice versa. If a positive trigger voltage is applied to the NPN base, the NPN is turned on and some current flows in the NPN emitter-collector circuit. Since the NPN collector feeds the PNP base, the PNP is also turned on, and the PNP collector output feeds into the NPN base, adding to the trigger voltage.

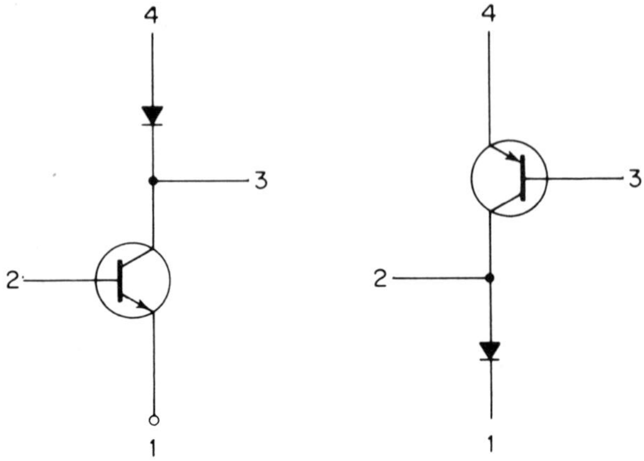


**FIGURE 3-24** Equivalent diode-transistor representation of an SCS

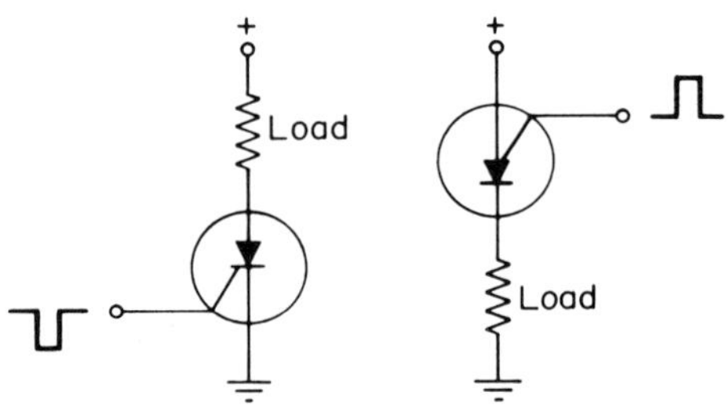


**FIGURE 3-25** Gate turn-off arrangement for SCSs

Load current then flows through the two transistors from the NPN emitter (or rectifier cathode) to the PNP emitter (or rectifier anode). Load current continues to flow even though the trigger is removed, since the current flow also keeps both transistors turned on. The same condition can be produced by a negative trigger applied to the PNP base. This turns on the PNP, which, in turn, turns on the NPN, until both transistors are fully conducting.

Normally, the currents do not stop until the load voltage is removed or, in the case of alternating current, the voltage drops to zero between cycles. However, if the load current is below a certain

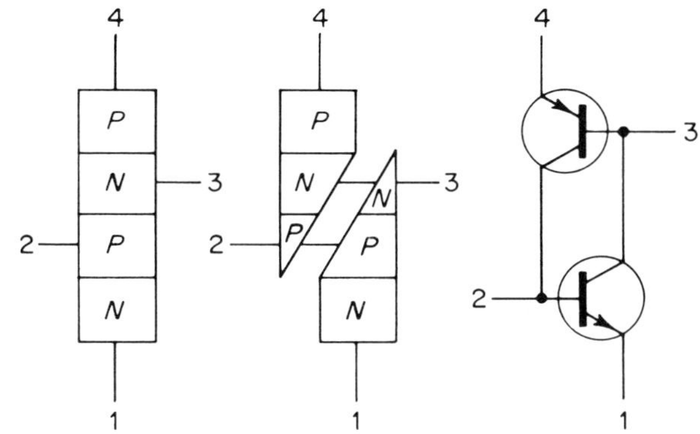


**FIGURE 3-26** Equivalent two-transistor representation of an SCS

level (different for each type of SCS), the SCS can be turned off by a trigger voltage. For example, if a negative trigger voltage is applied to the NPN base, less current flows in the NPN emitter-collector circuit. As a result, the PNP is turned on less, and the PNP collector output drops to aid the turn-off trigger voltage at the NPN base. The feedback process continues until the load current is completely stopped.

### *3-6.3 TRIACS AND DIACS*

The generally accepted definition for the term "triac" is bidirectional triode thyristor. However, the term (as coined by General Electric) is used to identify a "*tri*ode *ac* semiconductor switch" (or three-terminal switch). Like the SCR and SCS, the triac is triggered by a gate signal. Unlike either the SCR or SCS, the triac conducts in *both directions* and is therefore most useful for controlling devices operated by a-c power (such as a-c motors). Since an SCR or SCS is essentially a rectifier, two SCRs or SCSs must be connected back to back (in parallel or bridge) to control alternating current. (Otherwise, the a-c will be rectified into direct current.) The use of two or more SCRs or SCSs required elaborate control circuit (in most cases). Much of this can be eliminated by a triac circuit when a-c power is to be controlled.

Figure 3-27 shows the circuit symbol and outward physical appearance of a triac. Note that the symbol is essentially one SCR symbol combined with another complementary SCR symbol. Since the triac is not a rectifier (when turned on and conducting, current flows in

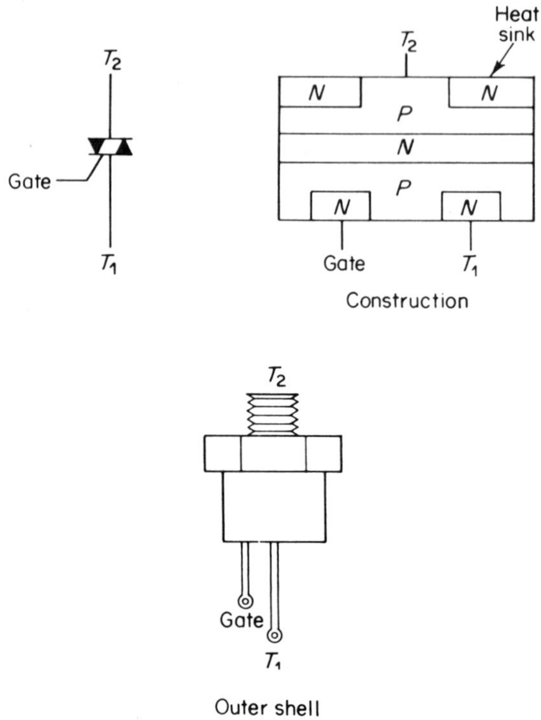


**FIGURE 3-27** Symbols, outer shell, and construction of a typical Triac

both directions), the terms "anode" and "cathode" do not apply. Instead, terminals are identified on the schematic by number. Terminal T1 is the reference point for measurement of voltages and currents at the gate terminal and at terminal T2. The area between terminals T1 and T2 is essentially a PNPN switch in parallel with an NPNP switch.

Like the SCS and SCR, the triac can be made to conduct when a "breakdown" or "breakover" voltage is applied across terminals T1 and T2, and when a trigger voltage is applied. Current continues to flow in one direction until that half-cycle of the a-c voltage (across T1 and T2) is complete. Current then flows in the opposite direction for the next half-cycle. The triac does not conduct on either half-cycle unless a gate-trigger voltage is present during the next half-cycle (unless the breakdown voltage is exceeded).

**Triac Trigger Sources.** Triacs can be triggered from many sources, as can SCRs and SCSs. One of the most common trigger sources is the *diac* shown in Fig. 3-28. Note that there are two types

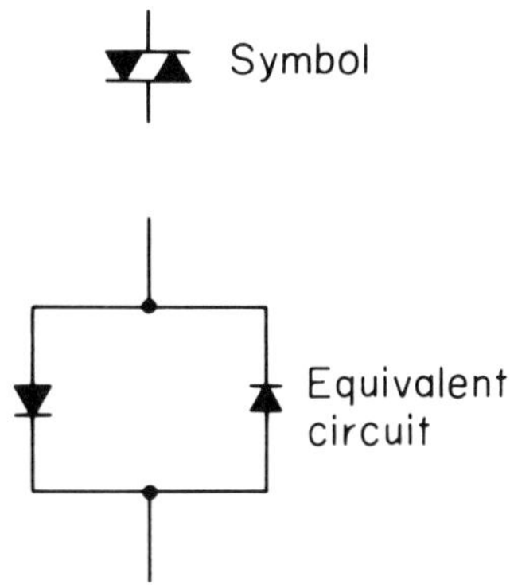


**FIGURE 3-28** Symbol and equivalent circuit of Diac

of diacs, diode and transistor. The term diac (officially bidirectional diode thyristor) was coined by General Electric to identify a "*di*ode *ac* semiconductor device." The diac is also known as a bilateral trigger diode on some schematics. No matter what term is used, the diac can be considered as a semiconductor device resembling a pair of diodes connected in complementary (parallel) form (just as the symbol appears on Fig. 3-28). The anode of one diode is connected to the cathode of the other diode, and vice versa.

Each diode passes current in one direction only, as in the case of the basic diode (Sec. 3-2). However, the diodes in a diac do not conduct in the forward direction until a certain "breakover" voltage is reached. For example, if a diac is designed for a breakover voltage of 3 V, and the diac is used in a circuit with less than 3 V, the diodes appear as a high resistance (no current flow). Both diodes conduct in their respective forward bias directions when the voltage is raised to any value over 3 V.

### *3-6.4 SUS AND SBS*

Figure 3-29 shows the symbols and equivalent circuits of the SUS and SBS. The SUS (silicon unilateral switch) is essentially a miniature SCR having an anode gate (instead of the usual cathode gate) and a built-in low-voltage avalanche (or zener, Sec. 3-2.1) diode between the gate and cathode. However, the SUS operates in a manner similar to the UJT (Sec. 3-5), but not exactly. The SUS switches at a fixed voltage, determined by the internal zener diode, rather than a fraction or percentage of the power supply voltage. Also, the switching current of the SUS is generally higher than that of the UJT.

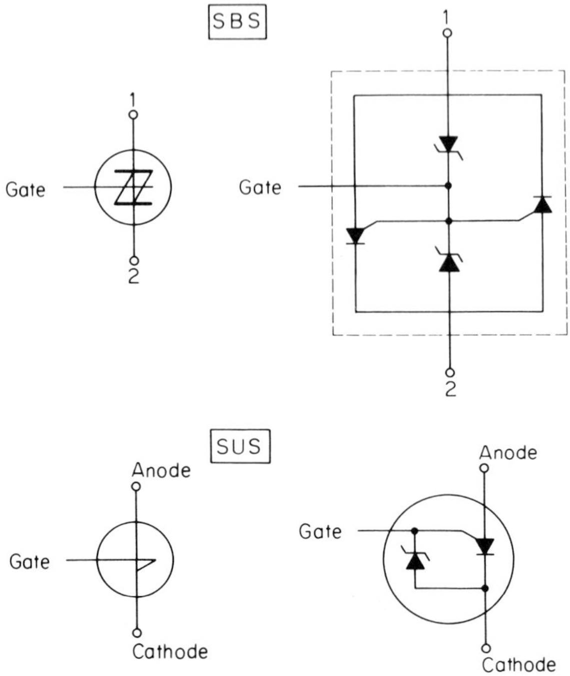


**FIGURE 3-29** Symbols and equivalent circuits for SBS and SUS

The SBS (silicon bilateral switch) is essentially two identical SUS structures arranged in inverse-parallel. Since the SBS operates as a switch with both polarities of applied voltages, the SBS is particularly useful for triggering triacs with alternate positive and negative gate pulses.

## 3-7 LIGHT-ACTIVATED DEVICES

Many of the active devices found on electronic schematics are *light-activated* or *light-sensitive*. Such devices are based on the *photocell* principle, so we will start by discussing this principle very briefly.

### *3-7.1 BASIC PHOTOCELL THEORY*

There are two basic types of photocells: *photovoltaic* and *photo-conductive*.

The *photovoltaic* cells produce an output voltage and current in the presence of light. For this reason, photovoltaic cells are often called *solar batteries* or *solar cells.* Both silicon and selenium are used in photovoltaic cells. However, silicon is the most effective. Photovoltaic cells are often used in sheets or panels, where many cells are connected together to act as a giant "sun battery."

The *photoconductive* cells are often called light-sensitive resistors or light-dependent resistors (LDRs), since they function as a resistor (rather than a battery) and do not generate an output voltage or current. Instead, photoconductive cells act as a resistance which varies in the presence of light, thus changing the amount of current being conducted through the circuit. The photoconductive cell is often made of cadmium sulfide.

### 3-7.2 PHOTOVOLTAIC CELLS

Figure 3-30 shows the photovoltaic cell symbol, which looks like a battery with arrows pointing toward the cell junction. The Greek letter lambda ($\lambda$) is supposed to be used in the circle surrounding the battery cell. However, the lambda is often omitted (sometimes the circle and the polarity symbol are also omitted). However, the arrows pointing toward the junction always indicate the light activation.

The main point to remember about photovoltaic cells is that when light strikes the cell, voltage and current (dc) are generated. Current flows from the negative terminal to the positive terminal, as in the case of conventional batteries (Chapter 2). Although selenium is not as efficient as silicon, selenium solar cells have characteristics more like the human eye and are used in applications to simulate the eye (such as camera exposure meters). Silicon is used when the cell must produce the maximum output for a minimum of light.

### 3-7.3 PHOTOCONDUCTIVE CELLS AND TRANSISTORS

Figure 3-31 shows the photoconductive cell symbol, which looks like a resistor with arrows pointing toward the resistance element. Again, the Greek letter lambda is supposed to be used in the circle. Fig. 3-31 also shows the basic construction of a typical photocell.

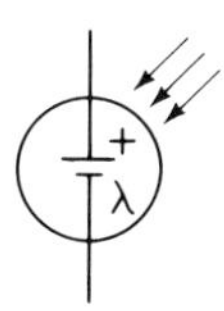


**FIGURE 3-30** Symbol for photovoltaic cell

The main point to remember about photoconductive cells is that an external voltage must be used, as shown in Fig. 3-31. When light strikes the photoconductive cell surface, current flow is increased. This is because cell resistance decreases in the presence of light. When light is removed or decreased, cell resistance increases and current flow decreases.

Figure 3-32 shows the symbol of a *photoconductive transistor* (also known as a *phototransistor*). Note that the symbol for a photoconductive transistor is similar to that of a conventional transistor. On many schematics, a standard transistor symbol (together with arrowheads) is used to indicate a photoconductive transistor. Most photoconductive transistors (usually germanium) operate essentially the same as conventional transistors, except that current flow is controlled by light striking the transistor junctions. An increase in light increases current flow, and vice versa.

Actually, most transistors are photoconductive to some extent. That is, current flow in the transistor is affected by light striking the transistor semiconductor material. However, conventional transistors are sealed so that no light can strike the semiconductor junction.

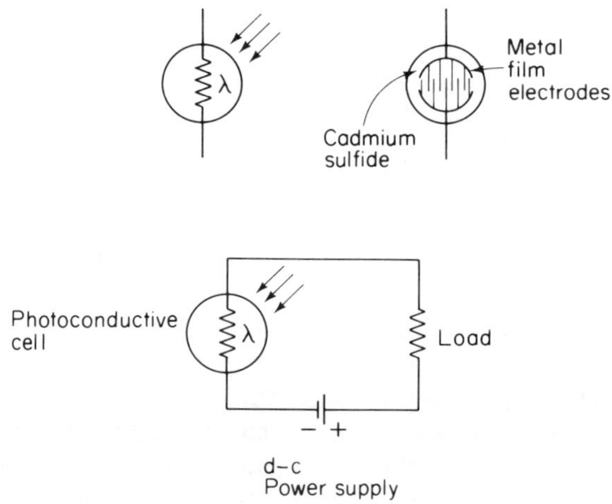


**FIGURE 3-31** Symbol and physical construction of photoconductive cell

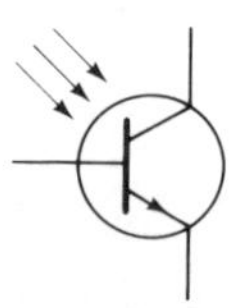

**FIGURE 3-32** Symbol for photoconductive transistor (phototransistor)

### *3-7.4 LIGHT-ACTIVATED CONTROL DEVICES*

When the photocell principles are applied to control devices, they are generally referred to as *light-activated semiconductors* (LAS). In PNPN control devices, such as the SCR or SCS (Sec. 3-6), the current increase caused by light striking the semiconductor junctions acts as a *trigger to turn on the device.* Conventional SCRs and SCSs are sealed to light. The junctions of LAS devices are exposed to light (by means of a window or opening in the enclosure). The most common LAS devices are the LASCR (light-activated SCR), LASCS (light-activated SCS), and the opto-coupler (or light-activated switch), all of which are discussed here.

**LASCR and LASCS.** Figure 3-33 shows the symbols and construction for the LASCR and LASCS. Note that the arrows indicate light or radiant energy striking the silicon chip. Both devices are similar in operation and characteristics to the corresponding SCR or SCS, except for the glass window on top of the can or enclosure. In a typical application, the gate or trigger terminal is connected to a fixed voltage in the circuit. The SCR or SCS is then triggered by light passing through the window onto the silicon chip area.

**Opto-coupler.** Figure 3-34 shows a typical light-activiated switch or opto-coupler. When a photoconductive cell is exposed to very strong light, the internal resistance drops to zero (or near zero). This

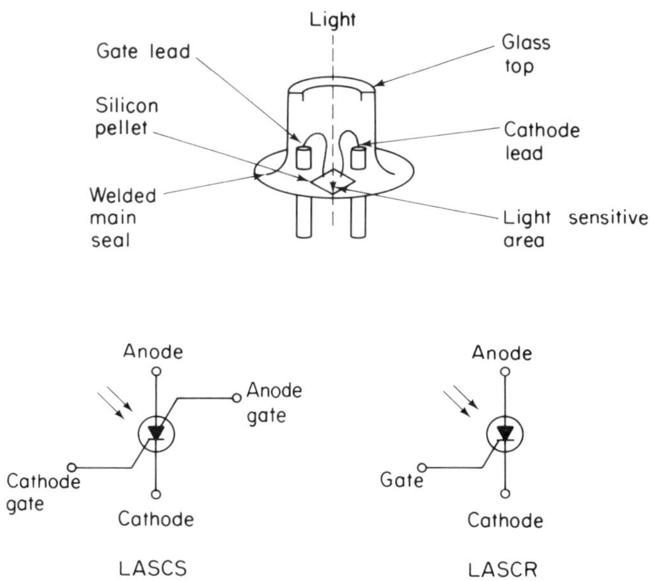


**FIGURE 3-33** Symbols and construction for LASCS and LASCR

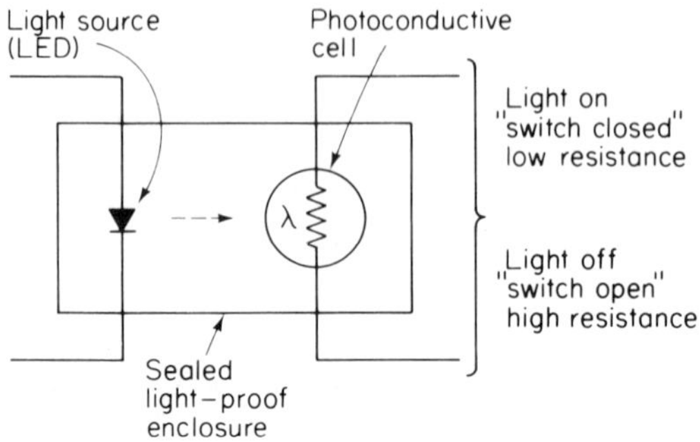


**FIGURE 3-34** Typical light activated switch or opto-coupler

has the same effect as closing the contacts of a switch. When all light is removed from the cell, the internal resistance increases to several million ohms. This has the same effect as opening the switch contacts.

In an opto-coupler, both the cell and a light source are sealed in a lightproof enclosure. When the light source is off, the photoconductive cell resistance is very high, and the "switch" is "open." When the light source is on, the light strikes the cell and causes the resistance to drop to zero. This "closes" the "switch." Such a switch is completely noise-free, and spark-free, which are important factors in many control applications. The light source can be an incandescent lamp. However, most present-day opto-couplers use a light-emitting diode (LED) similar to the LEDs used in digital readouts (as discussed in Chapter 6).

## 3-8 VOLTAGE VARIABLE CAPACITORS (VVCs)

The basic diode discussed in Chapter 2 has all the elements to form a capacitor. The anode and cathode act as conductors and form the capacitor "plates." The junction between anode and cathode acts as a dielectric (unless the junction is forward-biased). If the diode is reverse-biased, the junction is reinforced to further prevent any current flow. This makes the diode act more like a capacitor, where no current can flow between anode and cathode (except for possible leakage current).

The fact that a diode will act like a capacitor is usually an undesired characteristic in conventional diode operation. However, the characteristic can be put to good use when it is desired to have a

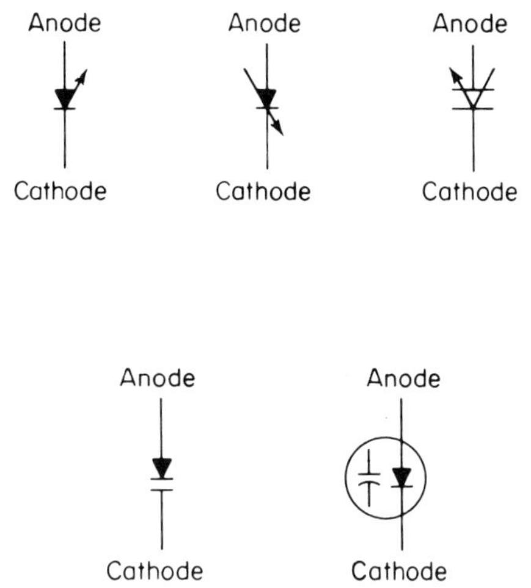


**FIGURE 3-35** Symbols for voltage variable capacitors (VVCs)

capacitor that will vary in value with changes in voltage. Several manufacturers produce voltage-variable capacitors (VVCs) under names such as Epicap, Capsil, Varactor, Voltacap, Paramp Diode Semicap, and Varicap. Similarly, there are a number of symbols used for diode capacitors, as shown in Fig. 3-35. Do not be surprised if you find any or all of these when reading schematics. Although any semiconductor material can be used for a VVC, silicon is used most often because of its high forward-to-reverse resistance and stable temperature characteristics.

You will most often find VVCs used in resonant circuits (Sec. 2-5). The capacitor value is controlled by a variable voltage. In turn, variations in capacitor value change the resonant frequency (or operating frequency) of the circuit. Thus, the circuit can be tuned by a variable voltage rather than by mechanical tuning of a variable capacitor.

## 3-9 SATURABLE REACTOR AND MAGNETIC AMPLIFIER

Saturable reactors and magnetic amplifiers operate on the principle of *variable reactance.* The basic circuit, shown in Fig. 3-36, looks like an ordinary transformer with an input and an output winding. The *control winding* has many turns of relative fine wire and carries little

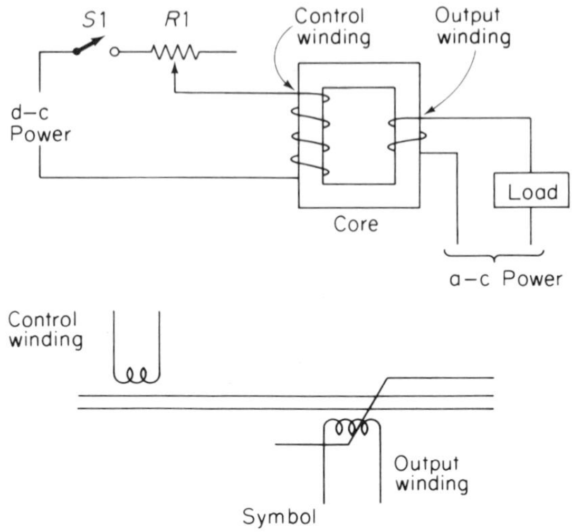


**FIGURE 3-36** Saturable reactor and magnetic amplifier

current. The *output winding* has fewer turns of heavier wire and carries a large current. In a typical circuit, the output winding is connected in series with the load or device to be controlled, and an a-c supply. The control winding has a variable resistance, switch, and a d-c supply.

Without going into any elaborate theory, the amount of direct current in the control winding determines the amount of alternating current in the output winding. Thus, it is possible to control a large alternating current with a small direct current. Since this is a form of amplification, the saturable reactor and the associated components are known as a *magnetic amplifier.* At one time, there was great hope that the magnetic amplifier (or mag amp) would be a substitute for vacuum-tube control systems. However, mag amps have higher levels of distortion than solid-state controllers and are generally limited to low-frequency operation (typically at the power-line frequency, 60 Hz or possibly up to 400 Hz).

Although the saturable reactor is not in common use as a magnetic amplifier, saturable reactors are very often found on the schematics of solid-state power supplies, as discussed in Chapter 4. As shown in Fig. 3-36, the symbol for a saturable reactor is different from that of a conventional inductance, and when a saturable reactor core is used in a transformer, the symbol is different from that of a conventional transformer.

## 3-10 INTEGRATED CIRCUITS (ICs)

Figure 3-37 shows the symbols used for integrated circuits. The most common symbol is a box with the terminals or connector pins (also called leads) identified by number. A triangle is often used when the symbol represents an IC amplifier. On some schematics, the IC is identified by reference designation and/or by manufacturer's part number.

Generally, parts within ICs are not shown on the schematic. One exception is where the IC is an *array* of parts such as shown in Fig. 3-38. An IC array provides several diodes, transistors, and other active devices within a single package. Because all the active devices are fabricated simultaneously and of the same material, they have (nearly) identical characteristics. This is particularly important in circuits where the characteristics must track each other with variations in temperature (such as a balanced two-channel amplifier). It is assumed that parts within IC arrays will be used together in some common circuits. Thus, the parts and their terminal or pin connections must be shown.

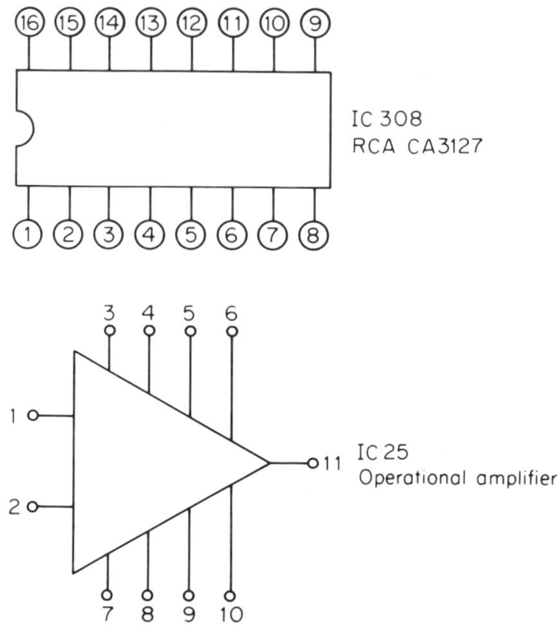


**FIGURE 3-37** Symbols for integrated circuits

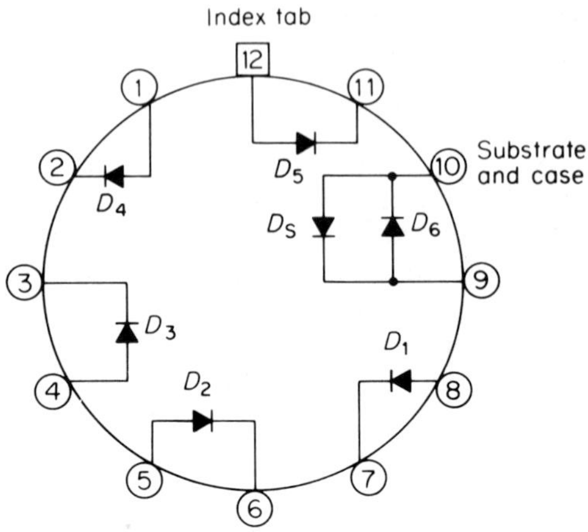


**FIGURE 3-38** Schematic for IC array

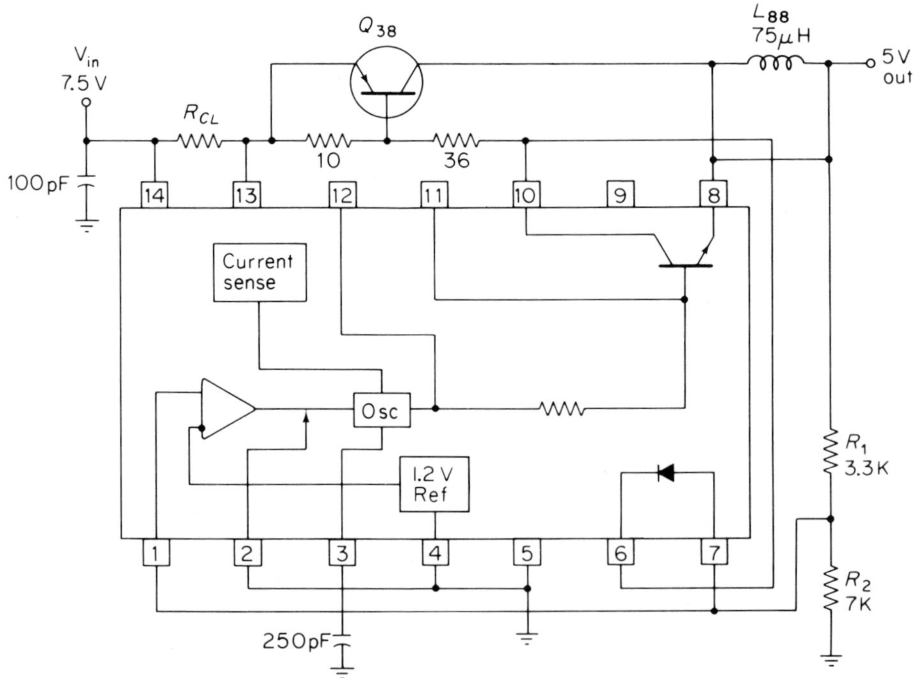


**FIGURE 3-39** Pinout diagram of a voltage regulator IC and external circuit parts

Figure 3-39 illustrates another case where IC parts are shown on the schematic. This is a *pinout* diagram of a voltage-regulator IC. All external parts and all key internal parts are shown to help the user understand operation of the circuit. Note that most of the internal parts are shown in block form.

## EXERCISES

3.1. Name one type of vacuum tube still in common use.

3.2. Name two types of electron-beam deflection used in CRTs.

3.3. What is the difference in the schematic symbols of a power diode and a signal diode?

3.4. Under what conditions will a zener diode pass current in the forward direction?

3.5. Under what conditions will a zener diode pass current in the reverse direction?

3.6. If you find a diode identified on a schematic as a 1N34A, what color code would you expect to find on the diode, and what is the maximum current capability of the diode? (Refer to Fig. 3-5.)

3.7. What are the three elements of a transistor shown on schematics?

3.8. What is the major difference between the NPN and PNP transistor symbol?

3.9. What is the preferred reference-designation letter for transistors?

3.10. What can you tell about the case style of a transistor from the schematic symbol?

3.11. Which element of a transistor is most often common or grounded?

3.12. In the schematic of Fig. 3-10, describe (briefly) the functions of C1, C2, R1, R2, RL, and RE.

3.13. What major change would you expect to find on the schematic of Fig. 3-10 if a PNP transistor is used in place of the NPN transistor shown?

3.14. Why is the output waveform shown in Fig. 3-10 not identical to the input waveform?

3.15. What are the input and output impedance characteristics of the circuit shown in Fig. 3-11?

3.16. What are the input and output impedance characteristics of the circuit shown in Fig. 3-12?

3.17. What is the typical or average voltage gain of the circuit shown in Fig. 3.12?

3.18. Name three functions of the bias network shown in Fig. 3-13.

**3.19.** What is the primary purpose of RE in Fig. 3-13?

**3.20.** What determines the approximate input and output impedances of the circuits shown in Fig. 3-14?

**3.21.** What is the major difference between the JFET and MOSFET transistor symbols?

**3.22.** Is the drain of an N-channel JFET (Fig. 3-15) normally connected to positive or negative on a schematic?

**3.23.** The gate, drain, and source elements of a FET correspond to the ____________, ____________ and ____________ of a two-junction transistor.

**3.24.** Is the dual-gate MOSFET shown in Fig. 3-15 N-channel or P-channel?

**3.25.** If you find a FET identified on a schematic as a 3N153, is it a JFET or a MOSFET, and what is the approximate gain or forward transconductance?

**3.26.** In the schematics of Fig. 3-18, describe (briefly) the functions of C1, C2, R1, R2, RL, and RS.

**3.27.** In a typical UJT circuit, the emitter is the ____________ and the base 1, base 2 circuit forms the ____________ .

**3.28.** What is the most common UJT circuit?

**3.29.** What is the main difference between the functions of a conventional diode rectifier and a controlled rectifier?

**3.30.** What is the main difference between the functions of an SCR and an SCS?

**3.31.** What is the main difference between the functions of an SCS and a triac?

**3.32.** Is the anode or the cathode of a triac (Fig. 3-27) connected to positive?

**3.33.** What is one of the most common devices used to trigger a traic?

**3.34.** The SUS equivalent circuit (Fig. 3-29) looks like an SCR with a zener, but the SUS operates more like a ____________.

**3.35.** The SBS is essentially two ____________ arranged inverse-parallel, and is particularly useful when the triggering pulses are ____________ .

**3.36.** What is the main difference between the functions of a photovoltaic cell and a photoconductive cell?

**3.37.** Photovoltaic cells are also called ____________ or ____________ .

**3.38.** Photoconductive cells are often called ____________ or ____________.

**3.39.** What identifies any light-activated device on a well-drawn schematic?

**3.40.** What is the main difference in the schematic symbol of a phototransistor and a conventional bipolar transistor?

**3.41.** What is the main difference in physical construction of a phototransistor and a conventional transistor?

**3.42.** Would you expect to find a photovoltaic cell or a photoconductive cell in an opto-coupler?

**3.43.** What is the main difference between the LASCR and the LASCS? How is this difference shown by the schematic symbol?

**3.44.** What controls the capacitance of a VVC?

**3.45.** What is the main difference in the schematic symbol of a saturable reactor and a conventional reactor (transformer, coil, inductor, etc.)?

**3.46.** What is the significance of the notch at the left end of IC308 shown on Fig. 3-37?

**3.47.** Why is IC25 on Fig. 3-37 shown as a triangle?

**3.48.** Why are parts of an IC array generally shown on the schematic?

**3.49.** What internal IC parts are shown on a pinout diagram?

**3.50.** Why are any internal parts shown on an IC pinout diagram?

# 4

# Reading Some Simple Circuit Schematics

In this chapter, we concentrate on reading the schematics of some very simple, but important, basic circuits. Such circuits are made up of the passive devices described in Chapter 2, and the active devices of Chapter 3. If you have been paying attention up to now, you should be able to recognize the symbols for the passive and active devices, and you should be able to describe (in general terms) their function. In effect, you should be able to see a schematic symbol and say "that is a diode, and it functions to pass current in one direction but not in the opposite direction."

By the time you have finished this chapter, you should be able to recognize complete circuits (amplifiers, oscillators, power supplies, waveform generators, control circuits, to name a few). You should also be able to describe the circuit functions. This will prepare you for Chapter 5, where we go on to see how the circuits are combined to form a complete piece of equipment (such as a TV set).

## 4-1 AUDIO AMPLIFIER CIRCUITS

Both two-junction transistors and FETs are used in audio amplifier circuits. FETs are usually limited to *voltage amplifier* applications. Two-junction transistors can be used in either voltage amplifier or *power amplifier* circuits. Since the number of different audio amplifier circuits is almost unlimited, it is impossible to cover all audio circuits here. Instead, we shall concentrate on a cross section of audio amplifier circuits. In reading audio circuits schematics, the primary points to remember are: the main function of an amplifier is to amplify signals applied to the input, and to deliver the amplified signals to the output. A secondary function of an audio amplifier circuit is to match impedances between input and output. The frequency of audio signals is generally considered anything up to about 20 kHz. However, many audio amplifiers will amplify signals at frequencies up to 100 kHz and higher.

### 4-1.1 BASIC AUDIO AMPLIFIER

Figure 4-1 is the schematic of a basic, single-stage, audio amplifier. Note that the basic audio circuit is similar to the transistor bias circuit discussed in Sec. 3-3 (and shown in Fig. 3-13). Input and output coupling capacitors C1 and C2 are added to prevent d-c flow to and from external circuits. Note that a *bypass capacitor* C3 is shown connected across the emitter resistor R3. Capacitor C3 is required only under certain conditions.

Input to the amplifier is applied between base and ground, across R2. Output is taken across the collector of Q1 and ground. The input signal adds to, or subtracts from, the bias voltage across R2. Variations in bias voltage cause corresponding variations in base current, collector current, and the voltage drop across collector resistor R4. Thus, the collector voltage (or circuit output) follows the input signal waveform, except that the output is inverted in phase. (If the input swings positive, the output swings negative, and vice versa.)

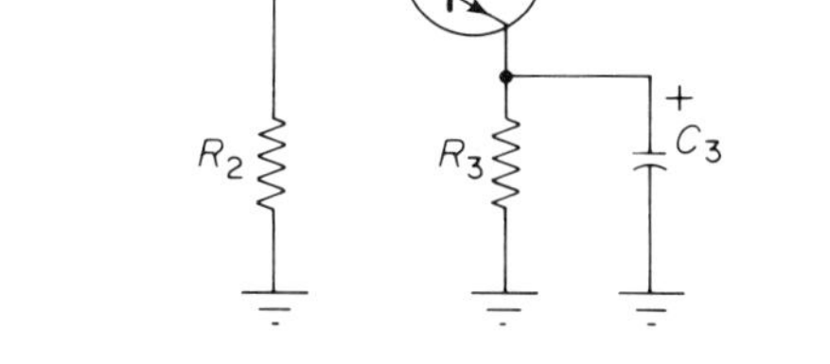


**FIGURE 4-1** Schematic of a basic, single-stage audio amplifier

Variations in collector current also cause variations in emitter current. This results in a change of voltage drop across the emitter resistor R3 and a change in the base-emitter bias relationship. The change in bias that results from the voltage drop across R3 tends to cancel the initial bias change caused by the input signal and serves as a form of negative feedback to increase stability (and limit gain). This form of emitter feedback (current feedback) is known as *stage feedback* or *local feedback*, since only one stage is involved. *Overall feedback*, or *loop feedback*, is sometimes used where several stages are involved.

The outstanding characteristic of the circuit in Fig. 4-1 is that circuit characteristics (gain, impedance, stability) are determined primarily by circuit values rather than transistor characteristics (such as beta).

### *4-1.2 BASIC AUDIO AMPLIFIER WITH EMITTER BYPASS*

Figure 4-1 shows a bypass capacitor C3 across emitter resistor R3. This arrangement permits R3 to be removed from the circuit as far as the signal is concerned, but leaves R3 in the circuit (in regard to direct current). An emitter bypass produces high signal gain, with good temperature stability. However, an emitter limits the low-frequency capability of the amplifier. Often, emitter bypass capacitors are electrolytic, so look for the polarity symbols on the schematic.

### *4-1.3 DIRECT-COUPLED VOLTAGE AMPLIFIER*

Figure 4-2 is a schematic of a two-state, direct-coupled amplifier. When stable voltage gains greater than about 20 are required, and it is not practical to bypass the emitter resistor of a single stage, two or

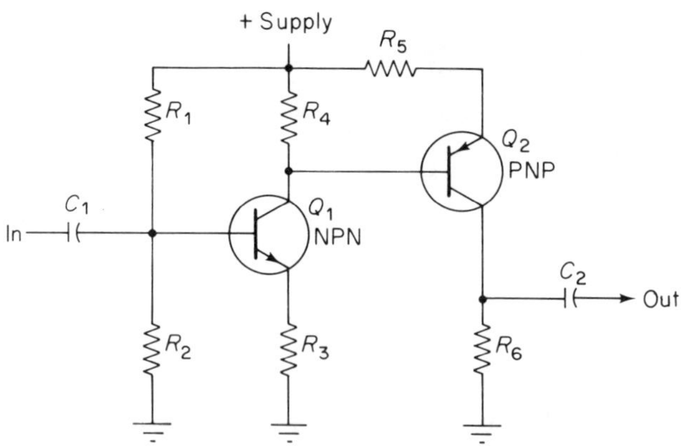


**FIGURE 4-2** Schematic of a two-stage, direct-coupled amplifier

more transistor amplifiers can be used. The output of one transistor is fed to the input of a second transistor. This is known as *cascading.* Generally, you will find no more than three direct-coupled transistors on a schematic, with two transistors adequate to do most jobs of voltage amplification. The stages can be coupled by means of capacitors. For example, two or three basic stages (such as shown in Fig. 4-1) can be connected together. More likely, you will find direct coupling, as in Fig. 4-2. This direct coupling eliminates the interstage coupling capacitors, as well as R1 and R2 for the second stage. Note that such direct-coupled pairs (often called *compounds*) use complementary NPN and PNP transistors, so that bias voltages are convenient.

### *4-1.4 TRANSFORMERLESS MULTISTAGE AUDIO AMPLIFIERS*

In addition to the direct-coupled amplifier, you will find a number of transformerless multistage circuits on schematics. These multistage circuits provide voltage, current, or power amplification of signals at audio frequencies. The circuits for the most important of these are the Darlington pair configurations (or Darlington compounds), the phase inverter or splitter, the emitter-coupled amplifier, the differential amplifier, the transformerless series output amplifier, the quasi-complementary amplifier, and the full-complementary amplifier. Each of these circuits, as well as various combinations of the circuits, is found on schematics in IC or hybrid packages. (Hybrid circuits are similar to ICs, except that ICs are usually completely functioning circuits, such as a complete audio amplifier, operational amplifier, etc.)

**Darlington Compounds.** Figure 4-3 shows the basic Darlington circuit (usually called the *Darlington compound*), together with two practical versions of the circuit, as you might find them on a schematic. As shown, the Darlington compound is an emitter follower (or common collector) driving a second emitter follower. As discussed in Chapter 3, an emitter follower provides no voltage gain but can provide considerable power gain.

The main reason for using a Darlington compound (especially in audio work) is to produce high current (and power) gain. For example, Darlington compounds are often used as audio drivers to raise the power of a signal from a voltage amplifier to a level suitable to drive a final power amplifier. Darlingtons are also used as a substitute for a driver section (or to eliminate the need for a separate driver).

**Multistage Darlingtons.** Darlington compounds need not be limited to two transistors. Three (and even four) transistors can be used in the Darlington circuit. A classic example of this is the General

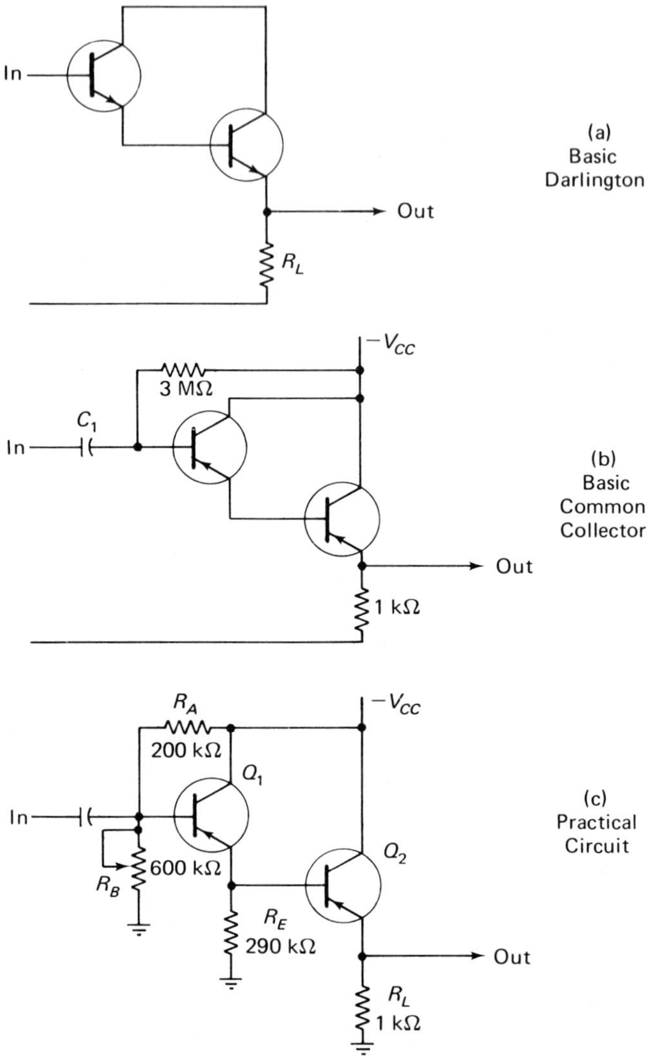


**FIGURE 4-3** Basic and practical Darlington compounds

Electric circuit of Fig. 4-4 (which is available in hybrid form). This circuit is essentially a common-collector and common-emitter Darlington, followed by a common-emitter amplifier.

**Phase Inverter or Splitter.** Figure 4-5 is the basic schematic of a phase inverter or splitter. As shown, the circuit is essentially a single-stage amplifier, with an output taken from the emitter as well as the collector. The values of R1 and R2 are usually equal, or near

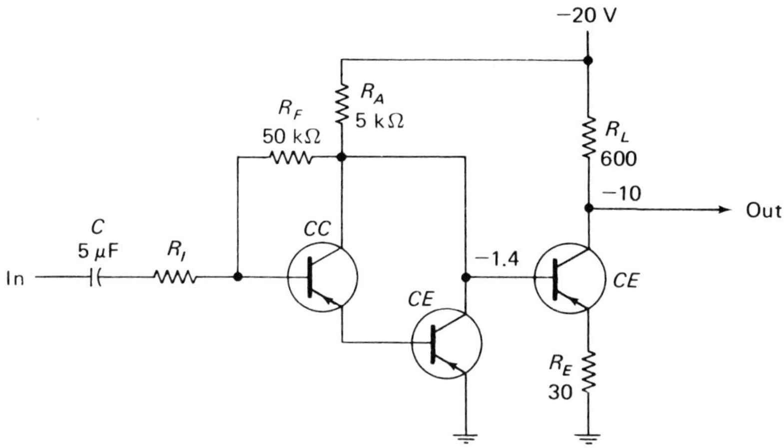


**FIGURE 4-4** General Electric CC-CE-CE multistage Darlington compound

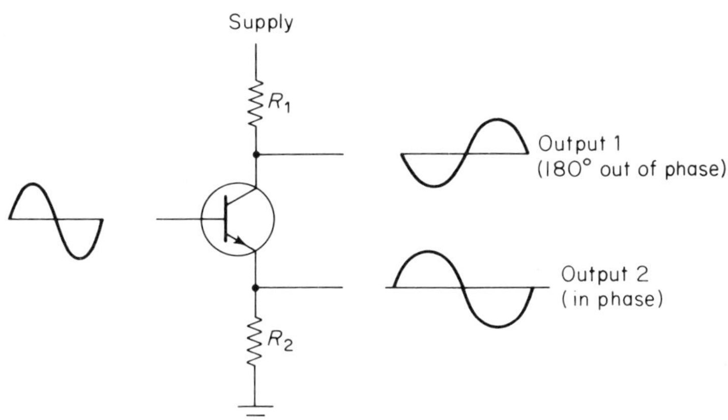


**FIGURE 4-5** Basic phase inverter or splitter

equal, so that there will be no voltage gain. The collector output is 180° out of phase with the input and, therefore, 180° out of phase with the emitter.

The main reason for using a phase inverter or splitter in audio work is to produce two 180° out-of-phase signals to drive a push-pull output stage. (This is the transformerless version of the *inter-stage transformer* discussed in Sec. 4-1.6.)

**Emitter-coupled Amplifier.** Figure 4-6 is the basic schematic of an emitter-coupled amplifier. This circuit is similar to the phase inverter in that two 180° out-of-phase outputs can be taken from

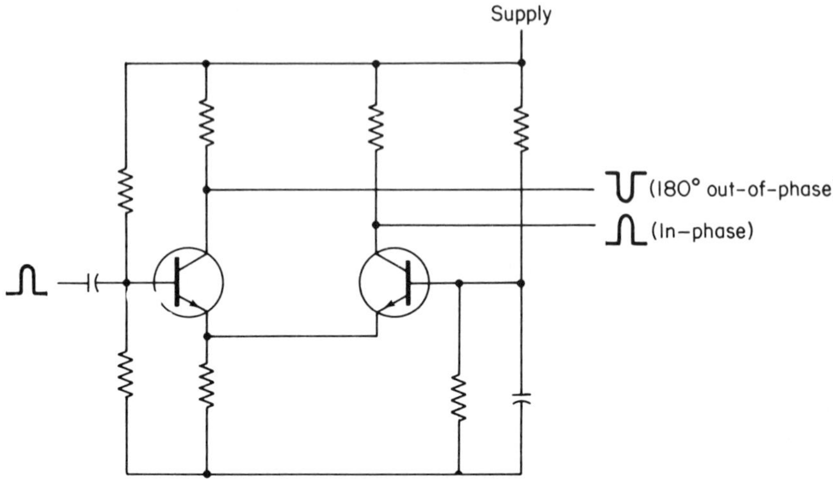


**FIGURE 4-6** Basic emitter-coupled amplifier

one input. Unlike the single-stage phase inverter, the emitter-coupled amplifier provides high gain. Therefore, an emitter-coupled amplifier can be used where a low-voltage input must be amplified to drive a push/pull output stage. (Push/pull is discussed in Sec. 4-1.6.)

**Differential Amplifier.** Figure 4-7 is the basic schematic of a differential amplifier. The differential amplifier is similar to the emitter-coupled amplifier, except that the two output signals are the result of a *signal difference* between the two inputs. An output is produced only when there is a difference in signals at the input. Differential amplifiers are often found on the schematics of oscilloscopes, electronic meters, and recording instrument amplifiers because signals common to both inputs (known as *common-mode signals*) are eliminated, or greatly reduced. A common-mode signal (such as power-line pickup) drives both bases in phase with equal amplified a-c voltages, and the amplifier behaves as though the transistors were in parallel. The emitter resistor produces emitter feedback, which reduces the common-mode signal gain, without reducing the differential signal gain.

**Transformerless Series Output Amplifier.** Figure 4-8 shows two typical transformerless series-output amplifiers found on audio circuit schematics. One configuration requires two power supplies but omits the coupling capacitor to the load. This circuit provides better low-frequency response (since there is no capacitor), but can be inconvenient because of the two power supplies. The circuit

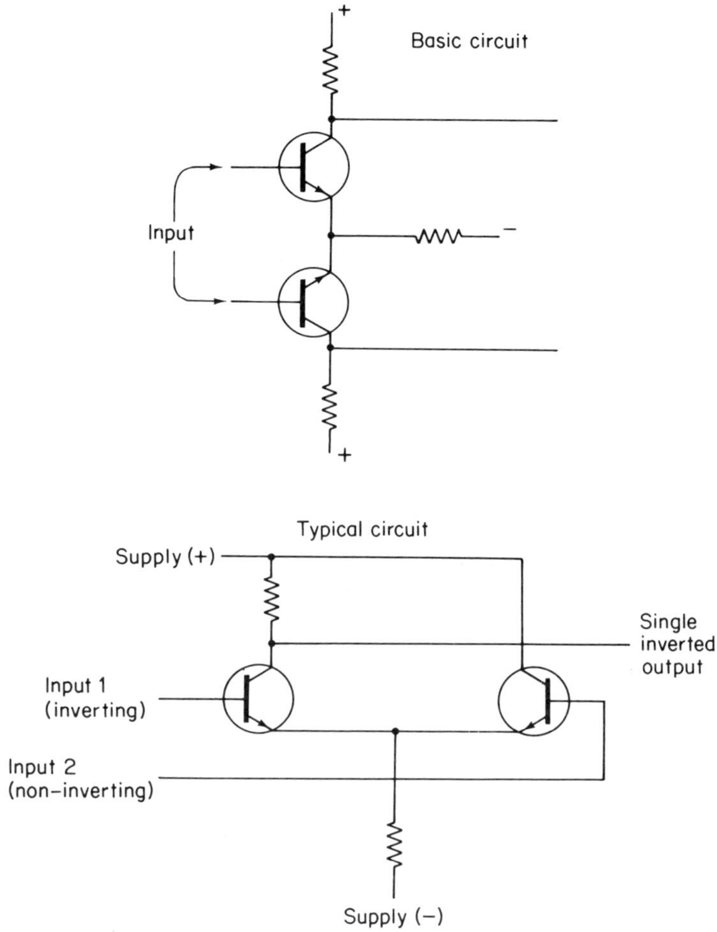


**FIGURE 4-7** Basic differential amplifier

with a single power supply has reduced low-frequency response because the capacitor forms a *high-pass* (or *low-cut*) *filter* with the load resistance. (This frequency filter effect is discussed further in Sec. 4-7.)

Either series output amplifier circuit has two drawbacks. A phase inverter is required to drive the series-output stage, even if gain is not required, since two input signals are required. Also, an additional driver stage may be necessary to bring the power up from the output of a voltage amplifier to a level required by the series-output stage. These problems are overcome by means of a *complementary amplifier*, either quasi-complementary or full-complementary.

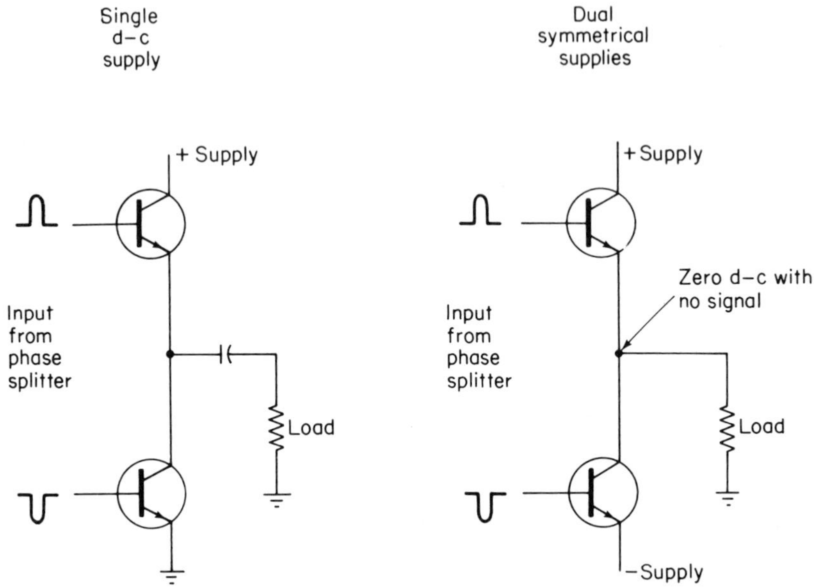


**FIGURE 4-8** Basic transformerless series output amplifier

**Quasi-complementary Amplifier.** Figure 4-9 is the schematic of a quasi-complementary amplifier output. This circuit consists of a Darlington compound using NPN transistors and a direct-coupled compound using an input PNP and an output NPN. Both base signals can be in phase (although are often at different voltage levels) so that a phase inverter is not needed.

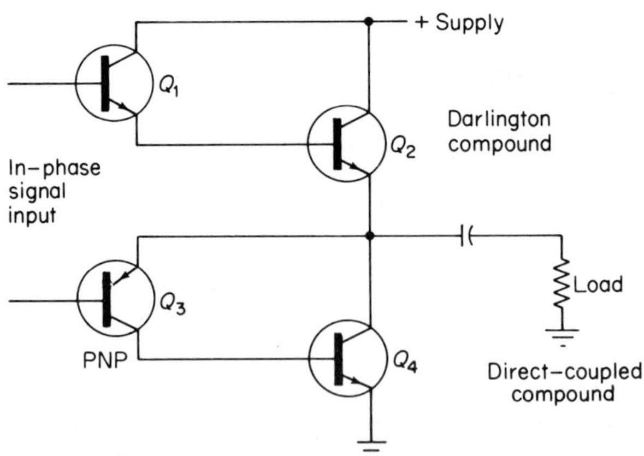


**FIGURE 4-9** Basic quasi-complementary amplifier

For example, if the input is positive-going, Q1 will be forward-biased, as is Q2. A positive-going input at Q3 will reverse bias Q3 since Q3 is PNP. This produces a negative output from Q3 to Q4, which is NPN, and reverse-biases Q4. When the input signal at the bases is negative-going, the condition reverses (Q3 reverse-biased, Q4 forward-biased).

**Full Complementary Amplifier.** Figure 4-10 shows two versions of the full-complementary amplifier output. Either version has an

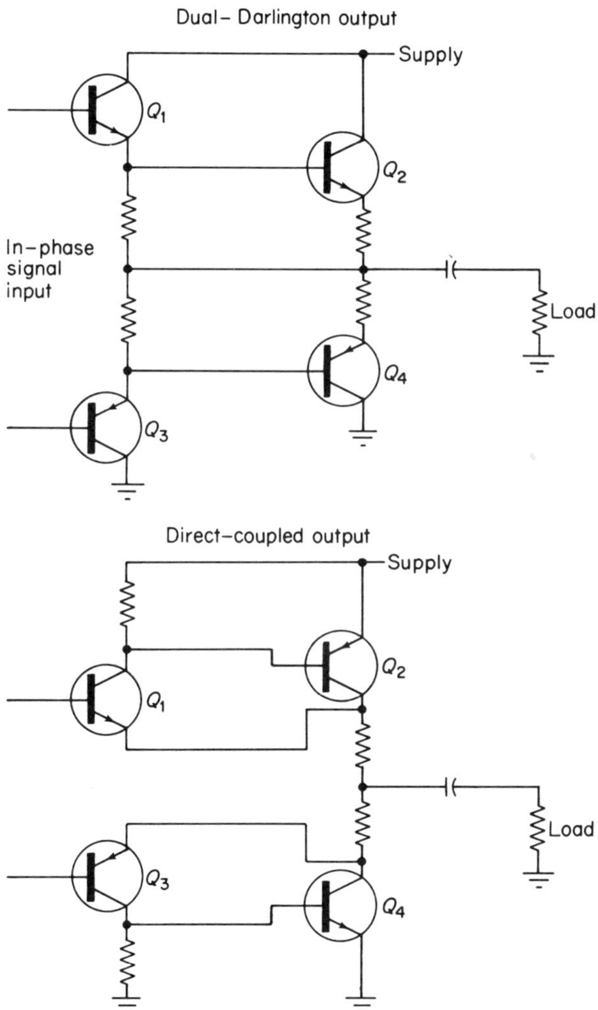


**FIGURE 4-10** Basic full-complementary amplifier

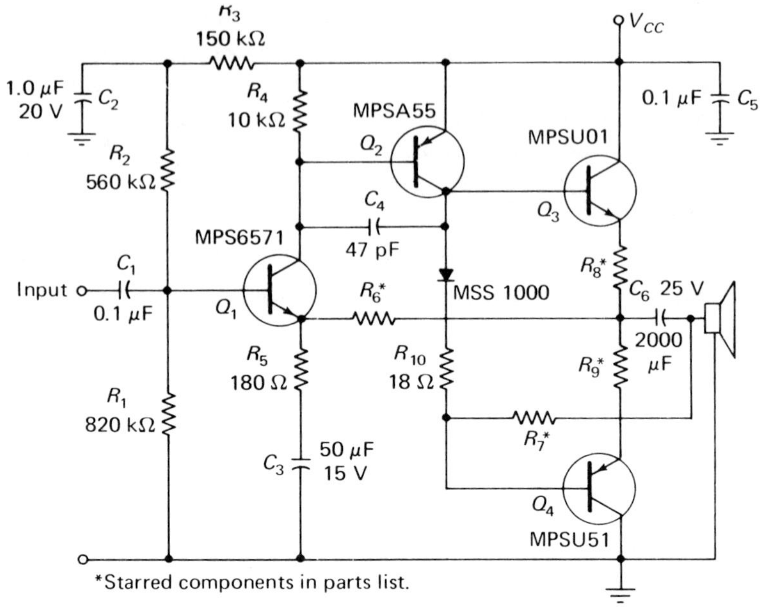


Parts List

| | 3 W | 5 W |
|---|---|---|
| $V_{CC}$ | 18 V | 22 V |
| $R_6$ | 3.9 kΩ | 4.7 kΩ |
| $R_7$ | 470 Ω | 560 Ω |
| $R_8$ | 0.75 Ω | 0.47 Ω |
| $R_9$ | 0.75 Ω | 0.47 Ω |

| | Amplifier Output | |
|---|---|---|
| | 3 W | 5 W |
| Idle current, nominal | 25 mA | 25 mA |
| Current drain at rated power output | 290 mA | 370 mA |
| Nominal sensitivity for rated power output | 0.25 V | 0.25 V |
| THD at rated maximum power output, 50 Hz to 15 kHz, nominal | 2% | 2% |
| THD at rated power output, nominal | 0.25 % | 0.25% |
| Input impedance, typical | 300 kΩ | 300 kΩ |
| Maximum power output at 5% THD, without current-limiting diodes | 4.5 W | 6.7 W |
| Maximum power output at 5% THD with current-limiting diodes* | 4.0 W | 6.0 W |

*See text.

**FIGURE 4-11** Four-transistor transformerless audio amplifier (courtesy Motorola)

advantage over the quasi-complementary in that both halves of the circuit are identical. This makes it easier to match both halves (for positive and negative signals) to minimize distortion that could be caused by uneven amplification of the signal.

One complementary circuit uses two Darlington compounds. This circuit is also known as a *dual-Darlington* output, and is used where power gain is needed. The other circuit uses two direct-coupled compounds, and is used where voltage gain is most needed. Phase inversion is not needed for either circuit. A positive-going input will forward-bias Q1 and Q2 and reverse-bias Q3 and Q4. A negative-going input will produce opposite results. As in the case of series-output circuits, the load-coupling capacitor can be omitted if two power supplies are used (one positive and one negative).

### *4-1.5 EXAMPLE OF TRANSFORMERLESS DIRECT-COUPLED AUDIO AMPLIFIERS*

Figure 4-11 is the schematic of an audio amplifier developed by Motorola using their complementary plastic transistors. The four-transistor circuit is capable of delivering 2 to 5 W of audio power with an approximately 0.25-V input signal. The circuit is essentially a series-output amplifier with a single power supply (of about 45 to 50 V).

### *4-1.6 MULTISTAGE AUDIO AMPLIFIER WITH TRANSFORMER COUPLING*

Audio amplifier stages are often coupled by means of transformers. There are advantages and disadvantages in using transformers in audio circuits. Transformer coupling is generally used where low-voltage operation is required or where impedances must be matched. This offsets the disadvantages of added weight and bulk of transformers, as well as frequency limitations. Below about 20 Hz, transformers act as a short across the amplifier. Above about 20 kHz, transformers have such high reactance that they appear as high resistances to signals passing through the amplifier.

Figure 4-12 is the schematic of a transformer-coupled audio amplifier. The circuit has a class A input or *driver* stage, and a class B push/pull output stage. The class A stage provides both voltage and power amplification as needed to raise the low input signal to a level suitable for the class B power output stage. The class A stage can be transformer-coupled or *RC*-coupled at the input as needed. Transformer coupling is used at the input, where a specific impedance-match problem must be considered in design. The class A input stage can be driven directly by the signal source or can be used with a

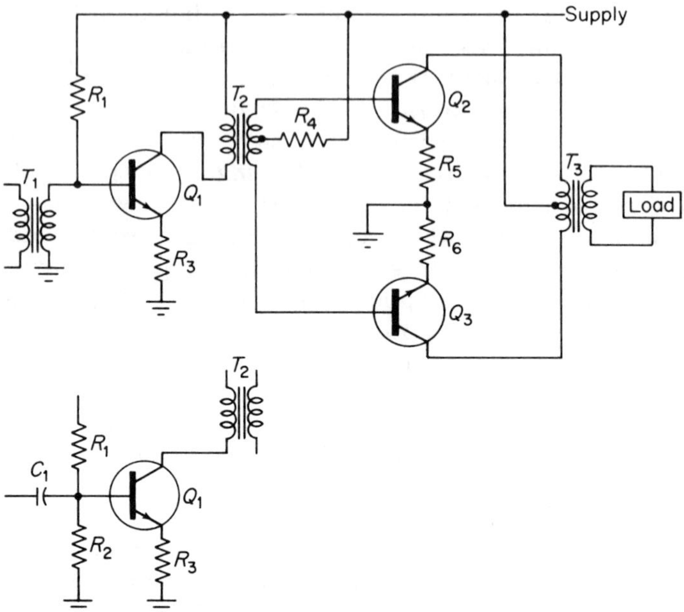


Alternate driver circuit with *RC* input

**FIGURE 4-12** Transformer-coupled audio amplifier (courtesy Motorola)

preamplifier for very low-level signals. When required, a high-gain voltage amplifier (such as a single stage with emitter bypass as described in Sec. 4-1.2) is used as a *preamplifier.*

The push/pull output stage may be operated as a class B amplifier. (The stage is called push/pull because one transistor is turned on, with the other transistor turned off, for each half-cycle of the signal. On one half-cycle, the polarity of the signal at one end of transformer T2 is positive, turning on the corresponding transistor. On the opposite half-cycle, the signal polarity at the output winding of T2 reverses, turning on the other transistor.) As discussed in Chapter 3, transistors are normally cut off and draw collector current only in the presence of an input signal when class B operation is used. Class B is the most efficient operating mode for audio amplifiers, since it draws the least amount of current (and no current where there is no signal). However, true class B operation can result in *crossover distortion.*

The effects of crossover distortion can be seen by comparing the input and output waveforms in Fig. 4-13. In true class B operation, the transistor remains cut off at very low signal inputs (because transistors have low current gain at cutoff) and turns on abruptly

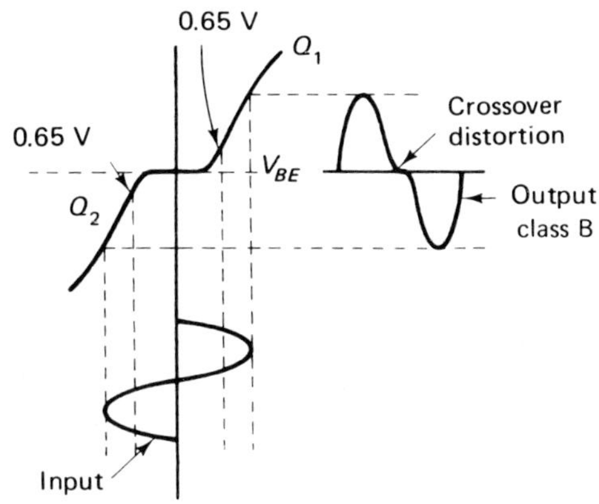


(a)

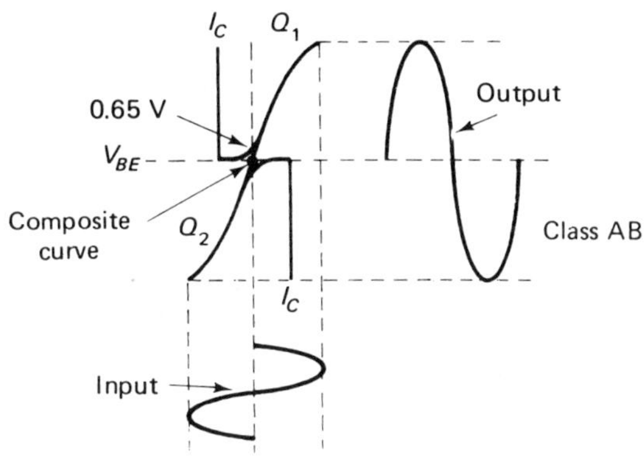


(b)

**FIGURE 4-13** Effects of crossover distortion, and how it is eliminated by forward bias of the base-emitter junction

with a large signal. As shown in Fig. 4-13a, there is no conduction when the base-emitter voltage ($V_{BE}$) is below about 0.65 V (for a silicon transistor). During the instantaneous pause when one transistor stops conducting and the other starts conducting, the output waveform is distorted.

Crossover distortion can be minimized by operating the output stage as class AB (or somewhere between B and AB). That is, the transistors are forward-biased (by a voltage drop across R4 in Fig. 4-12) just enough for a small amount of collector current ($I_C$) to flow with no signal. Some collector current is flowing at the lowest signal levels, and there is no abrupt change in current gain. The effects of this are shown in Fig. 4-13b. The combined collector currents result in a *composite curve* that is essentially linear at the crossover point. This produces an output that is faithful reproduction of the input, at least as far as the crossover point is concerned. Of course, class AB is less efficient than class B, since more current is used. Note that it is impossible to tell the class of operation from information on the schematic. However, it is generally a safe bet that any push/pull audio circuit is operated class B or AB. Single/ audio stages such as Q1 in Fig. 4-12 are usually operated class A.

### *4-1.7 AUDIO OPERATING AND ADJUSTMENT CONTROLS*

The most common operating controls for audio circuits used with music or voice reproduction equipment (hi-fi, stereo, public address, etc.) are the *volume* or *loudness* control, the *treble* control, and the *bass* control. The other most common audio control is the *gain* control (found on such circuits as the operational amplifier, industrial control amplifier, etc.). The gain and volume controls are often confused, since they both affect output of the amplifier circuit. A true gain control sets the *gain* of one stage in the amplifier, and thus sets the overall gain of the complete amplifier. A true volume control set the *level of the signal* passing through the amplifier, without affecting the gain of any or all stages. A gain control is usually part of a stage, while a volume control is usually found between stages or at the input to the first stage.

**Volume Control.** Figure 4-14 is the schematic of a basic volume control, which is a variable resistance or potentiometer connected as a voltage divider. The voltage output (or signal level) is dependent upon the volume control setting. Generally, the volume control is located at a low-signal-level point in the amplifier. The most common location for a volume control is at the amplifier input stage, or between the first and second stages. Capacitors C1 and C2 act as coupling capacitors for the signal and serve to isolate the volume control from other circuit elements (as far as dc is concerned). Note that C1 forms a high-pass (low-cut) *RC* filter with volume control R1, and C2 forms another high-pass filter with the input resistance of the following stage. (Filters are discussed in Sec. 4-7.)

**Gain Control.** Figure 4-15 is the schematic of a basic gain control, which is a variable resistance serving as the emitter resistance of

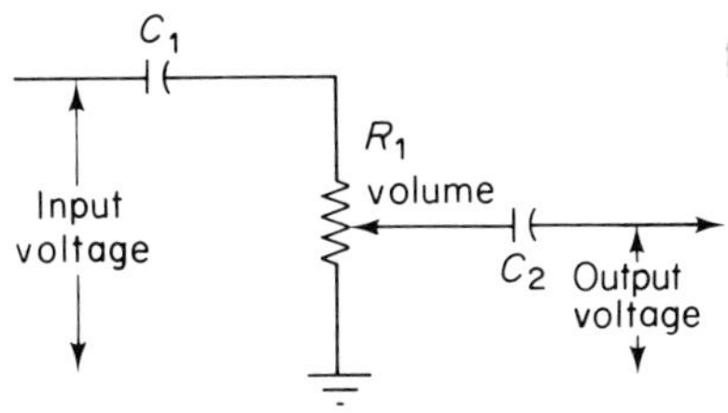


**FIGURE 4-14** Basic volume control circuit

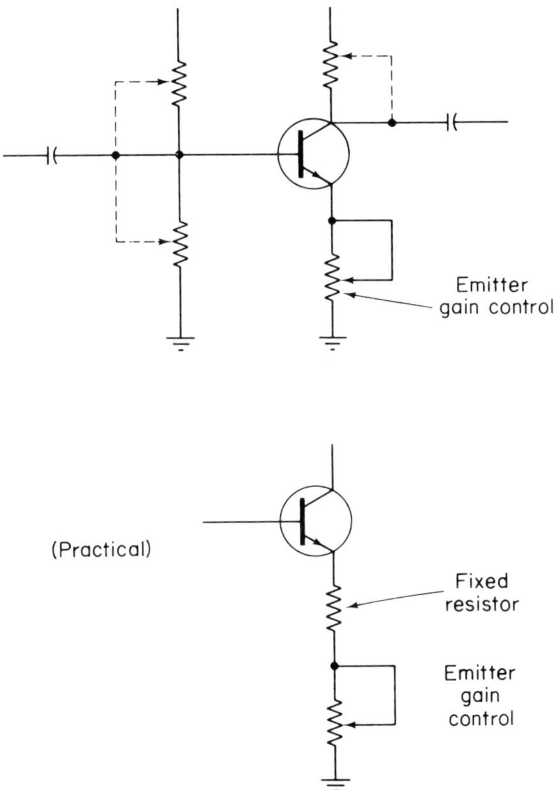


**FIGURE 4-15** Basic gain control circuit

an amplifier. Any of the three resistors could be used as the gain control, since stage gain is related to each resistance value (all other factors remaining constant). However, the emitter resistance is the most logical choice for a gain control. With all other factors remaining constant, a decrease in emitter resistance raises both the current gain and voltage gain (since there is less negative feedback). An increase in emitter resistance lowers stage gain.

**Treble Control.** Figure 4-16 is the schematic of a basic treble control. Such controls provide a means of adjusting the high-frequency response of an audio amplifier. These adjustments may be necessary because of variations in response of the human ear or to correct frequency response of a particular recording. As shown in Fig. 4-16, a typical treble control circuit consists of a variable resistance and a fixed capacitor. These two components form a *feedback circuit* between two stages. At any given frequency, the amount of feedback (and thus the frequency response) is set by adjustment of the variable resistance. As frequency increases, the capacitor reactance decreases. This also results in a change of feedback (and frequency response). Although the feedback circuit has some effect on voltage gain at all frequencies, the greatest effect is on the high-frequency end of the audio range.

**Bass Control.** Figure 4-17 is the schematic of a basic bass control. Such controls provide a means of adjusting the low-frequency response of an audio amplifier. These adjustments may be necessary because of variations in response of the human ear or because of other circuit conditions. As shown in Fig. 4-17, a typical bass control consists of a variable resistance and a fixed inductance. These two components form a bypass circuit around the volume-control coupling capacitor. The bass control could be placed elsewhere in the circuit, around another coupling capacitor. However, by placing the bass circuit at the volume control, it is impossible to boost low-frequency signals more when the volume is low.

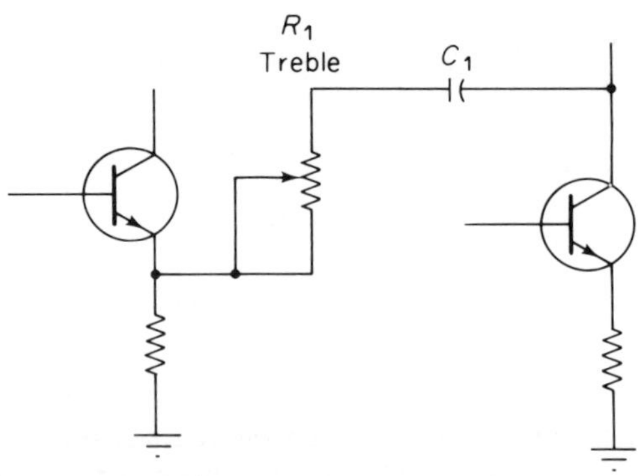


**FIGURE 4-16** Basic treble control circuit

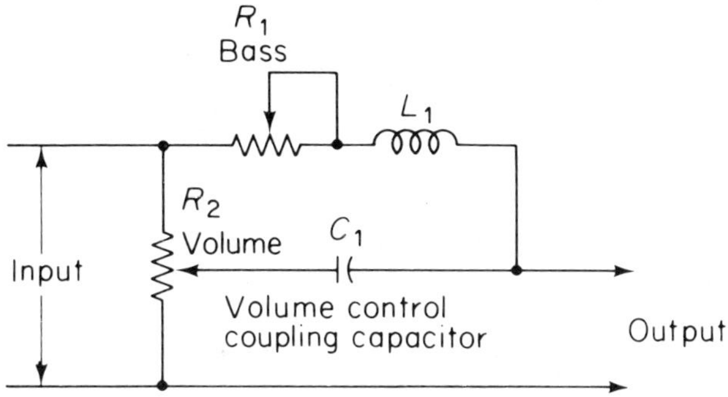


**FIGURE 4-17** Basic bass control circuit

As the frequency is lowered, the inductance of L1 decreases, bypassing more of the signal around the coupling capacitor C1. At any frequency, the amount of bypass (and thus the frequency response) is set by adjustment of the bass potentiometer R1. Since the upper end of the volume control R2 is bypassed by L1 and R1, the amount of bypass increases as the volume control is set to low level. Thus, the bypassed low-frequency signals will be stronger than the high-frequency signals at low-volume levels.

## 4-2 RADIO-FREQUENCY AND INTERMEDIATE-FREQUENCY AMPLIFIER CIRCUITS

Radio-frequency (RF) amplifiers are found mostly in radio receivers and transmitters. Thus, to fully understand operation of *RF* circuits, it is necessary to know how such circuits are used in receivers and transmitters, and how radio signals (including TV signals) are transmitted and received. We discuss all of this in Chapter 5. In this section, we will concentrate on the *RF* circuits themselves, so that you will be able to recognize them on schematics.

As in the case of audio amplifiers, both two-junction and FET transistors can be used in RF and IF (intermediate-frequency) amplifiers. Again, FETs are used for voltage amplifiers, whereas two-junction transistors are used for both voltage and power amplification. Note that we use the term IF along with the term RF. An IF amplifier is essentially a special-purpose RF amplifier used in radio and TV receivers. Typically, an IF amplifier is tuned over a very narrow band of signal frequencies, as discussed in Chapter 5.

Operation of RF and IF amplifiers is based on the use of *resonant circuits* (sometimes called *tank circuits*), consisting of a capacitor and a coil (inductance) connected in series or parallel, as discussed in Sec. 2-5. At the resonant frequency, the inductive and capacitive reactances are equal, and the circuit acts as a high impedance (if it is a parallel circuit) or a low impedance (if it is a series circuit). In either case, any combination of capacitance and inductance will have some resonant frequency.

Either (or both) the capacitance or inductance can be variable to permit tuning of the resonant circuit over a given frequency range. When the inductance is variable, tuning is usually done by means of a metal slug inside the coil. The metal slug is screwdriver-adjusted to change the inductance (and thus the inductive reactance) as required. Typical RF and IF circuits used in receivers (AM, FM, TV, communications, etc.) often include two resonant circuits in the form of a transformer (RF or IF transformer, etc.). Either the capacitance or inductance can be variable in such transformers.

### *4-2.1 TRANSMITTER RF CIRCUITS*

As discussed in Chapter 5, a typical transmitter consists of three stages: oscillator, multiplier, and final or power amplifier. RF oscillators are discussed in Sec. 4-3. Multipliers and power amplifiers are discussed here.

Figure 4-18 shows the schematics of typical RF power amplifiers. The same basic circuits can be used as frequency multipliers. However, in a multiplier circuit, the output must be tuned to a multiple of the input. A multiplier may or may not provide amplification. Usually, most of the amplification is supplied by the final amplifier stage, which is not operated as a multiplier. That is, the input and output of the final stage (generally connected to the transmitting antenna) are at the same frequency.

Note that the circuit of Fig. 4-18a has two tuning controls (variable capacitors) in the output network, while the network circuit of Fig. 4-18b has only one adjustment control. The circuit of Fig. 4-18a is typical for power amplifiers, where the output must be tuned to the resonant frequency by one control and adjusted for proper impedance match (usually to the transmitting antenna) by the other control (often known as the *loading* control). In practice, both controls affect tuning and loading (impedance matching). The circuit of Fig. 4-18b is typical for multipliers or intermediate amplifiers, where the main concern is tuning to the resonant frequency. The fixed parallel capacitor makes it possible to use a smaller variable capacitor, and prevents damage to the transistor when the variable capacitor is set to minimum.

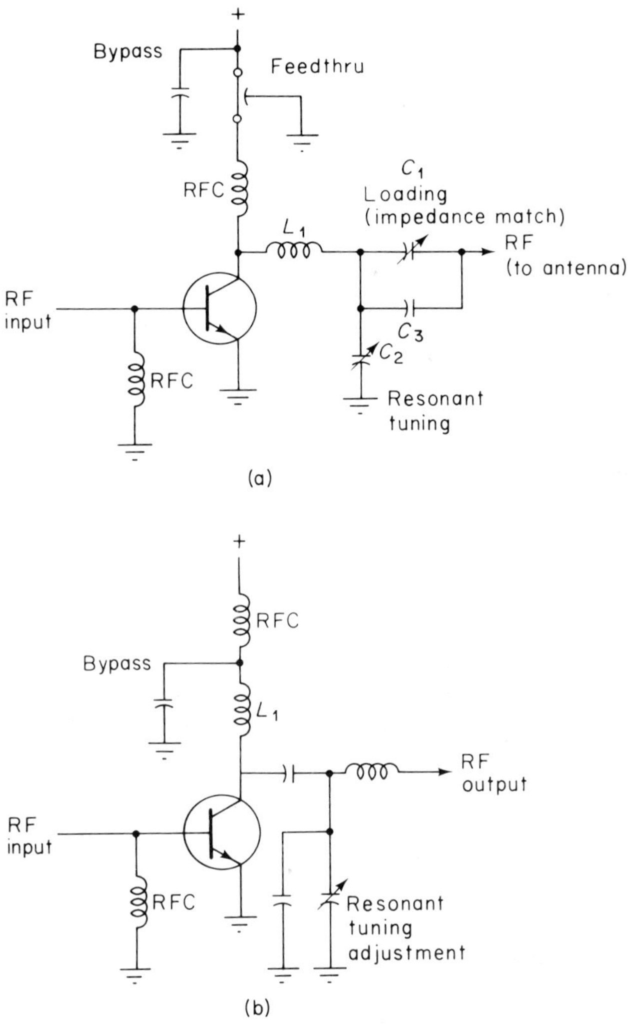


**FIGURE 4-18** Typical transmitter RF power amplifiers

RF amplifiers are usually operated class C (where the transistor conducts only in the presence of an input signal, either on the positive half or negative half, depending on the transistor type, NPN or PNP). Therefore, no fixed bias voltage is required for most RF amplifiers. The emitter is connected directly to ground, and the direct current path to the base is completed by a radio-frequency choke (RFC). The transistor collector is connected to the supply voltage through an RFC and (in some cases) through the coil portion of the resonant network. The RFC isolates the RF signal from the power supply. The power supply is bypassed to ground with either feedthrough capacitors or conventional capacitors. Either way, the RF signals are passed to ground rather than back to the power supply. Feedthrough capacitors are used at higher frequencies where the RF circuits are physically shielded from the power supply and other circuits.

### *4-2.2 RF VOLTAGE AMPLIFIERS*

Figure 4-19 is the schematic of an RF voltage amplifier. Such circuits are used primarily in receivers and receiver-type circuits. An IF amplifier or IF limiter amplifier are examples of voltage amplifiers. The input or first stage of a receiver may include a separate RF voltage amplifier (such as with some communications receivers). However, most solid-state receivers combine the RF voltage amplifier function with that of the local oscillator. Such circuits are discussed in Sec. 4-2.3 and in Chapter 5.

The circuit of Fig. 4-19 can be used as an IF amplifier, IF limiter, or separate RF amplifier, with few modifications. Both the input and output are tuned to the desired operating frequency by means of the resonant circuits. In this case, the resonant circuits are composed of transformers with a capacitor across the primary. The capacitors could be variable but are usually fixed. The resonant circuit is tuned by an adjusting slug between the windings.

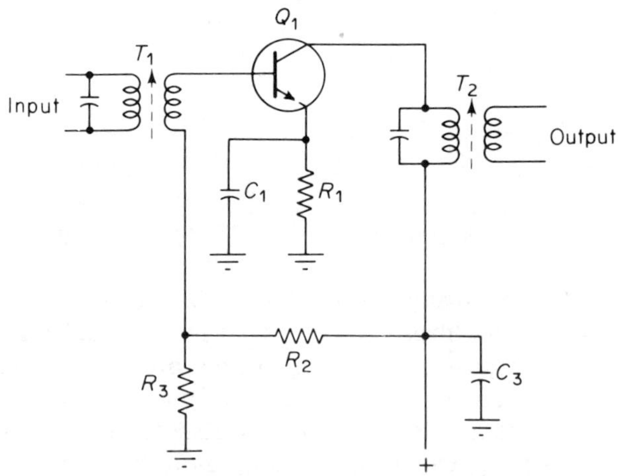


**FIGURE 4-19** Typical receiver RF voltage amplifier

The circuit is provided with a fixed bias through voltage divider resistors (R2, R3) and an emitter resistor (R1). This makes for a stable circuit with emitter feedback. Emitter resistor R3 is bypassed with capacitor C1 to provide high voltage gain. In some circuits, the fixed bias is eliminated (the secondary winding of T1 is returned directly to ground, and R2, R3 are omitted).

### 4-2.3 FREQUENCY MIXERS AND CONVERTERS

Figure 4-20 is the schematic of a typical frequency mixer and converter. Such a circuit is actually a combination of an RF voltage amplifier and an RF oscillator (Sec. 4-3). The individual outputs of the two sections are combined to produce an IF output, as discussed in Chapter 5. Usually, the RF oscillator operates at a frequency above the RF amplifier, with the difference in frequency being the intermediate frequency. The resonant circuit of T1 is tuned to the incoming RF signal, T2 is tuned to the oscillator frequency (RF + IF), and T3 is tuned to the intermediate frequency (IF). The resonant circuits of T1 and T2 are usually tuned by means of variable capacitors, ganged together so that both the oscillator and RF amplifier will remain at the same frequency relationship over the entire tuning range. For example, if T1 is tuned from 550 to 1600

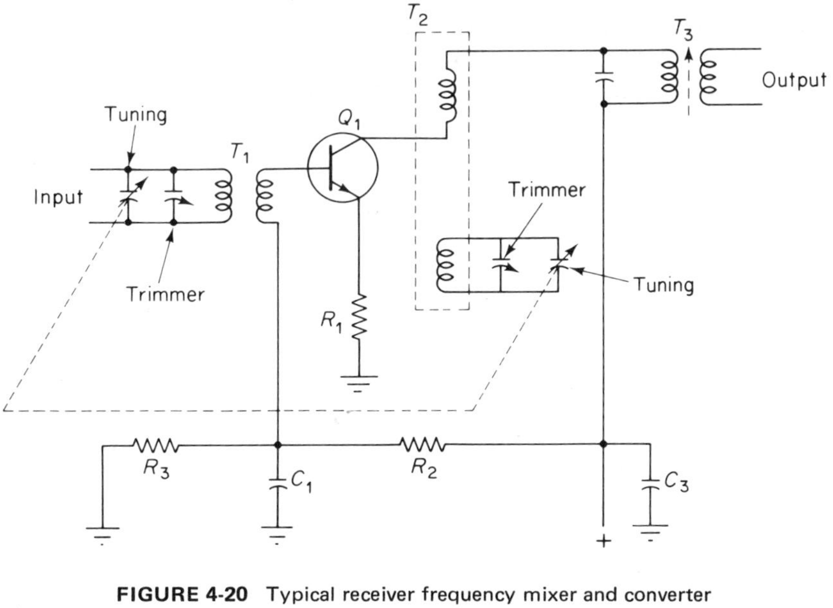


**FIGURE 4-20** Typical receiver frequency mixer and converter

kHz (the broadcast band), T3 is at a fixed intermediate frequency of 455 kHz, T2 tunes from 1005 to 2055 kHz. Usually, trimmer capacitors are placed in parallel with the variable capacitors to permit adjustment over the tuning range.

### *4-2.4 AM (AMPLITUDE-MODULATION) DETECTORS*

Figure 4-21 shows the schematics for two typical diode detectors, such as are found in AM radio and TV receivers. These detectors are essentially voltage detectors or rectifiers used to convert RF signal voltages into a d-c voltage. As discussed in Chapter 5, if there is audio-frequency information present on the RF signal (in the form of amplitude modulation), the detector functions to convert the audio voltage into pulsating d-c voltages.

Usually, the detector stage follows one or more IF stages. The output of the detector stage (pulsating d-c voltages at audio frequencies) is applied to the receiver audio section. In the circuit of Fig. 4-21a, the detector output is applied to a volume control in the audio circuits. In Fig. 4-21b, the detector load resistor acts as the volume control, eliminating the need for further controls in the audio circuits. Capacitor C1 provides a path for RF signals around resistance R1. These RF signals are converted to direct current by the detector CR1. This direct current passes through load resistance R1. If there is audio-frequency modulation present in the RF signals, the direct current passing through R1 varies (or pulsates) at the audio frequency. The output of the detector circuit is thus a pulsating direct current which can be amplified by the audio stages of the receiver.

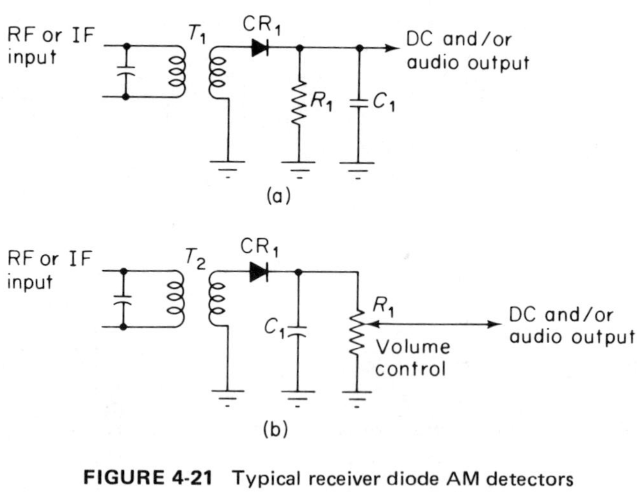


**FIGURE 4-21** Typical receiver diode AM detectors

### 4-2.5 *FM (FREQUENCY-MODULATION) DISCRIMINATOR AND RATIO DETECTOR*

Figure 4-22 shows the schematics for a discriminator and ratio detector, respectively. Discriminators and ratio detectors are used with FM radio receivers. Both circuits function to remove audio-frequency information present on the RF signal (in the form of frequency modulation) and to convert the audio voltages into pulsating d-c voltages. In either circuit, a diode is used as the basic detector. A ratio detector has an advantage in that it can be driven directly by an IF stage. A discriminator usually requires a limiter stage between the IF amplifier output and the discriminator input.

In both types of FM detector, the input transformer T1 shows an extra pickup winding in addition to the primary and secondary windings. A portion of the primary signal (from the IF amplifier or limiter stages) is sampled by the pickup coil and injected into the electrical center of the circuit. Not all discriminator and ratio detector transformers are so constructed. An alternative method is to obtain the sampled portion by eliminating the pickup coil and

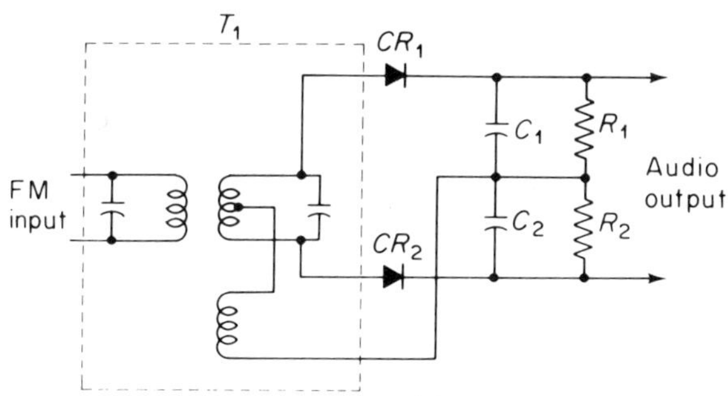


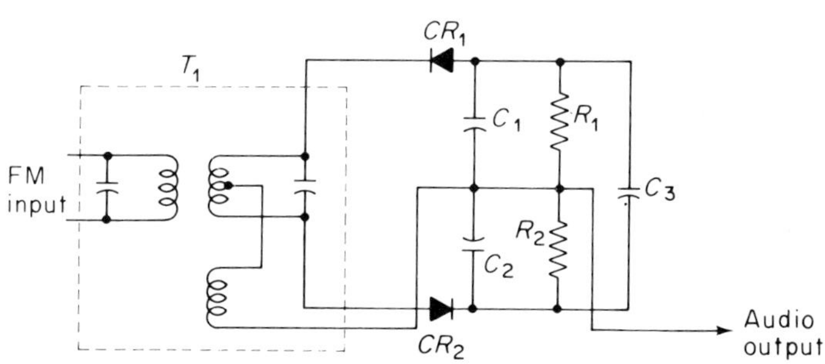


**FIGURE 4-22** Typical receiver FM **(a)** discriminator and **(b)** ratio detector

coupling this line to the primary by a capacitor. Discriminator and ratio detectors used in broadcast receivers are tuned to the intermediate frequency of 10.7 MHz.

With either FM detector, pulsating audio-frequency voltages are developed across load resistors R1 and R2. If there is no frequency modulation (the incoming signals remain at some center frequency), the voltage across R1 equals that across R2, but the polarities of the voltages are opposite. Thus, the voltages cancel and there is no output. If there is some frequency modulation (the incoming signals vary above and/or below the center frequency), the voltage across R1 is greater (or lesser) than that across R2, and the voltages do not totally cancel. Thus, there is a pulsating d-c output that corresponds to the amount of deviation above and/or below the center frequency.

### *4-2.6 AVC-AGC CIRCUITS*

Figure 4-23 shows the schematic of two AGC systems that are common to broadcast and communications receivers. Most receivers have some form of AVC-AGC (automatic volume control/automatic gain control) circuit. The terms AVC and AGC are used interchangeably on schematics. AGC is a more accurate term since the circuits involved control the gain of an IF or RF stage rather than

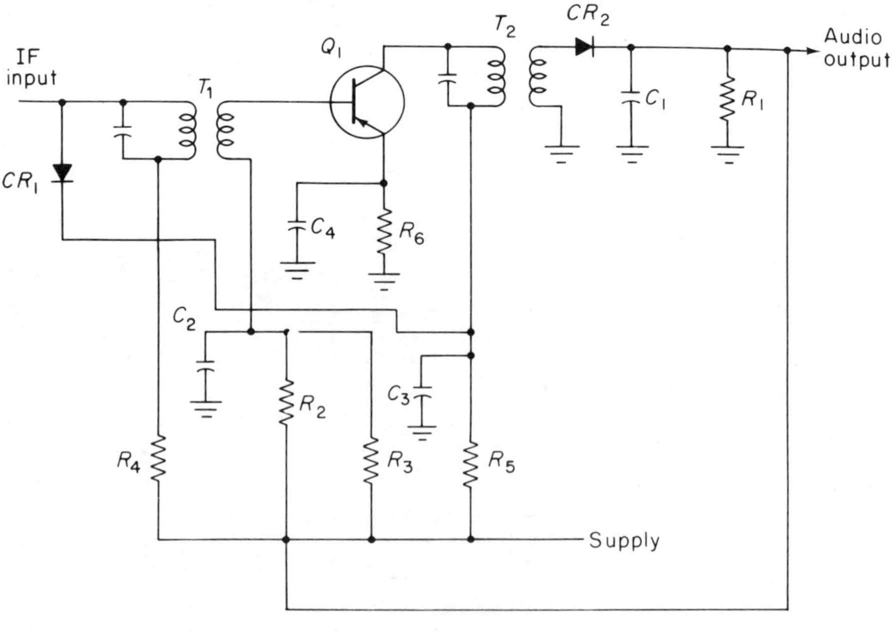


**FIGURE 4-23** Typical AVC-AGC circuits

the volume of an audio signal in an AF stage. However, in a broadcast receiver, the net result is an automatic control of volume. Either way, the circuit functions to provide a constant output despite variations in signal strength. An increased signal reduces gain, and vice versa.

In the circuit of Fig. 4-23, diode CR1 acts as a variable shunt resistance across the input of the IF stages. Diode CR2 functions as the detector and AGC as the bias source. Under no-signal conditions or in the presence of a weak signal, diode CR1 is reverse-biased and has no effect on the circuit. In the presence of a very large signal, CR1 is forward-biased and acts as a shunt resistance to reduce gain.

The output voltage of CR2 is developed across resistor R1 and applied to the audio stages. Resistor R1 also forms part of the bias network for the IF-stage transistor, along with resistors R2 and R3. The combined fixed bias (from the network) and variable bias (from the detector) is applied to the IF-stage base-emitter circuit. The detector bias varies with signal strength and is of a polarity that opposes variations in signal. That is, if the signal increases, the detector bias will be more positive (or less negative) for the base of a PNP transistor, and vice versa for an NPN transistor.

## 4-3 OSCILLATOR AND WAVE-SHAPING CIRCUITS

Oscillators are a form of *signal generator* circuit. That is, oscillators produce signals at frequencies from the very low audio range on up through the very high RF range. The signals can be sine waves, square waves, pulses, or sawtooth waves, depending on the oscillator circuit used. Similarly, the frequency of the signals can be very precisely fixed, or variable, depending on circuit design. In many cases, the signals produced must be modified or shaped. For example, square waves are often *integrated* or *differentiated.* Similarly, sine waves are often *clamped* and *clipped.*

It is impossible to cover every type of oscillator and signal wave-shaping circuit. However, the following paragraphs discuss the most common types of signal generating and shaping circuits found on schematics.

### *4-3.1 OSCILLATORS*

In general, transistor oscillators can be grouped into the following: *LC* (inductance-capacitance), crystal, *RC* (resistance-capacitance), blocking or relaxation, and multivibrator. All oscillators operate on the *feedback principle.* That is, part of the output signal produced by the circuit is fed back to the circuit input to sustain oscillation. However, the ways in which feedback is obtained, and the ways in which the frequency of the signals produced is controlled, vary considerably with the type of oscillator circuit.

**LC and Crystal-controlled Oscillators.** *LC* oscillators are those that use inductances (coils) and capacitors as the frequency-determining components. Typically, the coils and capacitors are connected in series- or parallel-resonant circuits and adjusted to the desired operating frequency. Either the coil or capacitor can be variable. *LC* oscillators are used at higher frequencies (RF). Both the Hartley and Colpitts oscillators can be designed with transistors. (The Hartley and Colpitts are classic vacuum-tube oscillator circuits, in common use for many years.) However, the Colpitts is generally more popular.

Transistor *LC* oscillators can be crystal-controlled. That is, a quartz crystal can be used to set the frequency of operation, with an adjustable *LC* circuit used to "trim" the oscillator output to an exact frequency. In addition to the Hartley and Colpitts, there are a number of other crystal oscillator circuits suitable for transistors. These include the Pierce oscillator, harmonic or overtone oscillators, and oscillators that use two transistors to provide the necessary feedback required for oscillation.

**Basic Hartley Oscillator.** Figure 4-24 is the schematic of the basic Hartley oscillator circuit. The frequency of the output signal is determined by the resonant frequency of C1 and L1. Typically, L1 is fixed and C1 is variable. Feedback is obtained by means of the

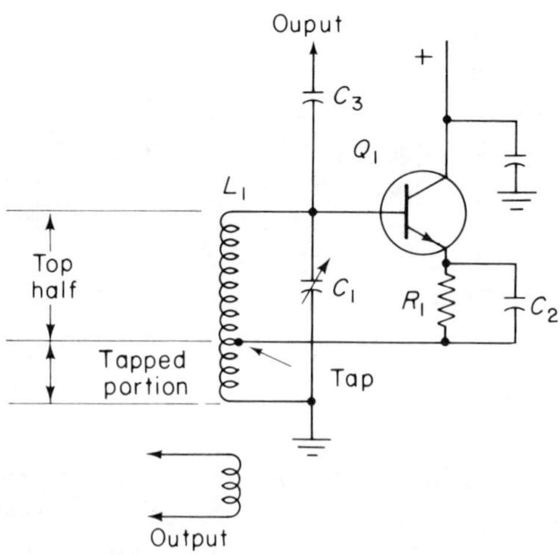


**FIGURE 4-24** Basic Hartley oscillator circuit

tap on L1. When current flows through the tapped portion of L1 to the emitter and collector of Q1, a magnetic field is set up around the tapped portion. This field produces a voltage in the top half of L1. This induced voltage is applied to the base of Q1 to increase the current flow. This action continues until C1 is fully charged. At that point, C1 discharged through L1 in the opposite direction, and the current through Q1 is decreased. Thus, current flow through Q1 is controlled at a frequency determined by the combination of C1 and L1 (the resonant frequency). Output from the Hartley circuit can be obtained through inductive coupling to L1, or by capacitive coupling to the base of Q1, as shown.

**Basic Colpitts Oscillator.** Figure 4-25 is the schematic of the basic Colpitts oscillator circuit. The frequency of the output signal is determined by the resonant frequency of C1 and L1. Typically, L1 is fixed and C1 is variable. Colpitts circuits are more commonly used in RF schematics than the Hartley circuit because of the mechanical difficulty involved in making the tapped coils required at high frequencies. Feedback is controlled in the Colpitts oscillator by the ratio of capacitance C2 to C3. Note that the *LC* circuits in Fig. 4-25 are in the base circuits, and feedback is between emitter and base, through C2 and C3. In practical Colpitts circuits, the *LC* components are in the collector, and feedback is between collector and emitter.

**Basic Crystal-Controlled Oscillator.** Figure 4-26 is the schematic of a Pierce oscillator. This type of oscillator is very popular because of its simplicity and minimum number of components. No *LC* circuits are required for frequency control. Instead, the frequency is

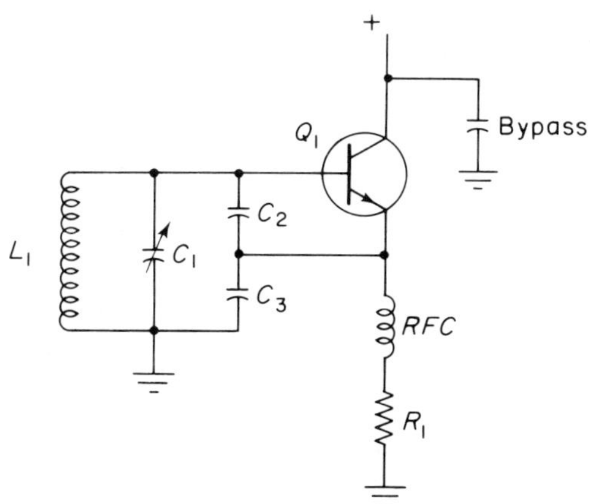


**FIGURE 4-25** Basic Colpitts oscillator circuit

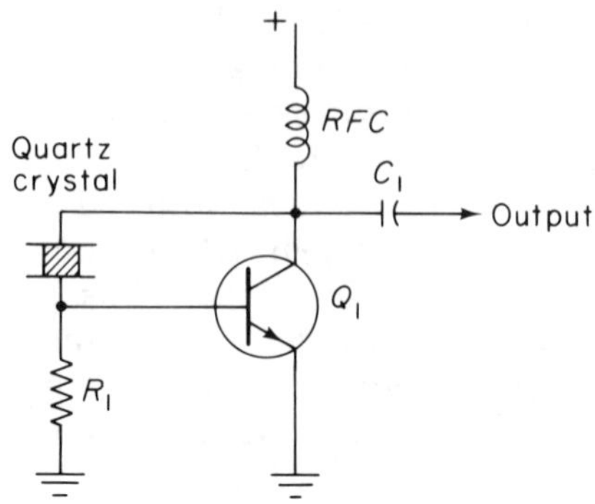


**FIGURE 4-26** Basic Pierce oscillator circuit

set by the quartz crystal. As discussed in Sec. 2-8.4, a quartz crystal produces an alternating current when placed between two plates and a voltage is applied to the plates. In the circuit of Fig. 4-26, the same voltage applied to the collector of Q1 is also applied to the quartz crystal. As a result, the crystal produces alternating current signals at some specific frequency (determined by natural characteristics of the crystal, such as thickness, cut, etc.) These a-c signals serve to lock the oscillator circuit at the same frequency. The transistor, in effect, serves to amplify the signals. Output from the Pierce oscillator can be taken from the transistor collector by means of a coupling capacitor, as shown.

**Harmonic or Overtone Oscillator.** It is frequently desirable to operate crystals at their *harmonic or overtone frequencies* (which are multiples of the fundamental or natural crystal frequency). The practical limit of crystal fundamental resonance is about 25 MHz for a transistor crystal oscillator. From 20 to 60 MHz, third-overtone (or third-harmonic) crystals are used. Fifth-overtone crystals are used for frequencies between 60 and 120 MHz, and seventh-overtone crystals at frequencies above 120 MHz. Usually, 150 MHz is the absolute limit for a solid-state crystal oscillator. Multipliers (Sec. 4-2.1) are used above this frequency.

Figure 4-27 shows the schematics of two overtone oscillators, with additional feedback. This additional feedback is obtained by means of a capacitive divider (C2, C3), acting as the *LC* circuit bypass. Most third-overtone crystals operate satisfactorily without this additional feedback, but for the fifth and seventh harmonics, the extra feedback is required. In both circuits, the LC circuit is tuned to resonate at the desired crystal overtone, not at the fundamental frequency. Output from the overtone oscillator can be taken

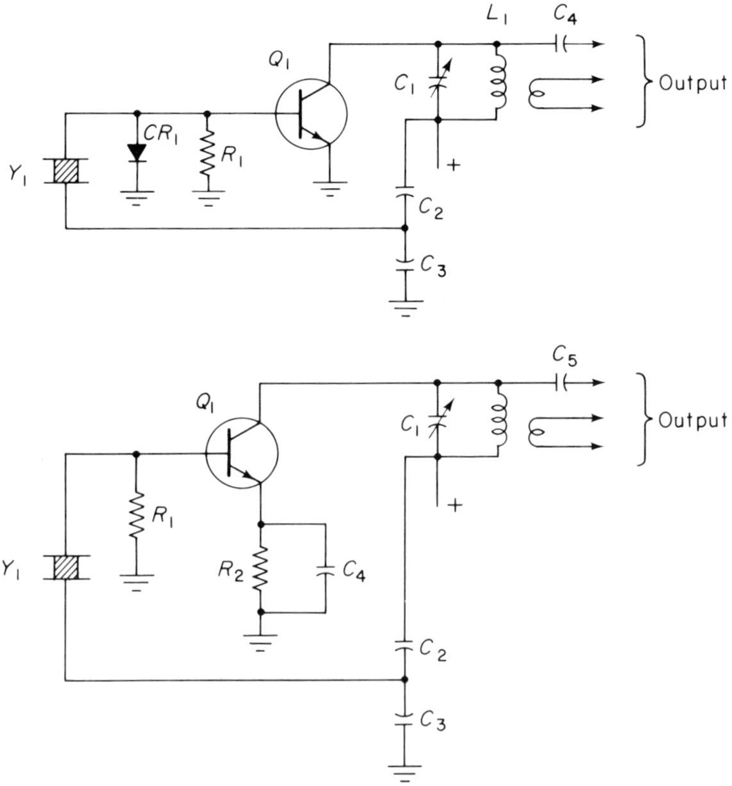


**FIGURE 4-27** Basic overtone oscillators

from the transistor collector by means of a coupling capacitor, or by means of inductive coupling, as shown.

**Practical Crystal-Controlled Oscillator Circuit.** Figure 4-28 is the schematic of a practical crystal-controlled solid-state oscillator. This circuit is one of the many variations of the Colpitts oscillator. However, the output frequency is fixed and controlled by the crystal. The circuit can be used over a narrow range of L1 (which is slug-tuned).

The resonant frequency of the oscillator circuit is set by the combination of $L$ and $C$ values. If the value of $C$ (the combined capacitance of C1 and C2) is made quite high (with a corresponding lower value of $L$), the resonant circuit is said to be "high-$C$" and usually results in sharper resonant tuning. A large value of $L$ (with a corresponding lower value of $C$) introduces more resistance into the resonant circuit, thus lowering the circuit $Q$ (Sec. 2-5.2) to produce broader resonant tuning.

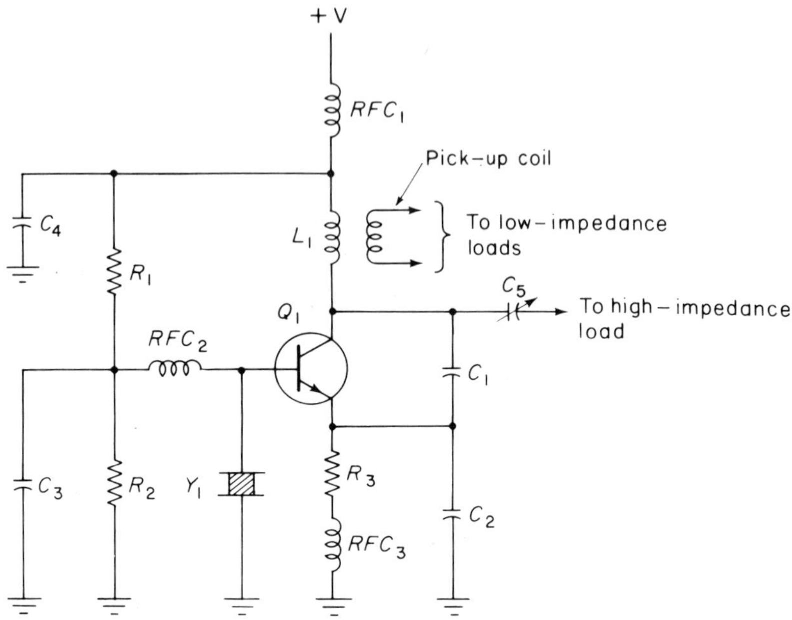


**FIGURE 4-28** Practical crystal-controlled oscillator circuit

Resistors R1 and R2 form a voltage divider and, in combination with R3, bias the transistor at a fixed operating point. Resistor R3 provides emitter feedback to maintain the transistor at the desired operating point. Capacitors C4 and C5 bypass RF signals to ground, whereas RFC1, 2 and 3 serve to keep RF signals out of the power supply.

The amount of signal feedback necessary for oscillation is set by the ratio of C1 and C2. For example, if C1 and C2 are of the same value, the feedback signal is one-half of the output signal appearing at the collector. This output (to the following stage) can be taken from L1 by means of a pickup coil (or low-impedance loads) or coupling capacitor C5 (for high-impedance loads). Generally, the most convenient output scheme is to use a coupling capacitor (C5) and make the capacitor variable. This makes it possible to couple the oscillator to a variable load (a load that changes impedance with changes in frequency).

**Practical Variable-Frequency Oscillator.** Figure 4-29 is the schematic of a practical variable-frequency oscillator capable of operating at frequencies up to about 0.5 MHz. This circuit is also one of the many variations of the Colpitts oscillator. Oscillation is sustained by feedback to the emitter from the junction of C1 and C2 (as is the case for the crystal oscillator). The circuit is tuned by

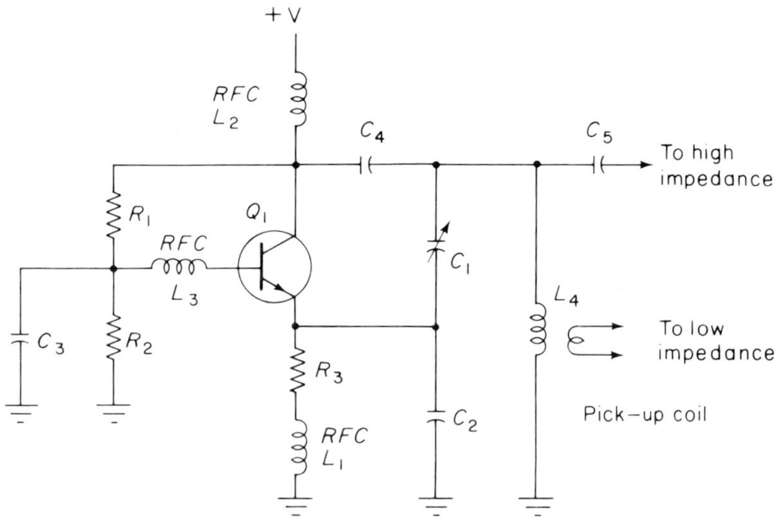


**FIGURE 4-29** Practical variable frequency oscillator

variable capacitor C1, which is in series with C2. Both C1 and C2 are in parallel with L4, and form a parallel resonant circuit. Capacitor C4 prevents direct current from passing through L4, but permits passage of RF output signals from the collector of Q1. This output can be taken from L4 by means of a pickup coil or coupling capacitor C5.

**RC Oscillator.** Figure 4-30 is the schematic of an *RC* oscillator. Such oscillators are used at audio frequencies instead of *LC* oscillators. *RC* oscillators avoid the use of inductances, which are not practical in the audio-frequency range. *RC* oscillators are usually operated class A, thus producing good waveforms. The feedback principle is also used in *RC* oscillators. The collector signal is fed back through an *RC* network to the base to sustain oscillation. The *time constants* of the *RC* network determine the oscillator output frequency. [The time constant of an *RC* circuit is discussed in much greater detail in Sec. 2-3.2.

**Blocking Oscillator.** Figure 4-31 shows the schematics of three blocking oscillators. Two versions of the circuit have tapped transformers. The blocking oscillator is one of the simplest solid-state oscillators and operates on the *relaxation* principle. The transistor is initially forward-biased to conduct through the transformer wiring. This causes a signal to be fed back to the base through C1. The base is driven very hard in the forward direction, so that C1 is rapidly charged through the forward-biased emitter-base junction. The output pulse at the T1 secondary is generated by the rapid turn-

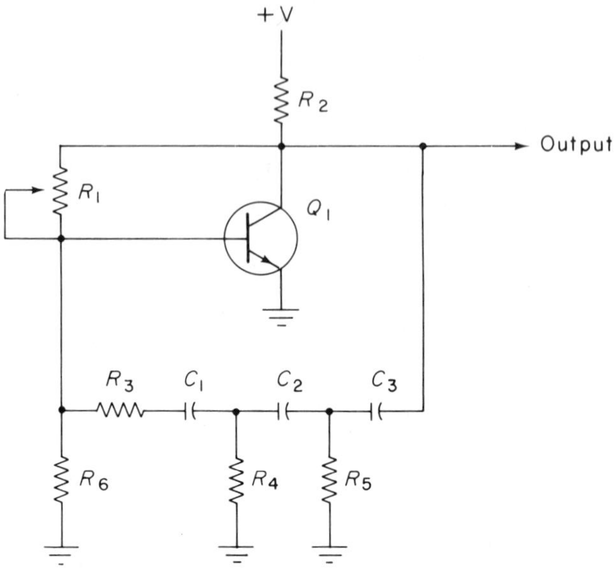


**FIGURE 4-30** RC oscillator circuit

on of the collector current, and the steepness of the pulse wavefront (shown in Fig. 4-31) is limited only by the leakage inductance of T1. The top of the pulse is flattened by collector-current saturation.

When saturation is reached, the feedback signal drops and is no longer able to forward-bias the transistor. Capacitor C1, in discharging, reverse-biases Q1 and discharges slowly through R1. The cycle is restarted by forward bias. The time constant of R1 and C1 determines the off-time between output pulses, and thus determines the oscillator frequency. Blocking oscillators are most commonly used in switching circuits.

Blocking oscillators can be used as driven oscillators, where a single output pulse is produced in response to a *trigger pulse.* This requires that the transistor emitter-base be reverse-biased as shown in Fig. 4-31. With this arrangement, the transistor is initially cut off so that a trigger signal switches the oscillator "on."

**Multivibrator.** Figure 4-32 is the schematic of a basic multivibrator. There are many types of multivibrators (or MVs), and they are often found in IC form. With any MV, the output of one-half of the circuit feeds into the input of the opposite half, and vice versa. For this reason, MVs are sometimes called *flip-flops.* One transistor is cut off while the other is being driven into saturation. The output of a multivibrator is a symmetrical square wave, as shown. That is, both positive and negative portions of the cycle are of the same

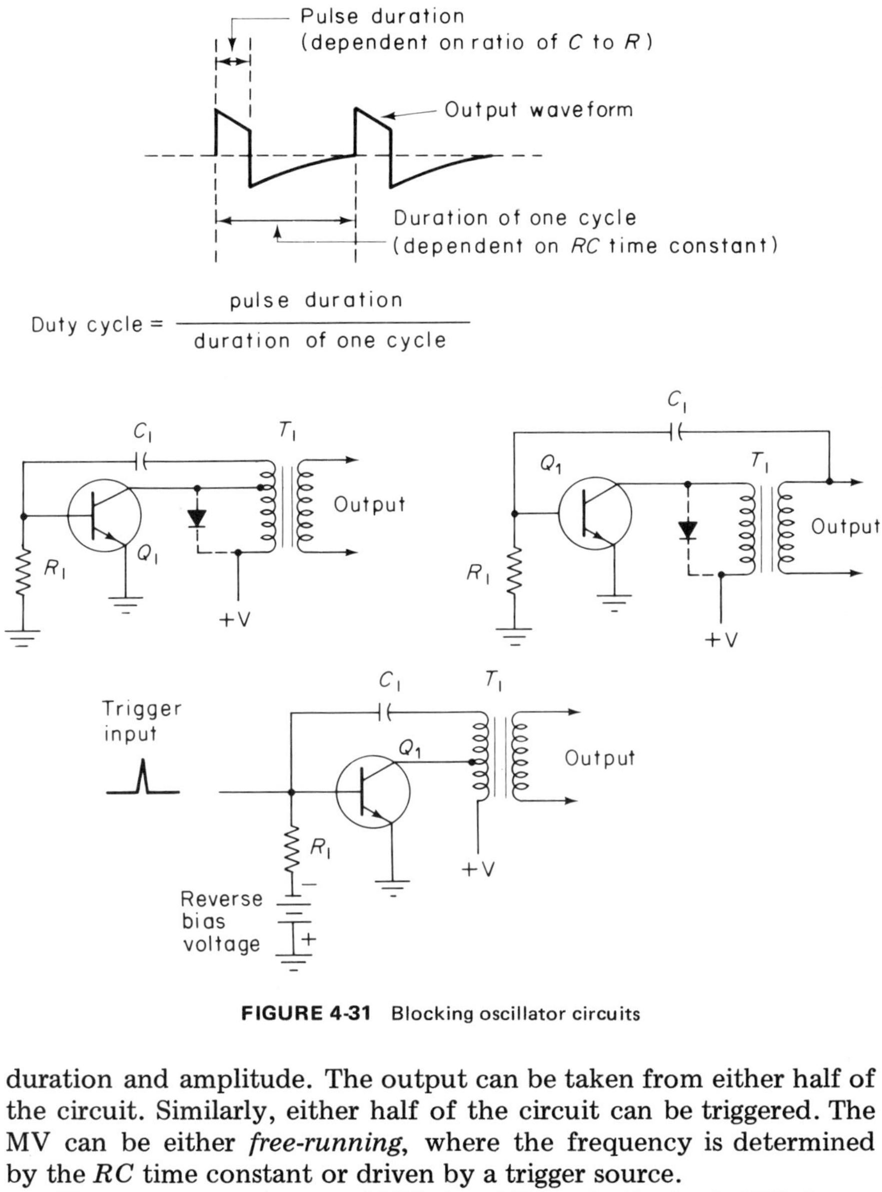


**FIGURE 4-31** Blocking oscillator circuits

duration and amplitude. The output can be taken from either half of the circuit. Similarly, either half of the circuit can be triggered. The MV can be either *free-running*, where the frequency is determined by the *RC* time constant or driven by a trigger source.

There are many types of MVs in addition to the basic MV shown in Fig. 4-32. Two of the most well-known MVs are the *bistable* and the *one-shot* (also known as a *Schmitt trigger*). With the bistable MV, each trigger pulse causes one transistor to saturate and the other to cut off, but two trigger pulses are required to return the circuit to the original state. With only one trigger pulse, the bistable MV will "flip" or "flop" to one stable state and remain in

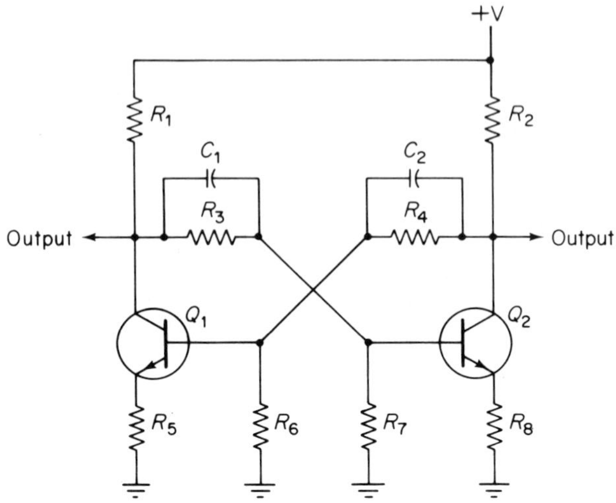


**FIGURE 4-32** Multivibrator circuit

that state. With the one-shot MV, each trigger pulse produces one complete output pulse. Typically, there is one output pulse of long duration for each input pulse. Thus, the one-shot MV (Schmitt trigger) is used as a *pulse stretcher*.

### *4-3.2 MODIFICATION OF SQUARE WAVES AND PULSES*

The outputs of MVs and similar pulse circuits are often modified to get a particular waveform. The most common wave-forming circuits used with square waves and pulses are the high-pass and low-pass *RC* filters described in Sec. 4-7.

When a square wave is applied to a low-pass *RC* filter, the output will be an *integrated waveform*, as shown in Fig. 4-33. In this case the *RC* network produces an output voltage proportional to the algebraic sum of the instantaneous values of the input voltage. Not

If the square-wave frequency is much lower than the cutoff frequency, both attenuation and modification will be at a minimum.

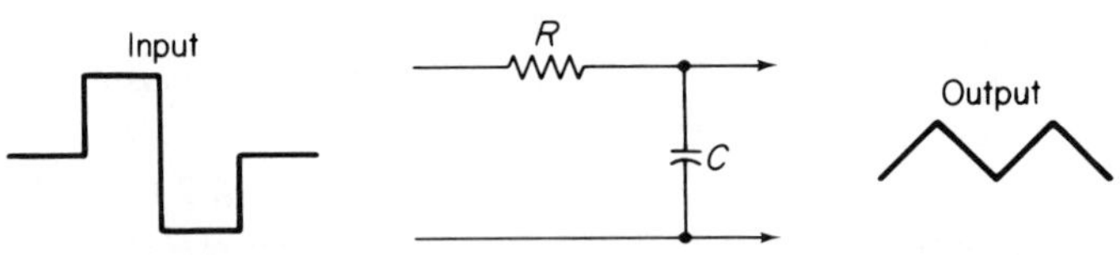


**FIGURE 4-33** Low-pass RC filter used to convert squarewaves into integrated waves

As the square-wave frequency increases, the output is a series of increasing and decreasing ramps. The ramps are linear, but only for a time that is shorter than about half the *RC* time constant. For very accurate integration of the square wave, the integration time should be about 0.1 of the *RC* time constant.
only will the amplitude be affected, but the waveform will change. The amount of attenuation, and the amount of waveform modification, are determined by the relation of the square-wave frequency to the *cutoff frequency*. *RC* filters have a cutoff frequency, at which the output drops to 0.707 of the input value (a 10-V input will drop to 7.07 V). The cutoff frequency can be approximated by the equation $1/6.28RC$, where $R$ is resistance in ohms and $C$ in capacitance in farads.

When a square wave is applied to a high-pass *RC* filter, the output will be a *differentiated waveform*, as shown in Fig. 4-34. In this case, the *RC* network produces an output voltage proportional to the time rate of change of the input voltage. As in the case of the low-pass (integrator) circuit, the amount and nature of the waveform modification of a high-pass (differentiator) circuit is determined by the relationship of cutoff frequency to square-wave frequency. The most common differentiation circuit is where the square-wave frequency is well below the cutoff frequency and the output becomes a series of "spikes," as shown.

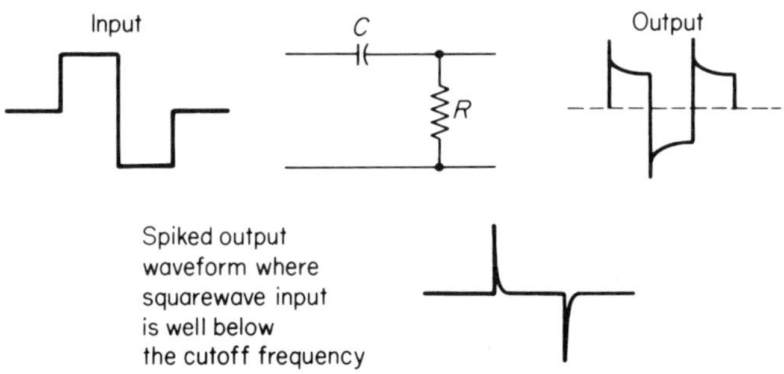


**FIGURE 4-34** High-pass RC filter used to convert squarewaves into differentiated waves

### 4-3.3 MODIFICATION OF SINE WAVES

There are many circuits used to modify and control sine waves. The most important of these circuits are clippers and clamps. These same clippers and clamps can be used to modify square waves and pulses.

**Series Diode Clippers.** A clipper circuit clips a particular waveform so that the amplitude of the output signal is proportional to those values of input waveform exceeding a certain critical value. A typical diode clipper is shown in Fig. 4-35. The diode CR1 is normally reverse-biased by the voltage applied to R1, and no current flows in the absence of an input signal.

If a positive input is applied, current will not flow until the input positive amplitude rises above the voltage applied to R1 (a bias voltage of 4.5 V). The output amplitude is then proportional to the amplitude of the input signal, which exceeds the 4.5 bias voltage. No current flows for a negative input signal, since the anode of CR1 is then negative. Therefore, the circuit clips those portions of the signal which are below the positive 4.5-V bias value. For example, if the input swings above and below zero by 12 V as shown in Fig. 4-35, and the bias is +4.5 V, the input will swing from 0 to +7.5 V.

**Shunt Diode Clippers.** Figure 4-36 shows the four basic types of shunt clippers. In all four circuits, the diode CR1 is forward-biased during a portion of the input waveform and will conduct. During conduction, the low forward resistance of the diode shunts (or shorts) the load resistance R1 and no output is produced. Thus, that portion of the input waveform is clipped. CR1 is reverse-biased during the remainder of the input, so the remainder of the waveform is reproduced across the output resistor R1.

In Fig. 4-36b-d, diode CR1 is reverse-biased by the fixed voltage (bias voltage). When the input signal exceeds this bias, CR1 conducts, and no output voltage is developed across R1. Thus, that portion of the input signal above the bias voltage is clipped, as shown. With negative input signals, the diode does not conduct, and the output is proportional to the input.

**Parallel Clippers or Slicers.** Two shunt clippers can be combined to produce a parallel clipper, or slicer, as shown in Fig. 4-37. The usual purpose of such a circuit is to clip the peaks of an a-c

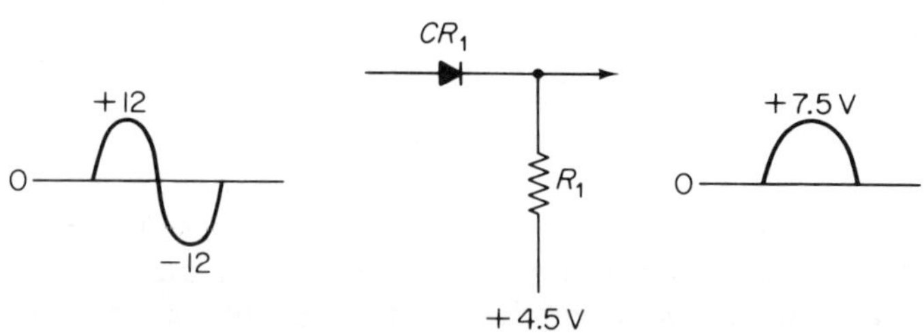


**FIGURE 4-35** Series diode clipper circuit

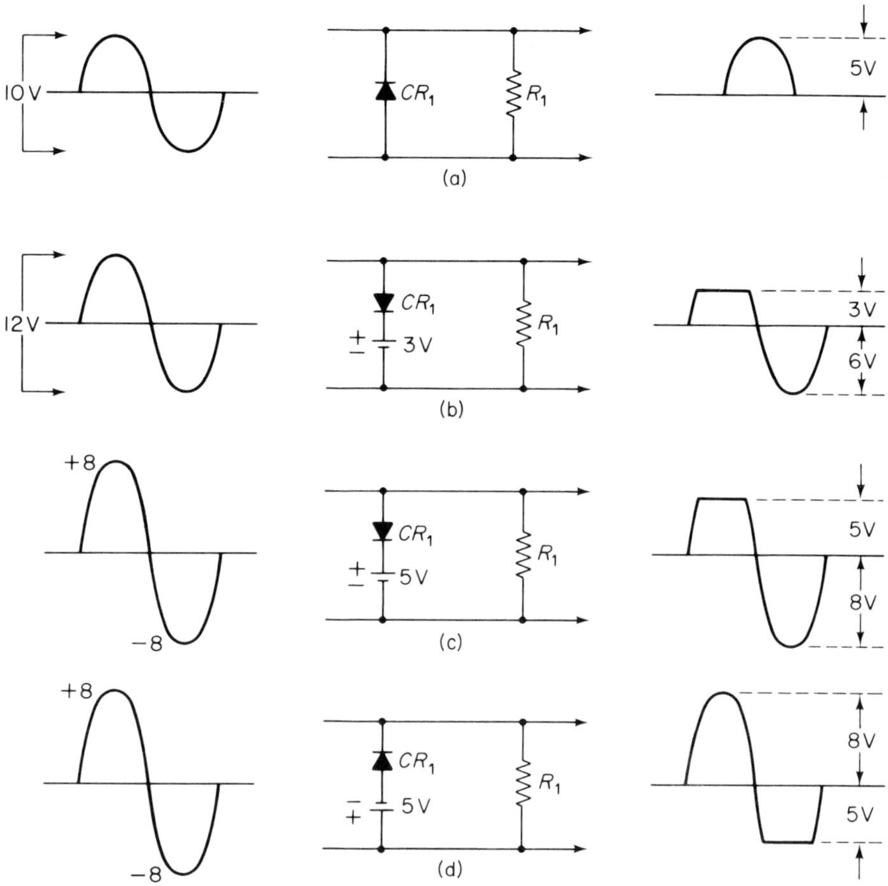


**FIGURE 4-36** Shunt diode clipper circuits

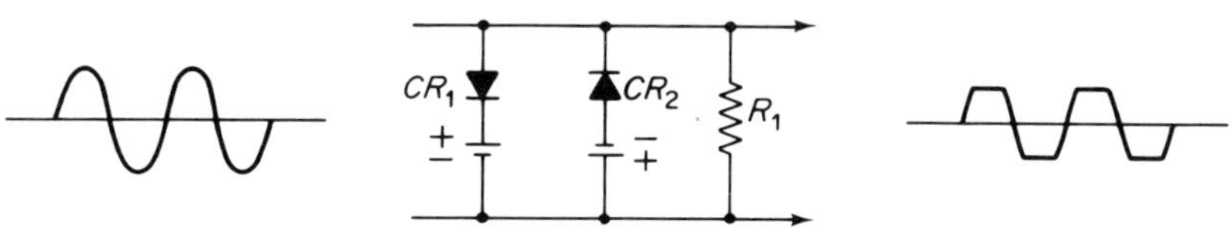


**FIGURE 4-37** Parallel clippers (slicer) circuit

input signal so that the output resembles a square wave (or pulse) that does not exceed a specific voltage value. The amount of clipping depends on the bias voltage in relation to the input signal. In effect, the output amplitude will equal the bias voltage. If no bias voltages are used, the output will be limited to the normal voltage drop of the diode (about 0.2 V for germanium and about 0.5 V for silicon).

**Shunt Limiter.** The biased shunt or parallel clipper can be used as a limiter. Such a circuit is shown in Fig. 4-38. Here, all input signals that exceed the bias voltage are limited to the value of the bias voltage. Diode CR1 is reverse-biased by the fixed voltage and will not conduct on any input pulses equal to, or below, the bias value. Any input pulses that exceed the bias value cause CR1 to conduct and remove the output. The circuit has no effect on signals of amplitudes below that of the bias voltage.

**Clamp or D-C Restorer.** When signals (pulses in particular) pass through coupling capacitors between stages, the d-c part of the signal is lost. For example, in the circuit of Fig. 4-39, if clamping diode CR1 was not present, the pulses on the output side of C1 would be of the same amplitude as the input pulses, except that the output pulses would be positive and negative with respect to ground. Assume that the input pulses have a 50-V peak value. Also assume that the steady-state value on the input side of C1 is +25 V (the collector of Q1 is normally +25 V with no signal). Between the pulses, C1 charges to the steady state (no signal) value. The input side of C1 is then positive with respect to ground, and the output side of C1 is negative with respect to ground.

When the pulses are present, C1 charges up to the 50-V peak value. On the input side, C1 goes from +25 to +75 V. On the out-

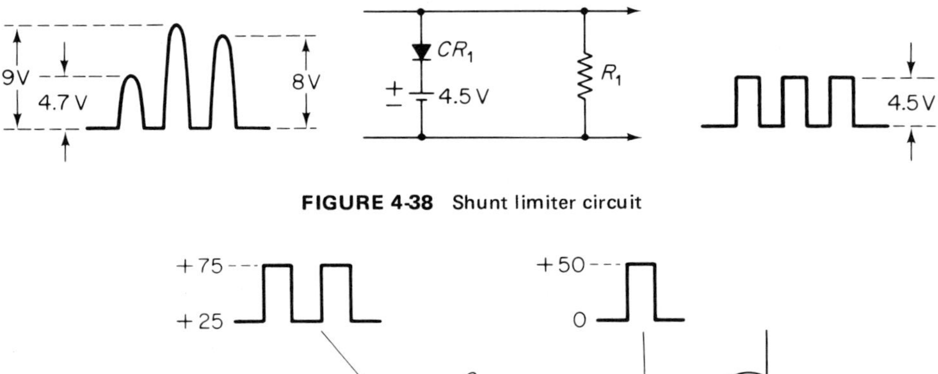


**FIGURE 4-38** Shunt limiter circuit

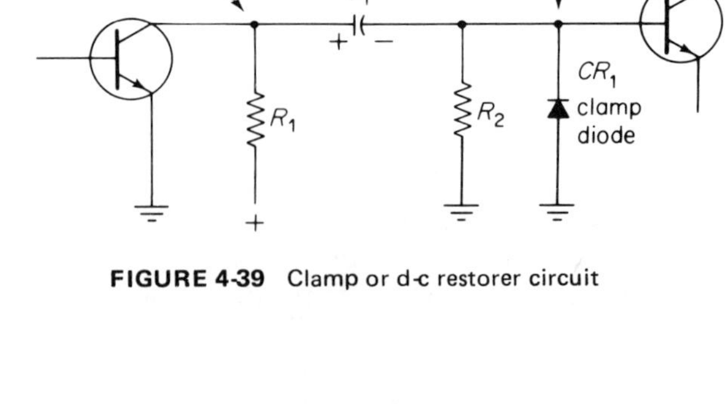


**FIGURE 4-39** Clamp or d-c restorer circuit

put side, C1 goes from −25 to +25. Therefore, the output pulses are positive and negative with respect to ground, even though the input pulses are always positive with respect to ground. Diode CR1 prevents this condition by discharging C1 between pulses (when the anode of CR1 goes positive). This makes the output side of C1 zero with respect to ground until the next pulse arrives. Then the output side of C1 rises to the peak value (in this case, 50 V).

### 4-3.4 UNIJUNCTION RELAXATION OSCILLATOR

The unijunction transistor (UJT) described in Sec. 3-5 is often used in a relaxation oscillator circuit. Such circuits are ideal for producing *sawtooth waveforms* or *sweep waveforms* where the output voltage rises from zero in a steady, linear fashion and then abruptly drops to zero. A typical UJT relaxation oscillator circuit is shown in Fig. 4-40. With this circuit, the positive supply voltage causes C1 to charge at a rate depending upon the time constants of R1, R2, and C1. When the charge reaches a level high enough to trigger the UJT, the impedance between the emitter and base 1 is lowered (negative resistance region discussed in Sec. 3-5). C1 then discharges, dropping the emitter into the reverse-bias region and stopping conduction. Capacitor C1 charges again, and the cycle is repeated. The oscillation frequency is dependent upon the R1, R2, and C1 time constants.

In some cases, a synchronization signal is applied to lock the oscillator to a given frequency. When synchronization (or sync) pulses are applied, the UJT circuit switches on and off at the frequency of the sync pulses. The sync pulse can be positive (applied at the

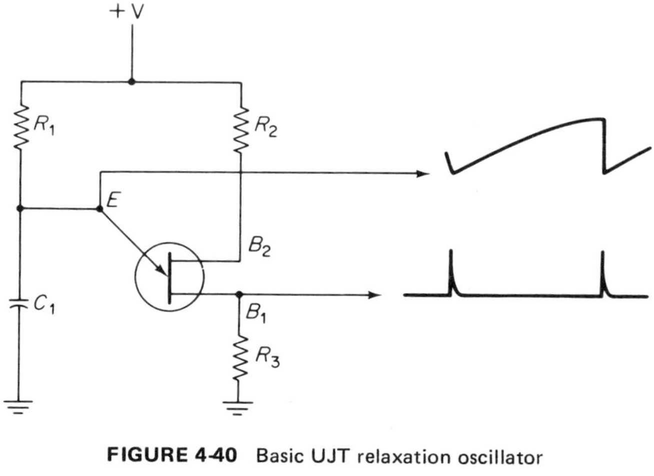


**FIGURE 4-40** Basic UJT relaxation oscillator

emitter) or negative (applied at base 2). Output from the UJT relaxation oscillator can be taken from the emitter or from base 1, as shown. The emitter output is a sawtooth, whereas the base 1 output is an irregular shaped pulse. In some circuits both outputs are used simultaneously.

## 4-4 OPERATIONAL AMPLIFIERS

The designation "operational amplifier" or "op-amp" was originally adopted from a series of high-performance, direct-coupled amplifiers. These op-amps were used to perform mathematical operations in analog computer (such as summation, scaling, subtraction, integration, etc.) Today, most op-amps are found in IC form, and are used as a replacement for any low-frequency low-power amplifier application, in addition to the computer applications.

### *4-4.1 OP-AMP SYMBOLS AND CIRCUITS*

Figure 4-41 shows the symbol used for op-amps, as well as the internal circuit. This is probably the last time you will ever see the circuit inside an IC op-amp. Even IC op-amp data sheets usually omit the full internal circuit details. Instead, they show the triangle symbol (indicating an amplifier), together with the external terminal or pin connections. So we will concentrate on external connections to the op-amp as you find them on a schematic. However, note that all stages of the op-amp are *direct-coupled* (no coupling capacitors). This direct coupling makes it possible to amplify very low frequency signals as well as changes in d-c levels.

### *4-4.2 OP-AMP INPUTS AND OUTPUTS*

Most IC op-amps have two inputs and one output. (On rare occasions, there will be two *differential* outputs. That is, when one output goes positive, the other output goes negative, and vice versa.) Virtually all IC op-amps have two differential inputs. One input is called the *inverting input* and is labeled with a minus (—) sign. When a signal is applied to the inverting input, the output will be inverted (when the input signal is positive, the output goes negative, and vice versa). The opposite occurs when a signal is applied to the noninverting (+) input (the output follows the input).

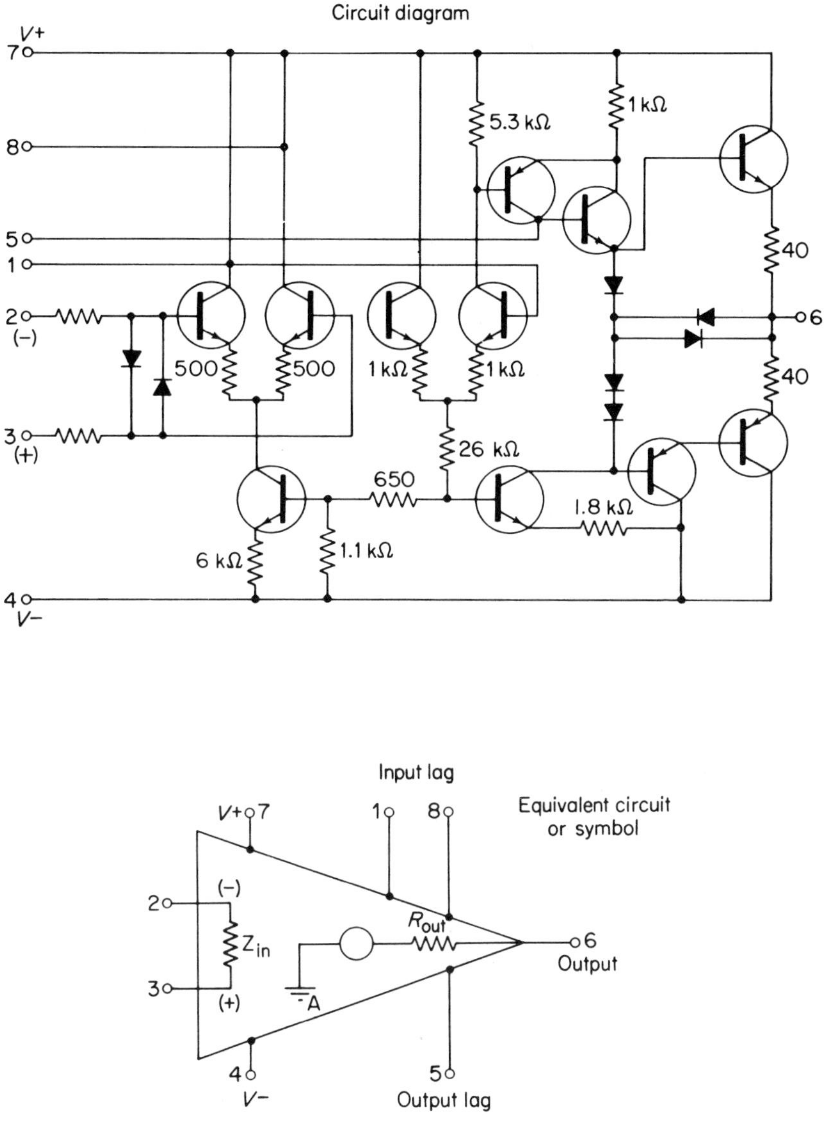


**FIGURE 4-41** Symbol and internal circuit of typical IC op-amp

In most circuits, only one input is used, with the other (inactive) input grounded (possibly through a resistor or other circuit parts). In practically all op-amp applications, the output is fed back to the input, usually to the inverting (—) input. This is negative feedback and makes it possible to control the gain of the op-amp. (Negative feedback tends to reduce gain; the more negative feedback the lower the gain.) If the output is fed back to the noninverting (+) input, this would be positive feedback (similar to that of an oscillator) and will cause the op-amp to act as an oscillator.

### *4-4.3 OP-AMP POWER SUPPLY*

Most op-amps require two power supplies, one positive and one negative (both with respect to ground). Generally, the power supplies are symmetrical (+6 V and −6 V; +12 V and −12 V), but not in all cases. Some op-amps are operated from a special single power supply circuit. Op-amps may or may not be grounded, depending on design. Both power supply terminals are generally bypassed to ground with capacitors.

### *4-4.4 OP-AMP COMPENSATION TERMINALS*

Most op-amps are provided with compensation terminals such as the *input lag* and *output lag* terminals shown in Fig. 4-41. External resistors and capacitors are connected at these terminals to set or change the characteristics of the op-amp for a particular application. For example, it may be necessary to alter the frequency-response characteristics of an op-amp to meet the requirements of a particular application. (Frequency response of an amplifier is the amount of gain to signals of a particular frequency or across a band of frequencies.) Op-amp data sheets describe how resistors and capacitors can be added at the compensation terminals to obtain the desired response. This is not usually shown on the schematic, but the parts (resistors/capacitors), their values, and the terminal connections are generally given.

### *4-4.5 TYPICAL OP-AMP APPLICATION*

Figure 4-42 is a schematic showing an op-amp connected in a typical *narrow-bandpass amplifier* (also known as a *tuned peaking amplifier*) circuit. Note that the output is fed back to the inverting (−) input through a parallel resonant circuit (L1, C1, R1). The greater the signal that is fed back to the inverting input, the lower will be the gain. However, as discussed in Sec. 2-5, a parallel resonant circuit provides high impedance at the resonant frequency. Thus, the output from the op-amp will peak at the resonant frequency. In effect, there is little or no feedback at the resonant frequency, and amplifier gain is maximum. At frequencies above and below the

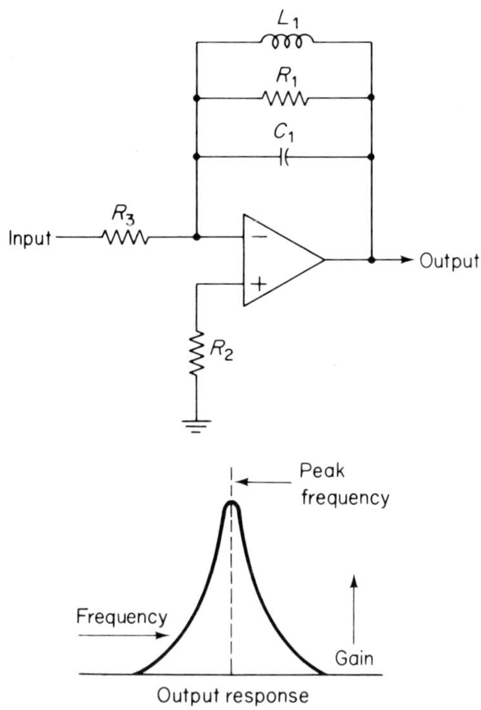


**FIGURE 4-42** Schematic showing an op-amp connected in a typical narrow-bandpass amplifier (tuned peaking amplifier) circuit

resonant frequency, there is some feedback. Therefore, gain is reduced by the negative feedback at all frequencies (but the resonant frequency). Note that the gain at frequencies other than the resonant frequency will be approximately equal to the ratio of R1/R3.

## 4-5 POWER SUPPLY CIRCUITS

The major function of any power supply is to provide a source of d-c voltage for operation of other circuits. Usually, this involves rectifying alternating current into pulsating direct current, filtering the pulsating dc into pure direct current, and then regulating the direct current to maintain a constant voltage and/or current. In the case of some equipment (usually portable), the process may also include converting direct current (from a battery) into alternating current so that it may be stepped up in voltage, then restored back to direct current at a higher voltage level.

### 4-5.1 *BASIC HALF-WAVE POWER SUPPLY*

The basic process of rectification by a diode is discussed in Chapters 1 and 3. When a single diode is used to convert ac into dc, the circuit is known as half-wave, since only one-half of each cycle is used. Usually, a filter circuit is required with a half-wave rectifier to smooth out the pulsating dc into pure dc or (more realistically) into dc with a slight "ripple." Usually, the peak-to-peak ripple voltage is expressed as a percentage of the total power supply output voltage. For example, if the power supply produces 100 volts of dc, and the ripple is 3 V peak to peak, there is a 3% ripple. The amount of ripple for any given type of power supply (half-wave, full-wave, etc.) depends on the type of filter.

A basic half-wave diode rectifier and filter circuit are shown in Fig. 4-43. The diode CR1 cathode is connected to a filter capacitor C1 (usually electrolytic) and to a filter choke L1. Another filter capacitor C2 is also used to smooth out the pulsating dc. (Filters are discussed further in Sec. 4-7.) A resistance R1 is placed across the power supply output. This resistance is not used in all circuits. Instead, the output is connected directly to the load circuit. If resistor R1 is used, it is known as a *bleeder* resistor, since it places a small current drain on the power supply and helps to stabilize the output. If resistance R1 is made up of a single *tapped resistor*, or a series of several resistors, it is possible to take several voltages from the bleeder network, as shown.

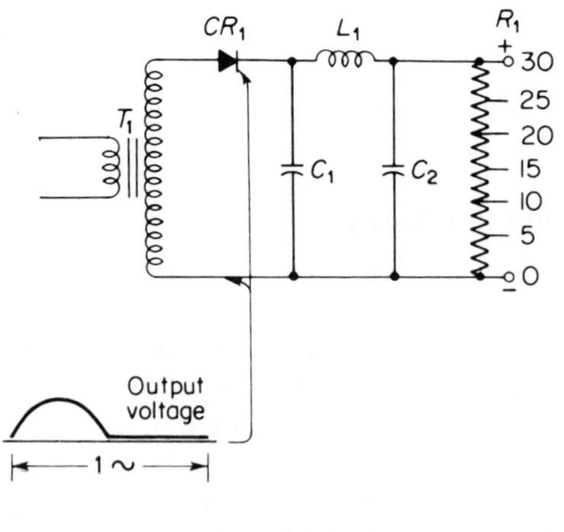


**FIGURE 4-43** Basic single-phase, half-wave rectifier circuit

### 4-5.2 *FULL-WAVE POWER SUPPLY*

Both positive and negative alternations of the a-c cycle are used in the full-wave supply of Fig. 4-44. For this reason, the full-wave supply is more efficient than the half-wave circuit. However, the full-wave circuit requires two diodes and a transformer with a center tap. As shown, one lead of the transformer secondary is connected to diode CR1, while the opposite end of the secondary is connected to a diode CR2. The center tap is connected to the common or ground circuit. The total voltage across the secondary of the transformer is about *twice* the voltage that appears at the power supply output.

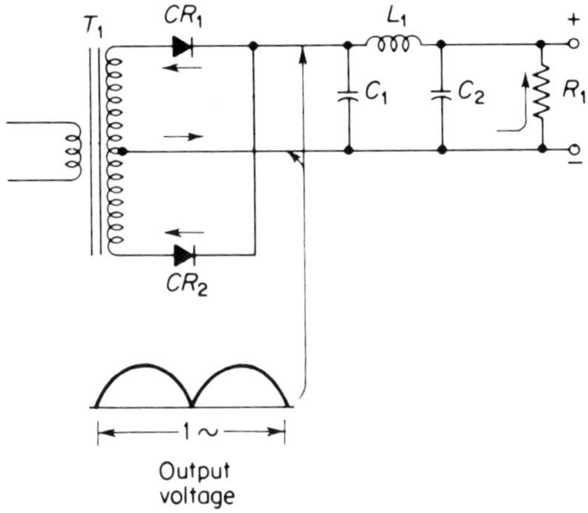


**FIGURE 4-44** Basic single-phase, full wave rectifier circuit

### 4-5.3 *FULL-WAVE BRIDGE POWER SUPPLY*

The full-wave bridge circuit is the most popular of the solid-state power supplies. As shown in Fig. 4-45, a bridge circuit makes it possible to have full-wave rectification using a transformer without a center tap. However, four rectifier diodes are required in the basic bridge circuit. When a positive alternation occurs, with the top end of the transformer secondary positive, current flows from the secondary bottom, through diode CR3, the load or bleeder resistor R1, and diode CR2, to the secondary top. On the next alternation, when the bottom end of the transformer secondary is positive, current

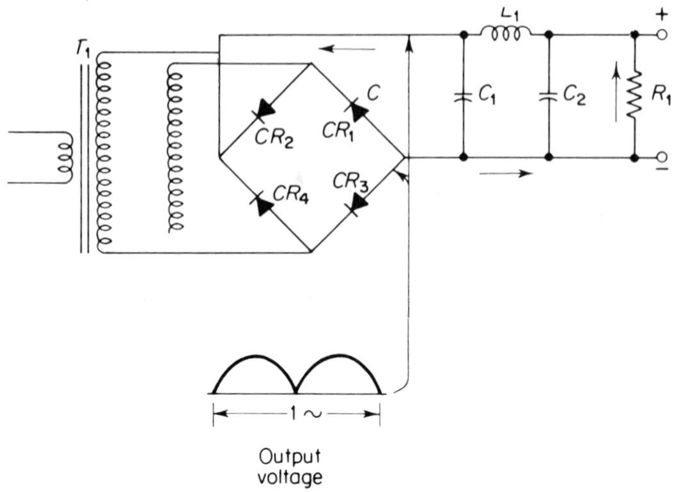


**FIGURE 4-45** Basic single-phase, full-wave, bridge rectifier circuit

flows from the secondary top, through diode CR1, the load or bleeder resistor R1, and diode CR4, to the secondary bottom. Full-wave rectification occurs since current flows through the bleeder resistance in the same direction on both half-cycles.

The filter capacitors C1 and C2 charge at a rate which is twice that for half-wave rectification, since current flows through the bleeder resistor in the same direction on both alternations or half-cycles. Consequently, the ripple frequence is twice that of half-wave rectification ripple frequency.

Since the discharge time between peaks of pulsating dc is only half that found in half-wave rectification, because there is a shorter time between peaks, the filter capacitors do not have as long to discharge. This makes it easier to maintain a relatively high charge and makes the output "smoother" than that of a half-wave supply.

### *4-5.4 THREE-PHASE POWER SUPPLIES*

The three-phase power supplies shown in Figs. 4-46 through 4-49 are similar to the basic half-wave and full-wave circuits, except that each phase is provided with its own diode (or half-wave) or diodes (full-wave). This reduces the amount of current through each diode (making it possible to use smaller diodes) and increases the ripple frequency (making it easier to filter).

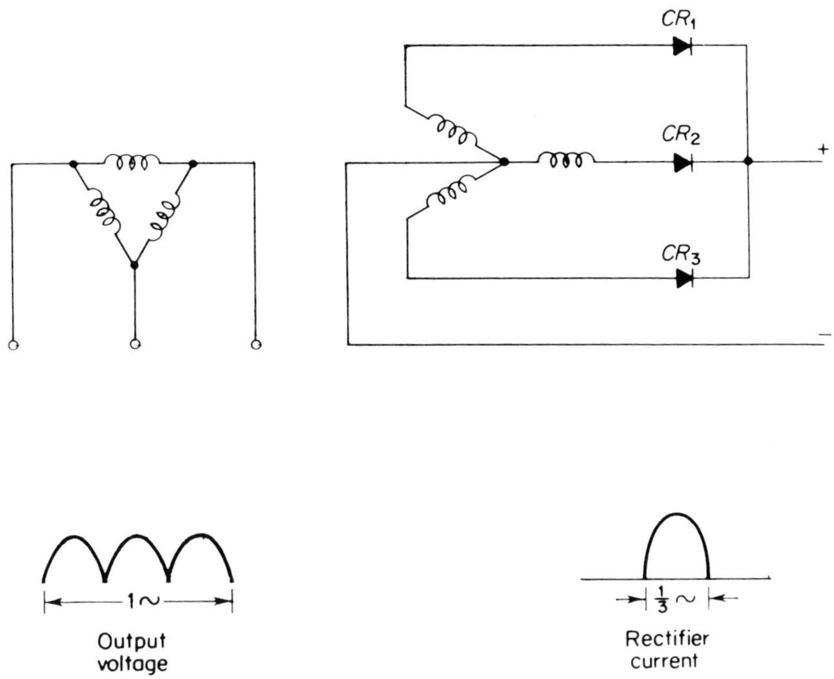


**FIGURE 4-46** Basic three-phase Y half-wave rectifier circuit

The three-phase Y half-wave circuit of Fig. 4-46 uses three rectifiers. One-third of the total output current flows through each rectifier, and the ripple frequency is three times that of a corresponding half-wave. That is, the ripple frequency is three times the line frequency.

The three-phase full-wave circuit of Fig. 4-47 uses six rectifiers. This circuit delivers twice as much voltage output as the circuit of Fig. 4-46 for the same transformer conditions. One-third of the total output current flows through each rectifier, and the ripple frequency is three times that of a corresponding full wave (or six times the line frequency).

The six-phase "star" circuit of Fig. 4-48 also uses six rectifiers. However, only one-sixth of the total output current flows through each rectifier.

The three-phase, double-Y, and interphase transformer circuit of Fig. 4-49 uses six half-wave rectifiers connected in parallel. This provides twice the output current of the half-wave circuit shown in Fig. 4-46, as well as a ripple frequency six times the line frequency.

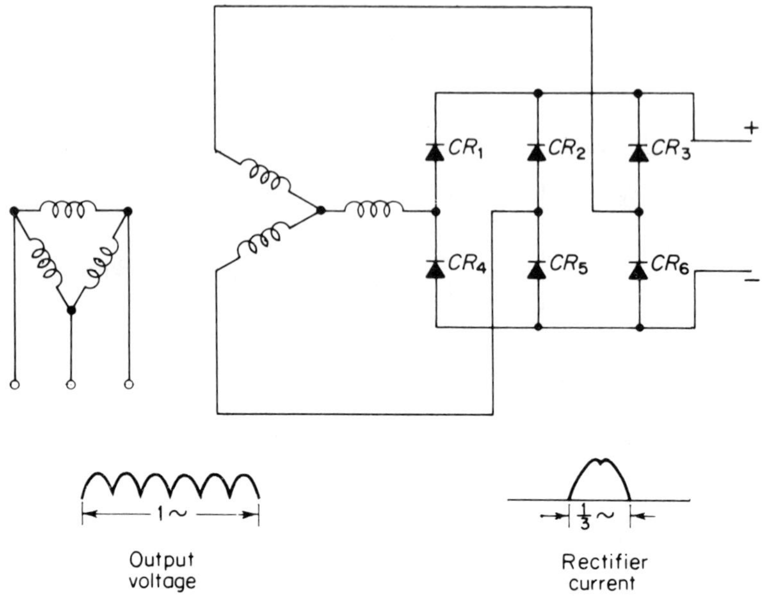


**FIGURE 4-47** Basic three-phase Y full-wave rectifier circuit

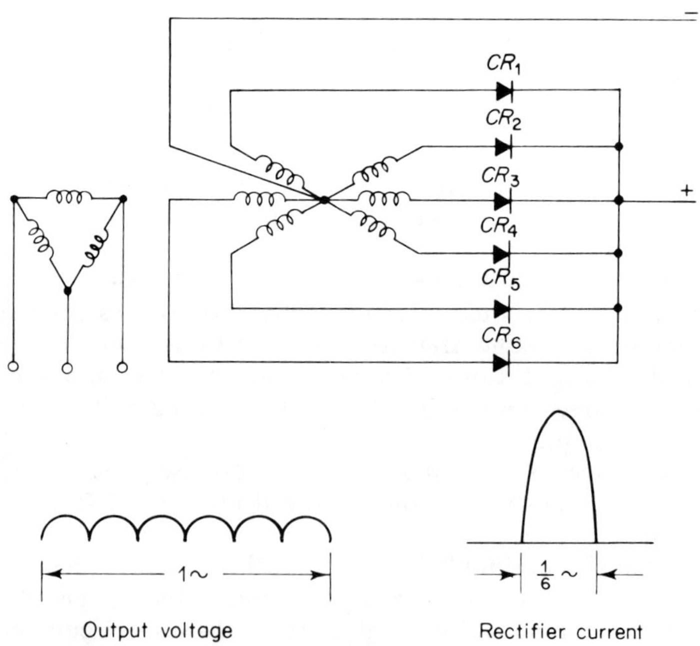


**FIGURE 4-48** Basic six-phase "star" rectifier circuit

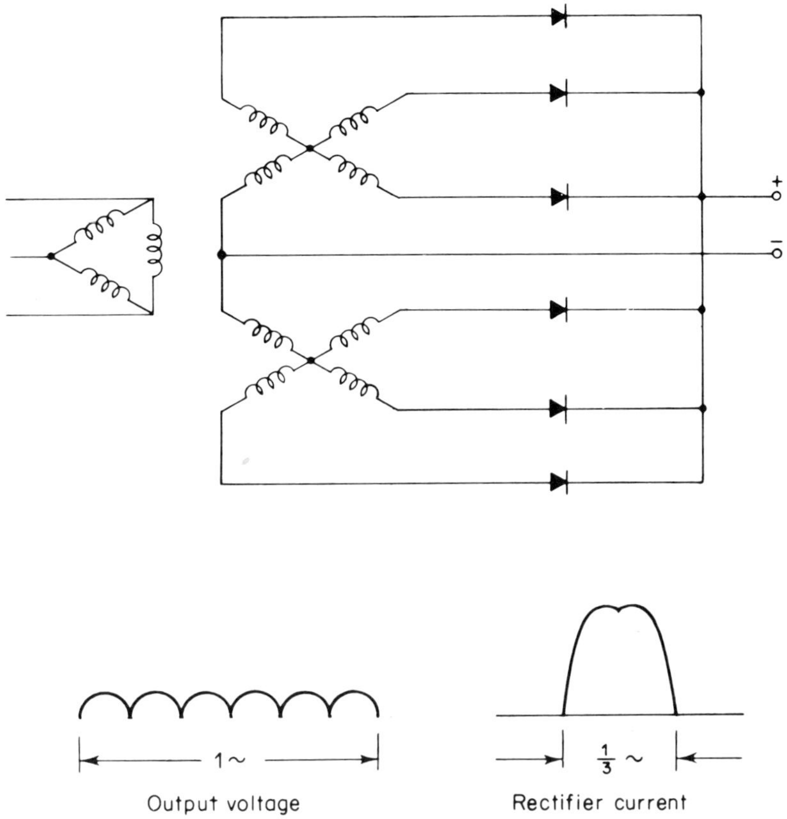


**FIGURE 4-49** Basic three-phase double-Y and inter-phase transformer circuit

### *4-5.5 VOLTAGE DOUBLING AND TRIPLING CIRCUITS*

Voltage doubling and tripling are used in power supply circuits where a higher voltage is required. Using such circuits, it is possible to increase an available a-c voltage *without* a transformer.

**Voltage Doubling Circuit.** As shown in Fig. 4-50, the basic voltage doubling circuit requires two rectifier diodes and two capacitors. When the a-c line alternation is such that terminal 1 is negative and terminal 2 is positive, as shown in Fig. 4-50a, current flows from terminal 1 in the direction shown by the arrows. The path of current from terminal 1 includes diode CR2 and capacitor C2, then returns to the positive terminal 2. Capacitor C2 is charged during this current flow with a polarity as shown.

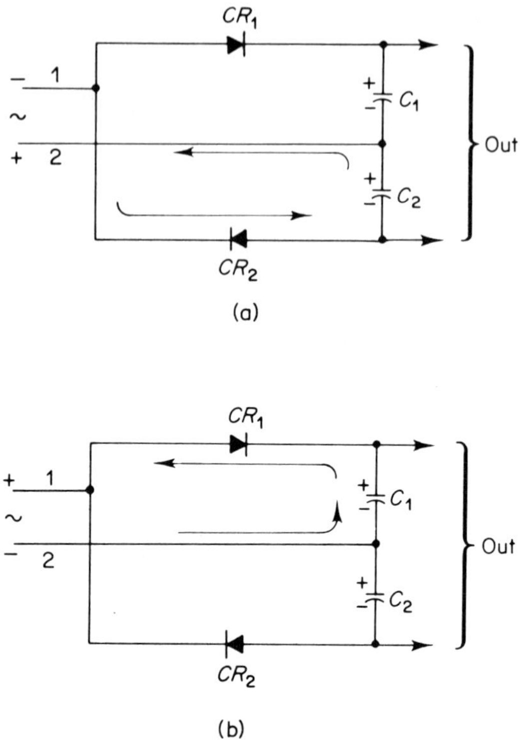


**FIGURE 4-50** Basic voltage doubling rectifier circuit

During the next alternation, when terminal 1 is positive and terminal 2 is negative, current flows as shown in Fig. 4-50b. As a result, current flows from terminal 2 toward C1, and charges C1 to the *peak* of the a-c voltage. Current continues through diode CR1 to the positive terminal 1.

Capacitors C1 and C2 are charged to the peak values of the a-c line voltage on alternate half-cycles. The d-c output voltage is taken from across the *two capacitors.* Therefore, the d-c voltage is approximately double that of the a-c voltage. The two capacitors are, in effect, in series, with the polarities adding to increase the total voltage.

**Voltage Tripling Circuit.** As shown in Fig. 4-51, the basic voltage-tripling circuit requires three rectifier diodes and three capacitors. When the a-c line alternation is such that terminal 1 is positive and terminal 2 is negative, current flows from terminal 2 through CR1 to terminal 1. Diode CR1 is forward-biased under this

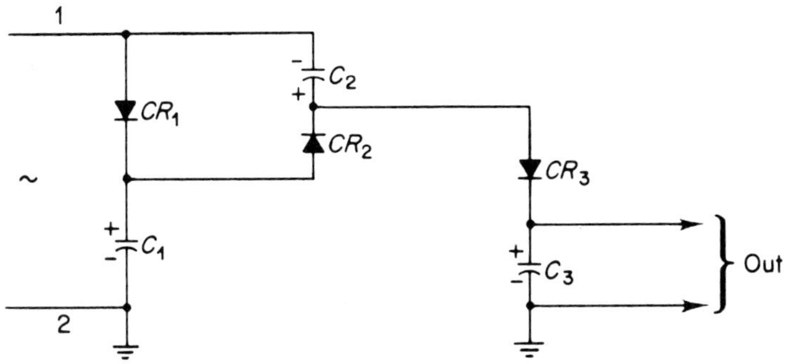


**FIGURE 4-51** Basic voltage tripling rectifier circuit

condition, permitting capacitor C1 to be charged to the a-c line voltage.

During the next alternation, when terminal 1 is negative and terminal 2 is positive, diode CR1 is reverse-biased and does not conduct. Instead, diode CR2 is forward-biased and permits current flow toward terminal 2, via capacitor C1. The charge already across C1 adds to the a-c line voltage and places a charge across C2. This charge is proportional to the line voltage added to the charge already across C1.

During the next alternation, when terminal 1 is again positive and terminal 2 is negative, capacitor C1 is again charged as before. But, at the same time, diode CR3 conducts and charges capacitor C3. When diode CR3 conducts, the charge across capacitor C3 is composed of the line voltage, plus the existing charge across C2. Since the C2 charge is already *double* voltage, this is added to the line voltage and produces a *triple* voltage across capacitor C3.

### *4-5.6 REGULATION CIRCUITS*

The amount of voltage that can be obtained from a power supply is determined by the voltage available at the source (line or transformer), less any voltage drop through the rectifiers and filter. This voltage drop is determined by the amount of current flowing through the power supply components (rectifiers, chokes, etc.). As more current is drawn from the power supply, the voltage drop across the power supply components increases and the output voltage is reduced. When less current is drawn from the power supply, the output is increased, since the power supply components have a minimum voltage drop.

The variation of voltage output with respect to the amount of current drawn from the power supply is known as *voltage regulation*, or simply *regulation*. Usually, regulation is expressed as a percentage. The value is determined by

$$\text{percentage of regulation} = \frac{(\text{no-load voltage})-(\text{full-load voltage})}{\text{full-load voltage}} \times 100$$

This equation applies to voltage regulation and takes into account the proportions of voltage increase and decrease with a change of load on a power supply. The smaller the difference between the full-load and no-load voltages, the better the regulation.

**Filter Circuit Regulation.** The filter components (choke coils and capacitors) provide some measure of voltage regulation. Should the output voltage drop, the capacitors will discharge and serve to keep the voltage constant. Similarly, a voltage drop will cause the magnetic field around the filter chokes to collapse (or decrease). The moving magnetic lines of force produce a current in the coil, but in the opposite direction to the current flow causing the magnetic lines. Thus, the filter choke sets up an opposition to any change in voltage amplitude.

The regulation provided by the filter network is sufficient for many applications. In other circumstances, the power supply output must be maintained at some critical voltage or current. A number of solid-state circuits have been developed to provide both voltage and current regulation. The following paragraphs summarize these circuits.

**Zener Diode Regulation.** The most common voltage regulation circuits found on solid-state schematics involves a zener diode. As discussed in Sec. 3-2.1, a zener diode maintains the voltage drop across its terminals at a constant value, no matter what current is being drawn. In its simplest form, the zener regulator consists of a series resistance R and a shunt-connected diode CR1, as shown in Fig. 4-52a. Such a circuit will maintain the output voltage at the rated value of the zener (within a given tolerance), even if the input voltage or output load change in value.

Sometimes it is necessary to regulate a voltage not available in standard zener diodes. This condition can be overcome by various circuit arrangements. Several zeners can be connected in series, as shown in Fig. 4-52b. The total regulated voltage is the sum of the individual zeners. A zener can also be connected as a series element, as shown in Fig. 4-52c. This circuit is used where only a small voltage drop is required. Several zeners can be connected in series to act as

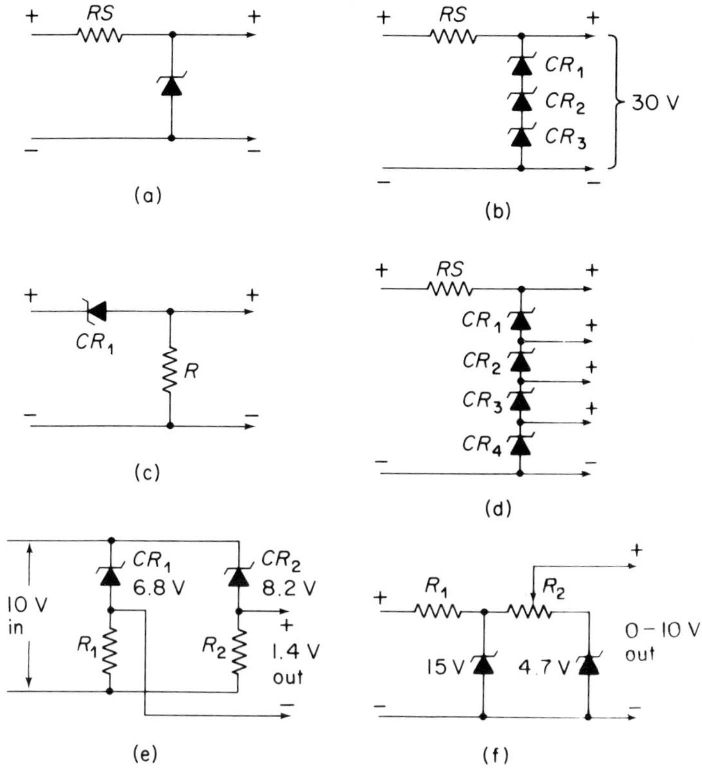


**FIGURE 4-52** Typical zener diode regulator circuits

a *divider* to obtain several regulated voltages, as shown in Fig. 4-52d. It is possible to arrange zeners so as to produce a regulated output at a voltage lower than that of the zener. Such a circuit is shown in Fig. 4-52e, where the output is the difference voltage between the two zeners. Zeners can also be used to provide an adjustable regulated voltage, as shown in Fig. 4-52f. Any combination of diodes can be used to obtain the desired range of output voltage.

**Extending Zener Regulation.** The voltage-control ability of zener diodes can be increased if the diode is used to control the operating point of a transistor or group of transistors. There are two basic types of transistor regulators: *shunt* and *series.* The shunt regulator is placed across the power supply output, while the series regulator is placed in series with the power supply output.

**Shunt Voltage Regulator.** The simplest form of shunt-transistor voltage regulator is shown in Fig. 4-53a. Transistor Q1 appears across the power supply output as a variable "bleeder resistor," with

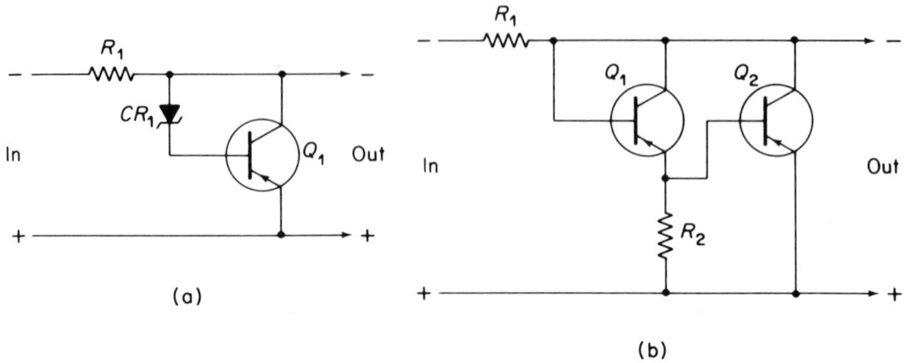


**FIGURE 4-53** Shunt voltage regulators

current flowing between emitter and collector. Base current flows through zener CR1. Both of these currents, as well as the load current, flow through the series resistor R1. When the forward bias of the Q1 base-emitter circuit is increased, the emitter-collector current is also increased (the emitter-collector resistance across the output is lowered). This lowers the output voltage. When the power supply is subject to large current changes, shunt regulators are often connected in cascade to increase their effectiveness, as shown in Fig. 4-53b.

**Series Voltage Regulator.** The basic series transistor voltage regulator is shown in Fig. 4-54. Transistor Q1 is placed in series with the power supply output, together with series resistor R1. Transistor Q1 acts as a variable "series resistor" with current flowing between the collector and emitter. Current also flows through resistor R2 and zener CR1 to establish a voltage at the base of Q1. The Q1 base voltage is determined by current flow through R2. The base voltage remains fixed in relation to the positive terminal of the power supply, but will vary in relation to the negative terminal. When the forward bias of the base-emitter circuit is increased, the emitter-collector

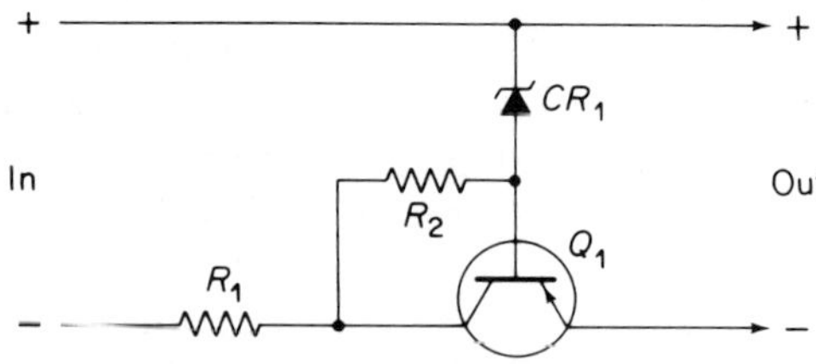


**FIGURE 4-54** Basic series voltage regulator circuit

"resistance" in series with the power supply output is lowered. This causes a lower voltage drop across the emitter-collector resistance and raises the power supply output voltage.

For example, if the power supply load should increase, more current will be drawn through the series resistor, as well as the Q1 emitter-collector resistance, and the power supply output voltage drops. Under these conditions, less current is drawn through CR1 and R2, increasing the forward bias on Q1. In turn, the decrease in Q1 emitter-collector resistance raises the power supply output voltage.

**Series Regulator with Feedback.** Series regulators are most often found in a feedback regulator circuit. Figure 4-55 shows one of the many variations of series feedback regulator. Note that the output is adjustable. In operation, the output voltage remains within about 0.5 V of the voltage at potentiometer R3 (at the base of Q3). Any variations in load (which might result in an undesired change in voltage across the load) cause corresponding changes in current through the base-emitter of Q3. Transistors Q1 and Q2 are connected as a Darlington pair or compound (Sec. 4-1.4). Any variations in output current (due to load changes) change the base current of Q2. These changes are amplified through Q1 and Q2. The change in emitter-collector current of Q1 opposes any changes in output current.

**Constant-Current Regulation.** Zeners can be used to regulate a power supply so as to produce a constant current rather than a constant voltage. The basic circuit is shown in Fig. 4-56. Transistor Q1 functions as a variable "series resistor" in the power supply output. Two parallel circuit paths exist. One path is through zener CR1

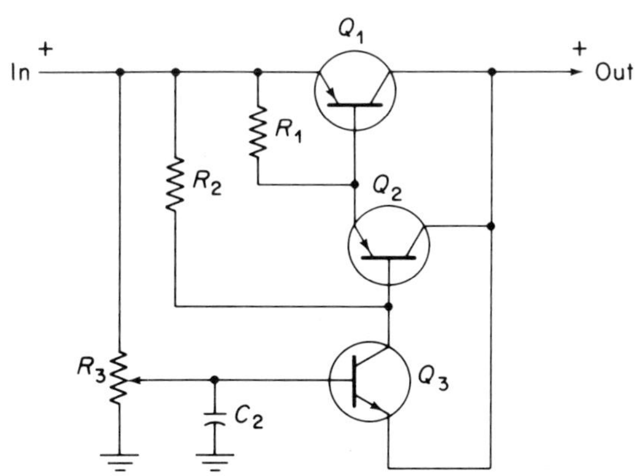


**FIGURE 4-55** Series voltage regulator with feedback

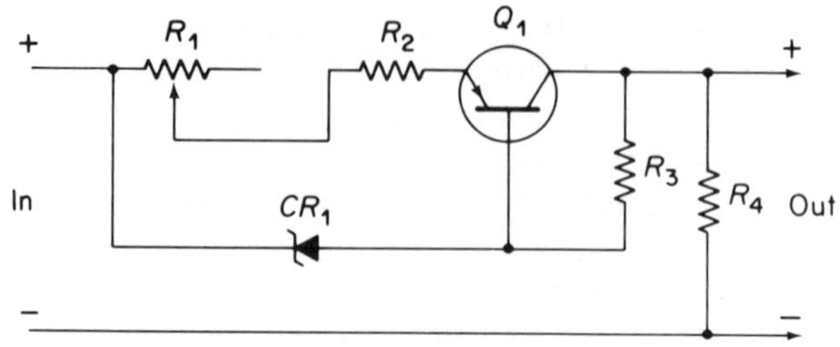


**FIGURE 4-56** Basic constant current regulation circuit

in series with bias resistor R3. The other path is through resistor R1 and transistor Q1.

Should there be any variation in current at the power supply output, the current through R3 will vary and change the forward bias on Q1. In turn, the emitter-collector resistance of Q1 varies to correct the current flow. The net result is that for every change in R3 current, there is an equal and opposite change in Q1 current. The current output of such a circuit is set by R1. The current remains constant (within limits) in spite of any load changes. However, the power supply output voltage will vary and the load changes.

### 4-5.7 POWER-CONVERSION CIRCUITS

Solid-state circuits have replaced the vibrator-type circuits for converters and inverters. Usually, the term *inverter* refers to a circuit used to change d-c power into a-c power. If the a-c output is rectified and filtered to provide direct current (say at a higher voltage), the overall circuit is referred to as a *converter*. The *push-pull switching inverter* is the most widely used type of power-conversion circuit. For inverter applications, the circuit provides a square-wave a-c output. When the inverter is used to provide d-c to d-c conversion, the square-wave voltage is usually applied to a bridge rectifier and filter.

Figure 4-57 shows the basic push-pull converter circuit using a *single transformer.* When the d-c input voltage is applied to the circuit, current tends to flow through both transistors Q1 and Q2. Because of the normal unbalance of the circuit, one transistor conducts more heavily than the other. Assuming that transistor Q1 conducts more heavily (initially), the rise in current through its collector inductance (one-half of the transformer winding) causes a voltage to be produced in the feedback windings of transformer T1 (which supply drive signals to the bases of Q1 and Q2). The base voltages thus produced are in the proper polarity to increase the current through Q1 and to decrease the current through Q2. Transistor

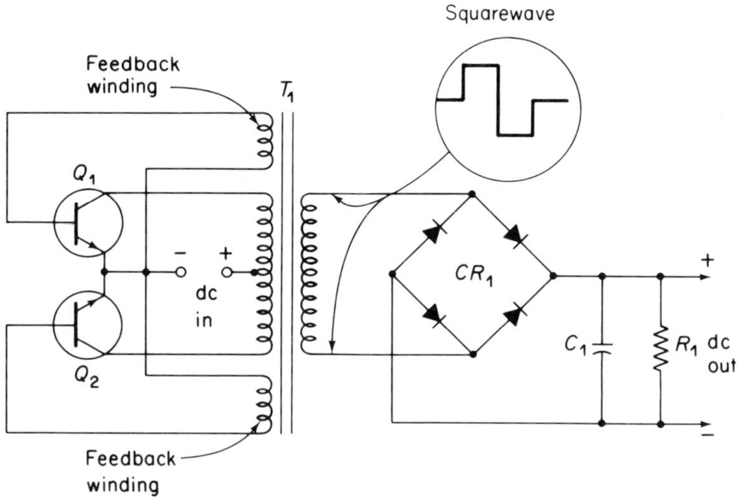


**FIGURE 4-57** Basic dc-dc push-pull converter circuit using a single transformer

Q1 is then quickly driven into saturation, with Q2 driven to cutoff.

When Q1 reaches saturation, the magnetic field around the collector is constant and no feedback voltage is applied to the transistor bases. With the cutoff base voltage removed, current is allowed to flow through transistor Q2. The increase in current through the collector inductance of Q2 causes voltages to be produced in the feedback windings, in the polarity that increases the current through Q2 and decreases the current through Q1. This effect is aided by the collapsing magnetic field about the collector inductance of Q1 (transformer winding) that results from the decrease in current through Q1. The feedback voltages produced by the collapsing field quickly drive Q1 beyond cutoff, and further increase the current flow through Q2 until saturation is reached. A new cycle is initiated each time one of the two transistors reaches saturation.

The expanding and collapsing magnetic fields around the primary of T1 produce an alternating current in the secondary of T1. This ac is rectified and filtered as described for the full-wave bridge supply in Sec. 4-5.3.

Figure 4-58 shows the basic push-pull converter circuit using *two transformers.* In this circuit, a small saturable transformer T1 provides the base drive for the switching transistors, and an output transformer T2 provides the coupling and desired output voltage (usually a step up) of the output delivered to the load circuit. Operation of the two-transformer circuit is essentially the same as that of the single transformer circuit.

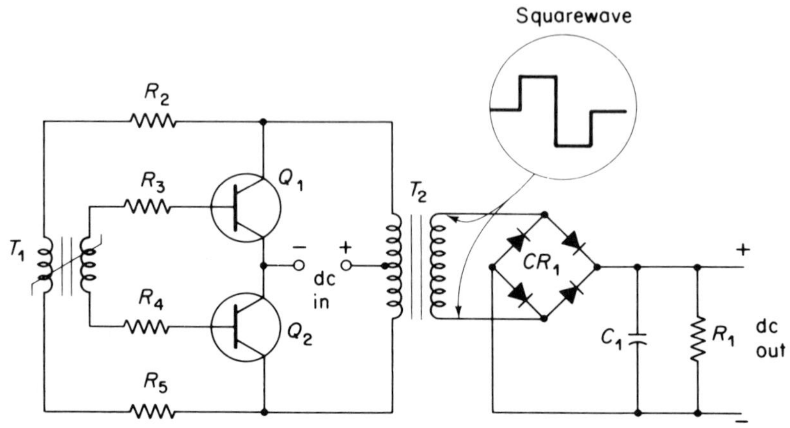


**FIGURE 4-58** Basic dc-dc push-pull converter circuit using two transformers

## 4-6 ATTENUATOR CIRCUITS

Attenuator circuits function to decrease or control signal strength. Most attenuators are resistive circuits or networks that decrease signal strength and match dissimilar impedances. In some cases, the attenuators are a group of fixed resistors in a circuit to provide a certain fixed amount of attenuation. In other cases, attenuators are variable resistances, similar to potentiometers or volume controls, with all the resistors contained in a single package.

Some attenuators are called *pads.* However, a pad is used primarily to match impedances, and may or may not provide some attenuation. A true attenuator provides attenuation, either fixed or variable.

Attenuators or pads may be either *asymmetrical,* where the input and output impedances are not the same, or *symmetrical,* with identical input and output impedances. In general, asymmetrical pads are used for impedance matching, but may also provide some attenuation. The symmetrical pads are used primarily as attenuators between two circuits with equal impedances.

Although there are any number of attenuator or pad circuits, they are generally versions of a few basic types. These basic types include the L, U, T, O, H, and pi types. In general, the L and U pads are asymmetrical (for impedance matching), and the T, O, H, and pi types are symmetrical (for attenuation).

### 4-6.1 ASYMMETRICAL L AND O PADS

The basic L-type pad shown in Fig. 4-59a is an asymmetrical pad used to match impedance of an input source to that of an output source. This provides a smooth transition between circuits of different characteristic impedances, with each circuit "seeing" a total impedance equal to its own characteristic impedance. Although a certain amount of signal attenuation usually occurs in

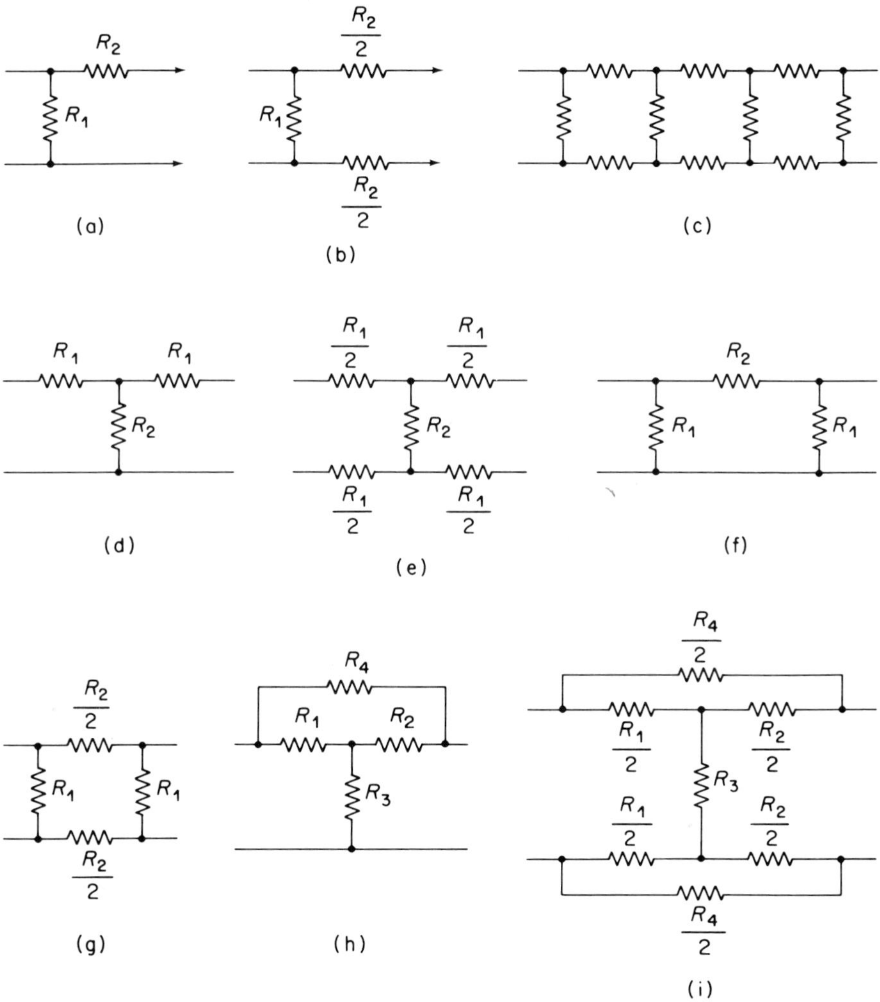


FIGURE 4-59 Typical attenuator and pad circuits

making this transition, the pad in Fig. 4-59a introduces a minimum of loss while performing its impedance-matching function.

A balanced arrangement of the L-type pad (known as the U pad) is shown in Fig. 4-59b. Here, each series resistor has one-half the value of R2 shown in Fig. 4-59a. When the pads are placed in cascade, as in the schematic of Fig. 4-59c, the network is sometimes referred to as a *ladder pad.*

### *4-6.2 SYMMETRICAL T AND H PADS (ATTENUATORS)*

The T pad shown in Fig. 4-59d is a symmetrical pad where the impedance of the input circuit matches that of the output circuit. The only function of the T pad is attenuation of the signal. Since impedance matching is not involved, the values of all R1 resistors are identical, and the values of R1 (total) and R2 are chosen to provide the degree of attenuation desired. The T pad of Fig. 4-59d is an unbalanced circuit. The balanced version (sometimes known as an H pad) is shown in Fig. 4-59e. In the balanced H pad, the R1 values are cut in half, as shown.

### *4-6.3 SYMMETRICAL PI AND O PADS (ATTENUATORS)*

The pi-type pad shown in Fig. 4-59f is a symmetrical unbalanced type. The balanced version (sometimes known as an O pad) is shown in Fig. 4-59g. Since the input impedance is the same as the output impedance, no impedance matching is involved. The resistor values are chosen to provide the degree of attenuation desired.

### *4-6.4 SYMMETRICAL BRIDGED T AND H PADS*

An additional resistor can be shunted across the series resistors of the T pad and H pad to form what is known as a *bridged pad.* The bridged T and H pads are shown in Fig. 4-59h and i, respectively.

## 4-7 FILTER CIRCUITS

The primary purpose of a filter is to discriminate against the passage of certain groups of frequencies while simultaneously passing other groups of frequencies. Although filter circuits range from the very simple to the very complex, there are only two types of filters, *active* and *passive.* The active filters described here are essentially a solid-state frequency-selective amplifier circuit. The passive filters are either *RC* (resistor-capacitor), used primarily for low- or audio-frequency applications, or *LC* (coil capacitor), for use at higher frequencies. Filters can also be classified by the frequencies that they are designed to pass or reject. The four most common filter classifications are as follows:

1. A *low-pass filter* is one that passes all frequencies below a selected value and attenuates higher frequencies. The low-pass filter is also known as a high-cut filter.
2. A *high-pass filter* is one that passes all frequencies above a selected value and attenuates lower frequencies. The high-pass filter is also known as a *low-cut filter.*
3. A *band-elimination filter,* also known as a *band-stop, band-rejection,* or *band-suppression* filter, is one that suppresses a selected band of frequencies while passing all lower and higher frequencies.
4. A *bandpass filter* is one that passes a selected band of frequencies while rejecting all lower and higher frequencies.

### 4-7.1 RC FILTERS

*RC* filters are the simplest form of audio-frequency filter. Low frequencies are removed by a series capacitor and shunt resistor (Fig. 4-60a), and high frequencies are filtered by interchanging the resistor and capacitor (Fig. 4-60b). The filters in Fig. 4-60 are essentially *RC* circuits with the following three factors: resistance, reactance, and impedance. The amount of attenuation produced by *RC* filters is dependent on the ratio of resistance or of reactance to impedance.

In the high-pass (low-frequency attenuation) circuit of Fig. 4-60a, input voltage is applied across both the resistor and capacitor. Output voltage appears across the resistance. As frequency decreases, capacitive reactance increases, as does the impedance. Since the input voltage remains constant, the current through the circuit decreases ($I = E/Z$) with an increase in impedance. Thus, current through the resistance decreases and the output voltage drops. In a high-pass *RC* circuit, the attenuation is calculated by

$$\text{output voltage} = \text{input voltage} \times \frac{\text{resistance } (R)}{\text{impedance } (Z)}$$

Since the value of *R* is fixed, for practical purposes, and the value of *Z* varies indirectly with frequency, the voltage output varies directly with frequency (output voltage drops with decreases in frequency).

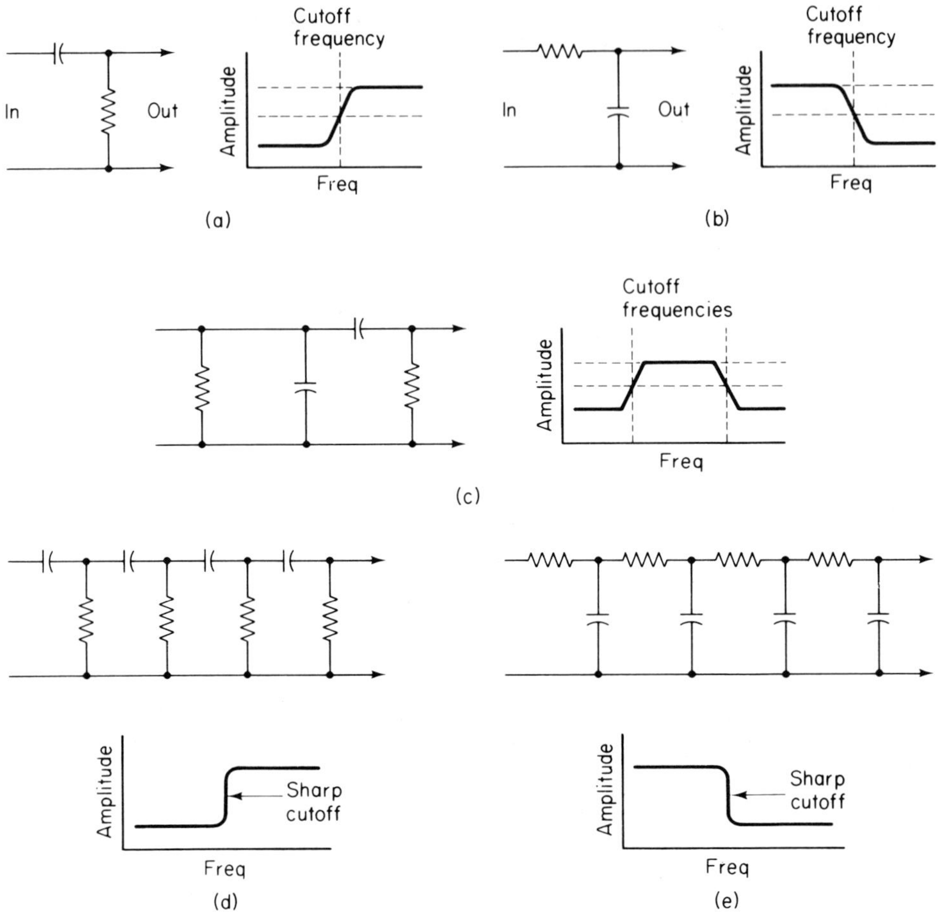


**FIGURE 4-60** Typical RC filter circuits

In the low-pass (high-frequency attenuation) circuit of Fig. 4-60b, input voltage is applied across both the resistor and capacitor. Output voltage is taken across the capacitor. As frequency increases, the capacitive reactance decreases, as does the impedance. However, since the impedance is composed of both reactance and fixed resistance, the impedance does not decrease as much as the reactance. Thus, the decrease in reactance (in relation to impedance) causes the voltage output to decrease. In a low-pass *RC* filter circuit, the attenuation is calculated by

$$\text{output voltage} = \text{input voltage} \times \frac{\text{reactance } (XC)}{\text{impedance } (Z)}$$

Since the reactance varies at a greater rate than the impedance, the voltage output varies directly with frequency (voltage drops with increases in frequency).

In the bandpass *RC* filter of Fig. 4-60c, both the high-pass (low-cut) and low-pass (high-cut) filter circuits are combined in a common circuit.

A single *RC* filter will provide a gradual transition from the passband to the cutoff region. If a rapid transition is required, two or more *RC* filters can be combined, as shown in Fig. 4-60d and e.

### *4-7.2 LC FILTERS*

The most precise filters are composed of inductors (coils) and capacitors. An *LC* filter can also include resistors. The input and output connections of most *LC* filters are terminated in source and load resistances or by impedances equal in value to the impedance of the filter. Except for special applications, it is not practical to use *LC* filters for audio frequencies. The inductances required at low frequencies are quite large, and therefore heavy and bulky.

**Low-Pass LC Filter.** Figure 4-61a shows a typical low-pass *LC* filter in its simplest form, the *basic L configuration.* All *LC* filters take advantage of the fact that capacitors and coils operate inversely in the presence of alternating current. That is, inductive reactance increases with frequency, and capacitive reactance decreases with frequency. Thus in the low-pass *LC* filter, the parallel unit (a capacitor across the line) has a decreasing reactance as frequency is increased. This acts to bypass high frequencies but has an increasing reactance to low frequencies. The series unit (a coil in the line) has an increasing reactance to high frequencies but passes the low frequencies.

The basic L configuration provides a gradual cutoff of signals at frequencies above a given point (cutoff frequency). When a sharper cutoff frequency is desired, an additional capacitor or inductance can be added to the circuit. If an inductor is added as shown in Fig. 4-61b, the filter is known as a *T type* (because of its appearance). If a capacitor is added as shown in Fig. 4-61c, the filter is known as a *pi type.*

All of the filters shown in Fig. 4-61 are known as *constant-K filters* because the impedance of the filter remains constant no matter what frequencies are involved. That is, if capacitive reactance goes down for a given increase in frequency, the inductive reactance goes up by a corresponding amount (and vice versa).

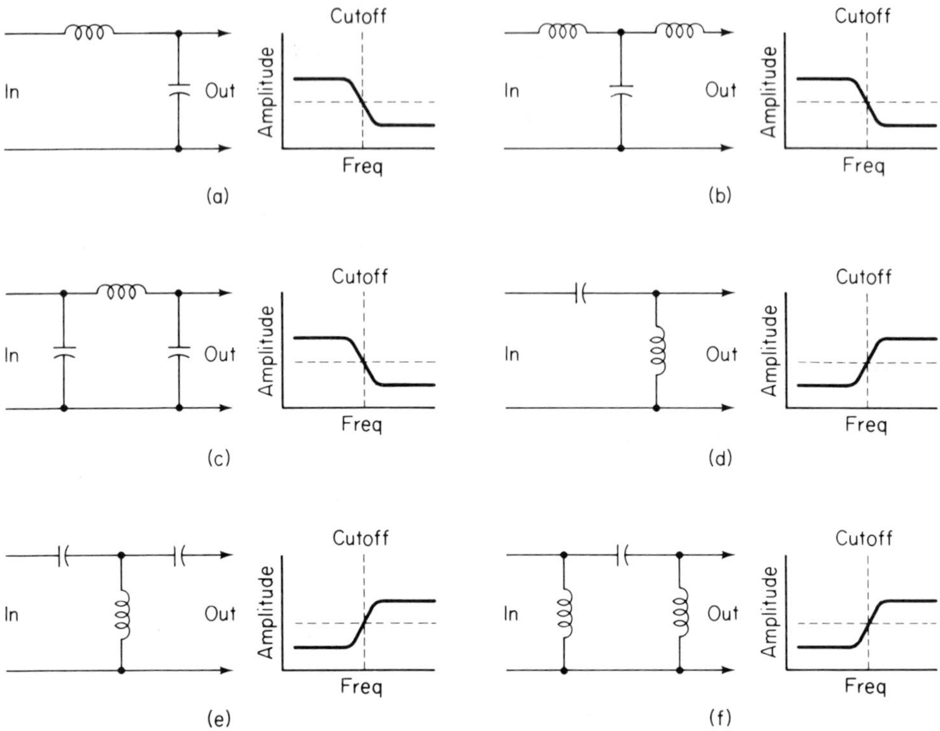


**FIGURE 4-61** Typical LC filter circuits

**High-Pass LC Filter.** Figure 4-61d shows a typical high-pass *LC* filter in the simplest form, the basic L configuration. High-pass *LC* filters operate on the same principles as the low-pass filters, except that the components are interchanged (coil across the line, capacitor in series). When a sharper cutoff frequency is required, additional capacitors or inductances are added, as shown in Fig. 4-61e (T-type) or f (pi-type).

**Band-Elimination LC Filter.** Two *LC* circuits can be combined as a band-elimination filter, shown in Fig. 4-62. This circuit takes advantage of the different impedance characteristics of series- and parallel-resonant *LC* circuits. A parallel *LC* circuit has maximum impedance at the resonant frequency. A series *LC* circuit has minimum impedance at the resonant frequency.

The series circuit of the filter has minimum impedance at the center frequency of the desired band. The impedance increases on either side of resonance. This acts to bypass the center frequency. The parallel circuit has maximum impedance at the center frequency,

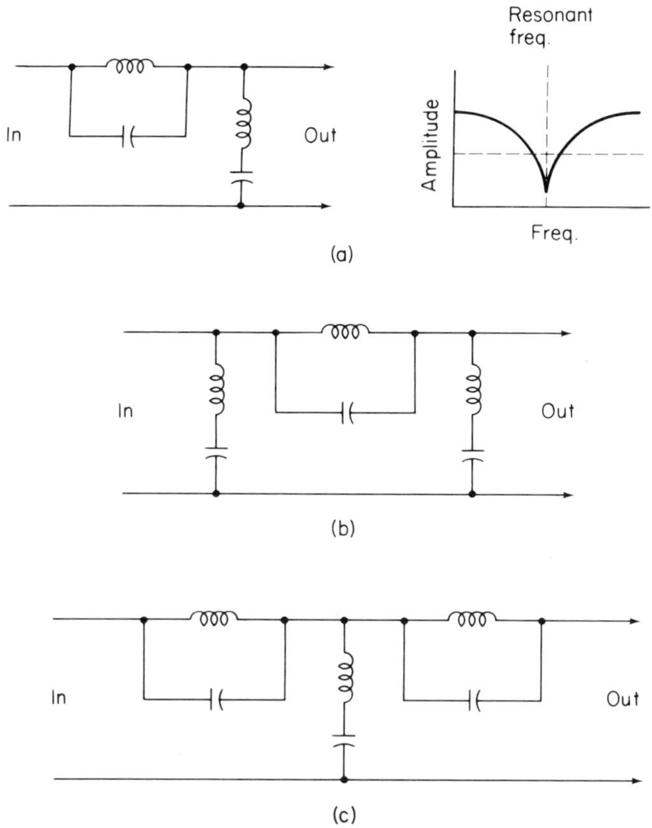


**FIGURE 4-62** Band-elimination LC filter circuits (a) Basic, (b) Pi-type, (c) T-type

with the impedance decreasing on either side of resonance. This prevents passage of the center frequency as well as a band of frequencies on either side. As shown in Fig. 4-62, band-elimination filters are also found in T-type and pi-type configurations.

**Bandpass LC Filter.** Two *LC* circuits can be combined as a bandpass filter shown in Fig. 4-63. Again, this circuit takes advantage of the different impedance characteristics of series- and parallel-resonant *LC* circuits. However, in the band-pass filter, the series circuit is placed in series with the line, and the parallel circuit is placed across the line. This acts to pass the center frequency and to reject or bypass frequencies above and below the center frequency. As shown in Fig. 4-63, bandpass filters are also found in T-type and pi-type configurations.

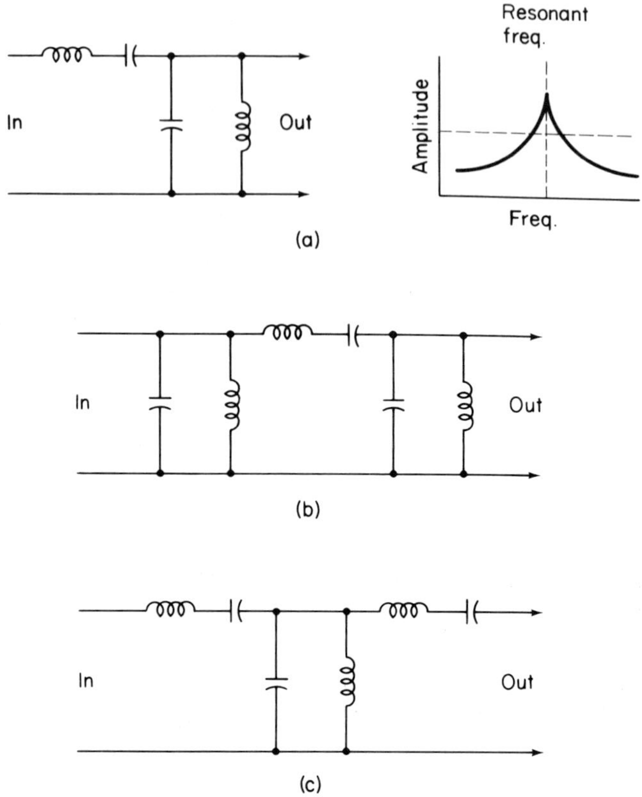


**FIGURE 4-63** Band-pass LC filter circuits (a) Basic, (b) Pi-type, (c) T-type

### *4-7.3 ACTIVE FILTERS*

In addition to passive filters (either *LC* or *RC*), it is possible to use amplifiers to form filters. There are two advantages in using these active filters. First, it is possible to obtain the equivalent of an inductive reactance without actually using a heavy and bulky inductance usually required for a typical *LC* audio filter. Second, the use of an active filter eliminates the signal loss normally associated with passive filters.

**Low-Pass Active Filter.** Figure 4-64 shows the basic circuit of an active low-pass filter. The amount of gain and the sharpness of the cutoff frequency are set by the amount of feedback in relation to

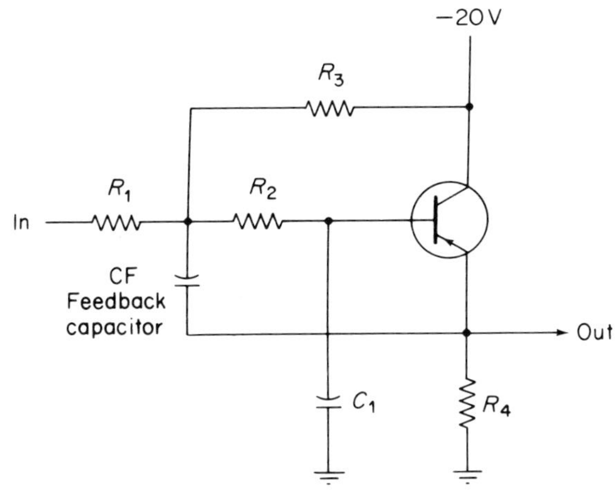


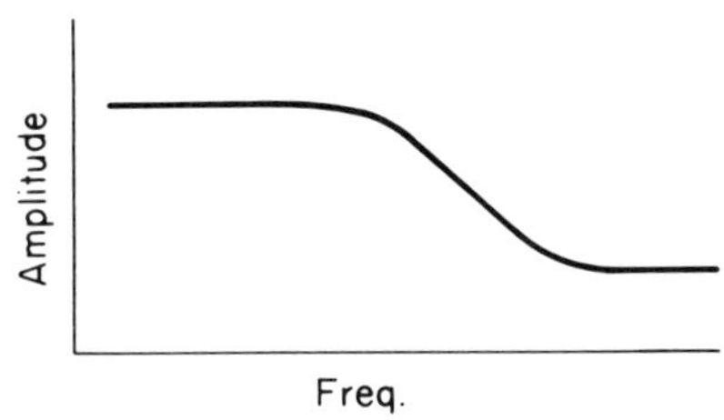


**FIGURE 4-64** Low-pass active filter circuit

signal (which, in turn, is set by component values). The feedback is positive and thus adds to the signal. However, the feedback amplitude (across the entire frequency range) is just below the point necessary for oscillation. The circuit in Fig. 4-64 is an emitter follower, which typically has no voltage gain.

**High-Pass Active Filter.** Figure 4-65 shows the basic circuit of an active high-pass filter. The circuit of Fig. 4-65 is the inverse of the Fig. 4-64 circuit. That is, the Fig. 4-65 circuit uses capacitors in series with the base, with feedback obtained through RF rather than CF. The gain and sharpness of cutoff are set by the amount of feedback (determined by circuit values).

**Active Bandpass Filter.** The circuits in Figs. 4-64 and 4-65 can be cascaded to provide a bandpass filter. In such an arrangement, the low-pass filter generally follows the high-pass filter.

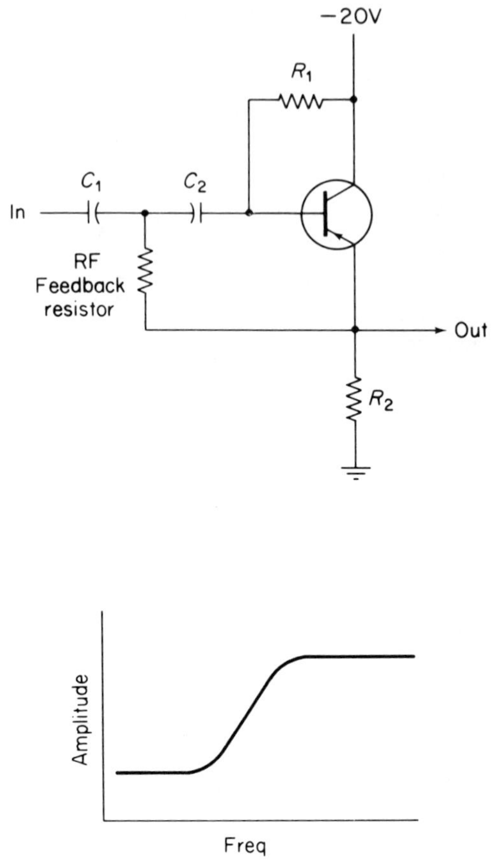


**FIGURE 4-65** High-pass active filter circuit

**Active Peaking Filter.** The circuits in Figs. 4-64 and 4-65 can be cascaded to provide a peaking filter with the proper selection of components. However, a single-stage tuned amplifier will produce the same results. Such a circuit is shown in Fig. 4-66.

## 4-8 ELECTRONIC CONTROL CIRCUITS

This section describes some basic control applications for the electronic devices covered in Chapter 3 (SCRs, Triacs, photocells, etc.). Keep in mind that these represent only a very small fraction of the electronic control circuits that you may find on schematics.

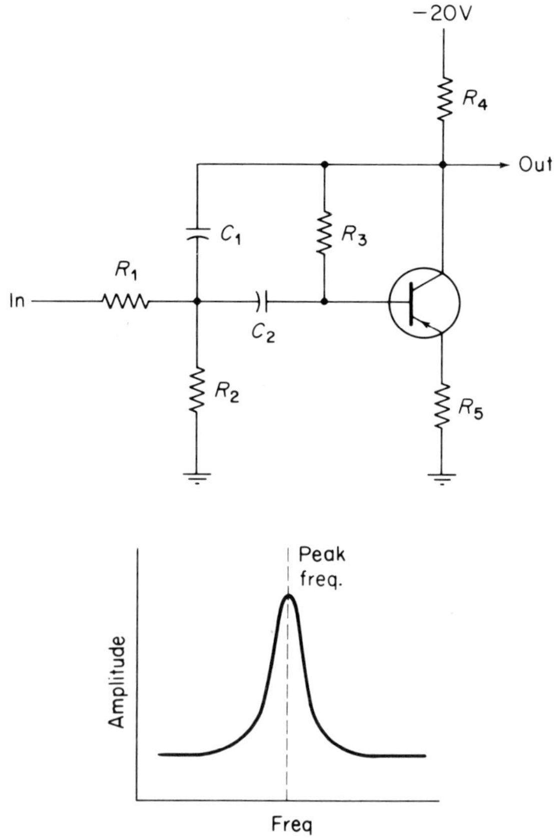


**FIGURE 4-66** Active peaking filter circuit

This is because there is an increasing use of electronic control devices in such applications as motor control, power control, and switching.

### *4-8.1 BASIC PHASE CONTROL*

Phase control is the process of rapid ON-OFF switching which connects an a-c power supply to a load for a controlled fraction of each cycle. This is a highly efficient means of controlling the *average power* to loads such as lamps, heaters, and motors. Control is accomplished by governing the *phase angle* of the a-c waveforms at which the control device is triggered. The control device then conducts for the remainder of the half-cycle. Figure 4-67 shows the basic forms of a-c phase control.

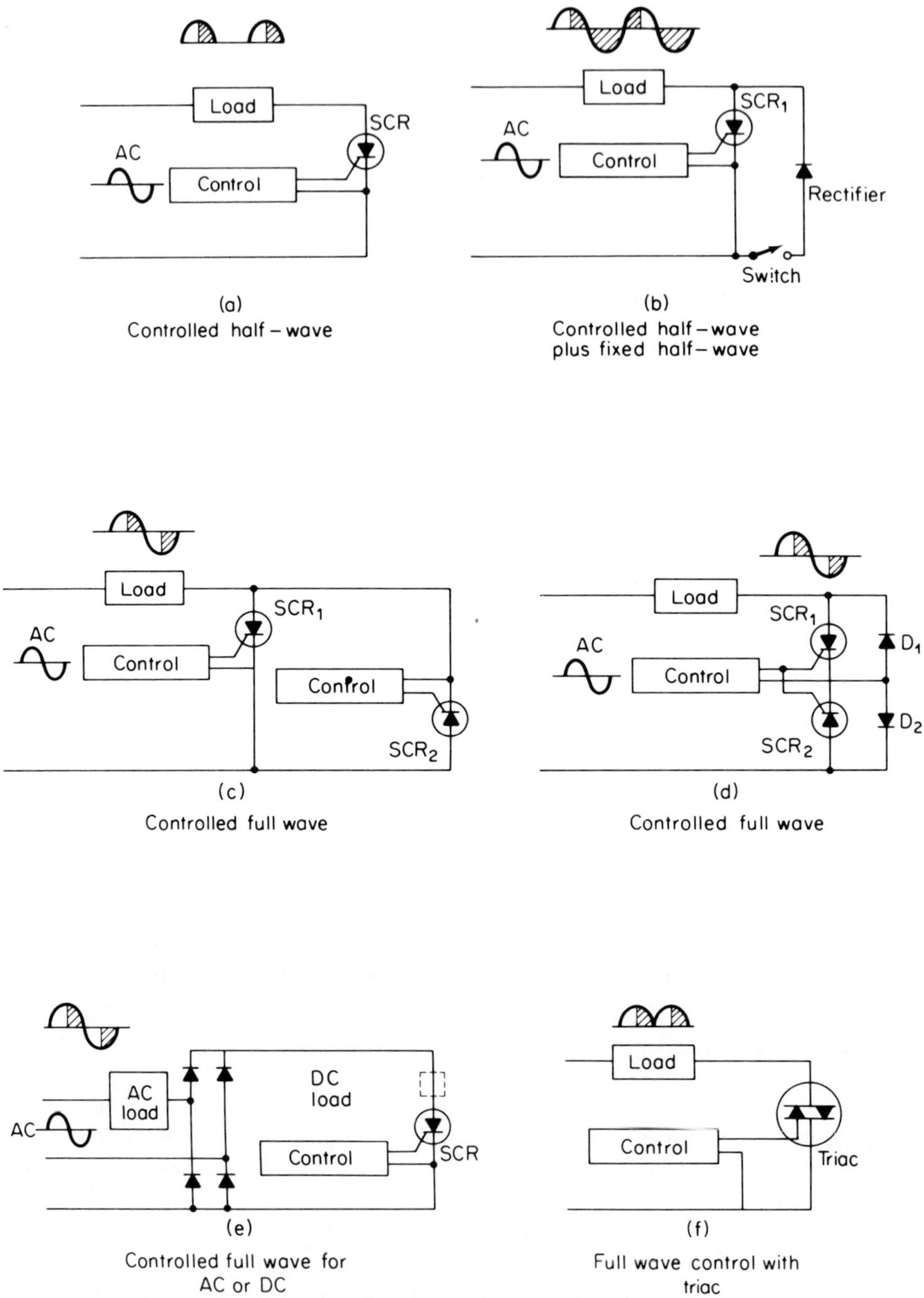


**FIGURE 4-67** Basic forms of a-c phase control

The simplest form of phase control is the half-wave control shown in Fig. 4-67a. This circuit uses one SCR for control of current flow in one direction only, and is used for loads that require power control from zero to one-half of full-wave maximum. The circuit is also useful where direct current is required (or permitted) for the load.

The addition of one rectifier, as shown in Fig. 4-67b, provides a fixed half-cycle of power which shifts the power control range to half-power minimum and full-power maximum. The use of two SCRs, as shown in Fig. 4-67c, controls from zero to full-power but requires two gate signals. A single gate or trigger signal is required for the zero to full-power control circuit of Fig. 4-67d. The most flexible circuit, Fig. 4-67e, uses one SCR inside a bridge rectifier and may be used for control of either a-c power or full-wave rectified d-c power.

Generally, the most simple, efficient, and reliable method of controlling a-c power is by the use of a triac, as shown in Fig. 4-67f. This circuit provides full-wave control, in both directions, using only one gate or trigger (often a diac). For that reason, the triac-diac combination has been by far the most popular a-c power control circuit.

**Basic Trigger Circuits for Phase Control.** Most phase control trigger circuits use some form of relaxation oscillator, such as the UJT oscillator discussed in Sec. 3-5. Since relaxation oscillators are simply *timing circuits*, they must be synchronized to the a-c power supply to provide proper phase control control. That is, their timing cycles must start simultaneously with each alternation of the a-c power. This is usually done by taking the oscillator input voltage from the a-c supply.

There are many ways of connecting the various versions of the basic oscillator circuit, using the different triggering devices, control devices, supply, and load circuits. Fig. 4-68 shows the basic half-wave and full-wave phase control circuits. With either circuit, the load, control device, and trigger device are all operated from the a-c power supply. Capacitor C is charged through adjustable resistor R on alternations of the a-c supply. In the half-wave circuit, C is charged once for each half-alternation or half-cycle. In the full-wave circuit, C is charged on both alternations of each cycle. When C is charged to a certain voltage, the trigger device turns on and provides a trigger signal to the control device. This switches the control device on for the remainder of the half-cycle, and power is applied through the load.

The time required to charge C to the trigger point is set by adjustment of R. In turn, this determines the portion of the half-

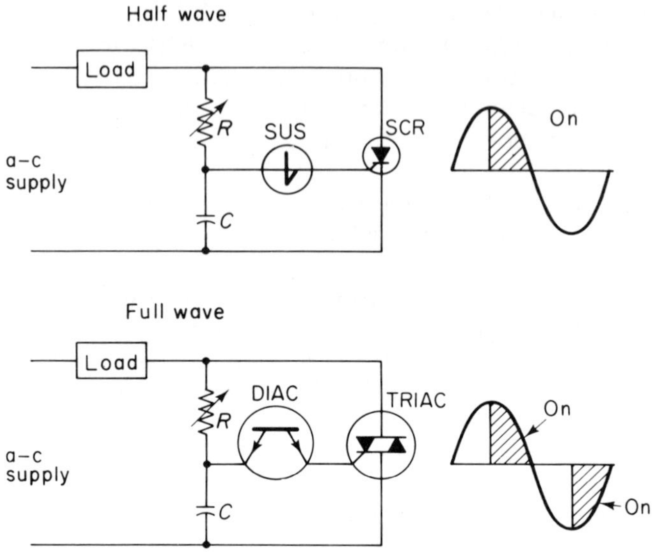


**FIGURE 4-68** Basic half-wave and full-wave phase control circuits

cycle in which the control device is on, and the amount of power applied to the load. For example, assume that the value of R is adjusted to minimum resistance. This causes C to charge quickly, and switch both the trigger and control devices on *early in the half-cycle.* Thus, maximum power is applied to the load. With R set at maximum resistance, C charges slowly, and the devices switch on late in the half-cycle, providing minimum power to the load. Note that both trigger and control devices are switched off when the a-c power goes to zero between each half-cycle.

### *4-8.2 STATIC SWITCHING CIRCUITS*

Electronic control devices are not limited to phase control. In fact, almost any static switching function otherwise performed by various mechanical and electromechanical switches can be accomplished with electronic controls. The following paragraphs describe an example.

**Electronic Proximity Switch.** Figure 4-69 is the schematic of an electronic proximity switch. This circuit is ideally suited for use in elevators (both for door safety controls and floor selector buttons), supermarket door control, bank-safe monitors, flow switch actuators, and conveyor belt counting systems.

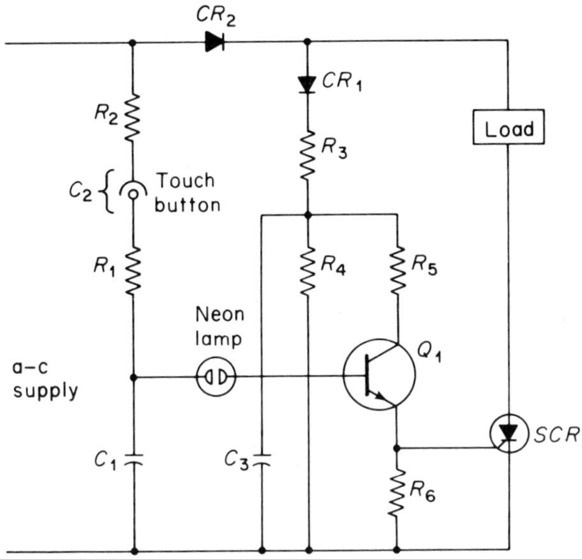


**FIGURE 4-69** Electronic proximity switch

Capacitor C1, resistors R1/R2, and the sensor plate "capacitor" C2 form a voltage divider across the a-c supply. The voltage across C1 depends on the ratio of C1/C2 and the supply voltage. The capacitance of C2, in turn, depends on the proximity to the sensor plate of any reasonably conductive and grounded object (metals, human body, etc.). As soon as the voltage across C1 exceeds the breakdown voltage of the neon lamp, capacitors C1 and C2 discharge through the base and emitter of transistor Q1. This discharge current is amplified by Q1 and turns the SCR on to energize the load.

The load remains energized as long as the sensor plate or "button" is touched. When the button is no longer touched, the load is de-energized. A latching action can be provided by replacing CR1 when a short circuit (so that the SCR remains triggered on both half-cycles). The circuit must then be reset by an auxiliary contact in series with the SCR.

### *4-8.3 ELECTRONIC MOTOR CONTROL CIRCUITS*

One of the most common electronic control applications is *phase control of electric motor speed.* There are many motor control systems, so we will discuss only the basics here. The main problem in motor speed control is sensing the motor speed and then providing feedback to offset changes in speed. If speed increases (due to a change in load or supply voltage), the feedback signal slows the motor down, and vice versa.

In d-c and universal motors, speed is usually sensed by monitoring the CEMF generated during the SCR on-time. (CEMF, or counter electromotive force, is a voltage or current generated by an electric motor when the motor armature rotates. This CEMF is in opposition to the motor supply voltage, and is approximately proportional to speed. A universal motor is one that will operate with either a-c or d-c power.) In the case of shunt and permanent-magnet motors, the CEMF is directly proportional to speed. If speed increases, the CEMF increases, and vice versa. Thus, by comparing the CEMF against an adjustable voltage, and using this comparison to control the SCR on-time, motor speed can be controlled within close tolerance. In the case of series motors, the CEMF is almost proportional to speed, so the same basic principle can be used.

In a-c induction motors, speed is often sensed by means of a tachometer generator. (Such tachometers consist of a small d-c generator mechanically coupled to the motor.) If motor speed increases, tachometer output voltage increases, and vice versa. This tachometer feedback signal is used to control SCR on-time.

**D-C and Universal Motor Speed Control.** Figure 4-70 is the schematic of the basic circuit for speed control of d-c and universal motors using an SCR. This phase-control circuit operates by comparing the motor CEMF voltage against a reference voltage (which is rectified by CR2, and set by adjustment of R2).

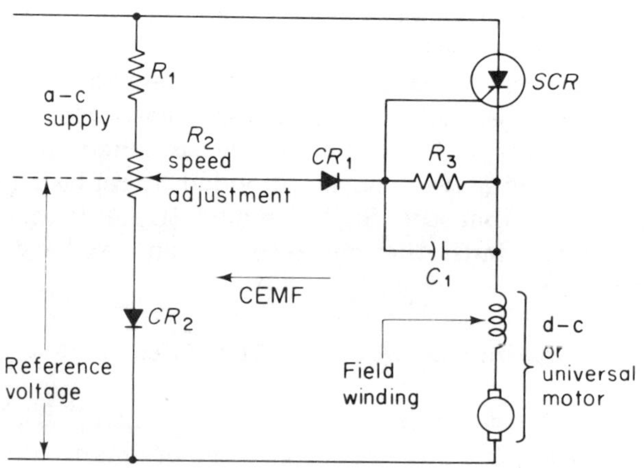


**FIGURE 4-70** Basic d-c and universal motor speed control circuit

If the CEMF is greater than the reference voltage (the motor is going faster than the selected speed), CR1 is reverse-biased (the cathode is more positive than the cathode), the SCR is not triggered, and no power is applied to the motor. When the motor slows down, the CEMF voltage drops and is less than the reference voltage, CR1 is forward-biased (anode more positive), the SCR is triggered, and power is applied to the motor. The speed at which this occurs may be varied by changing the magnitude of the reference voltage (by adjustment of R2).

**Induction Motor Control.** Figure 4-71 shows the basic circuit for speed control of a-c motors using a tachometer generator geared to the motor. The output of the tachometer generator (or feedback signal) is rectified by diodes CR1 through CR4 and applied through resistor R1 to amplifier transistor Q1. This feedback signal is inverted by Q1 and is used to charge capacitor C1 through Q2. The charge on C1 determines the SCR on-time.

If the motor speeds up, the tachometer generator produces a greater output, which is rectified into a positive d-c voltage by the diodes. This positive voltage output is inverted by Q1 to a negative voltage which charges C1 to a lower value. The SCR then triggers at a later point, and the motor slows down. The charge on C1 (and thus the motor speed) can also be set by adjustment of R1.

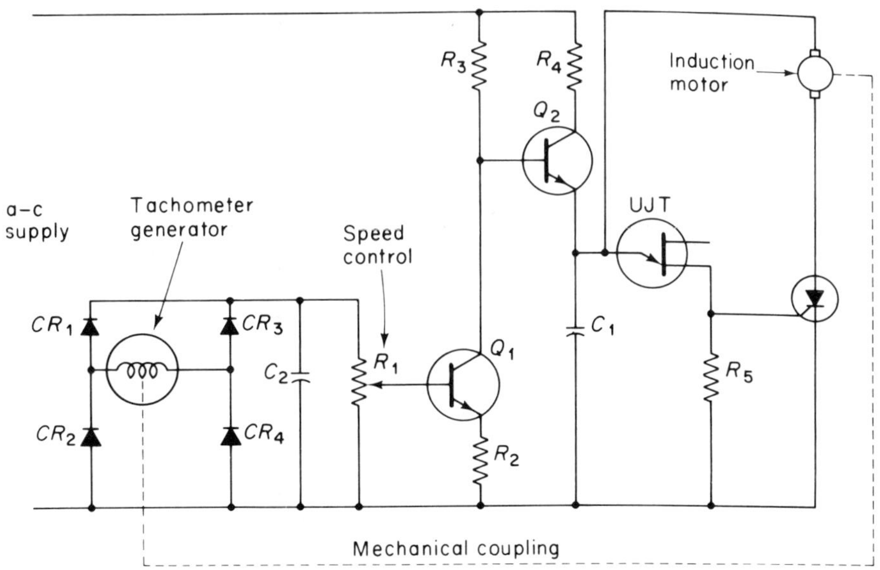


**FIGURE 4-71** Induction motor speed control circuit

### 4-8.4 LIGHT-ACTUATED CONTROL APPLICATIONS

Light-presence detectors and light-absence detectors are typical examples of light-actuated control applications. Figure 4-72 shows the basic circuits for both applications.

In the case of the *light-presence detector*, power to activate the load is switched on only when light is applied. Capacitor C starts to charge only when light strikes phototransistor Q1. The rate at which C charges is set by resistor R1. When C has charged sufficiently, the UJT fires and triggers the SCR, applying power to the load. When light is removed from Q1, capacitor C is disconnected from the power source and does not charge. Thus, the UJT and SCR do not trigger, and power is removed from the load. A typical application

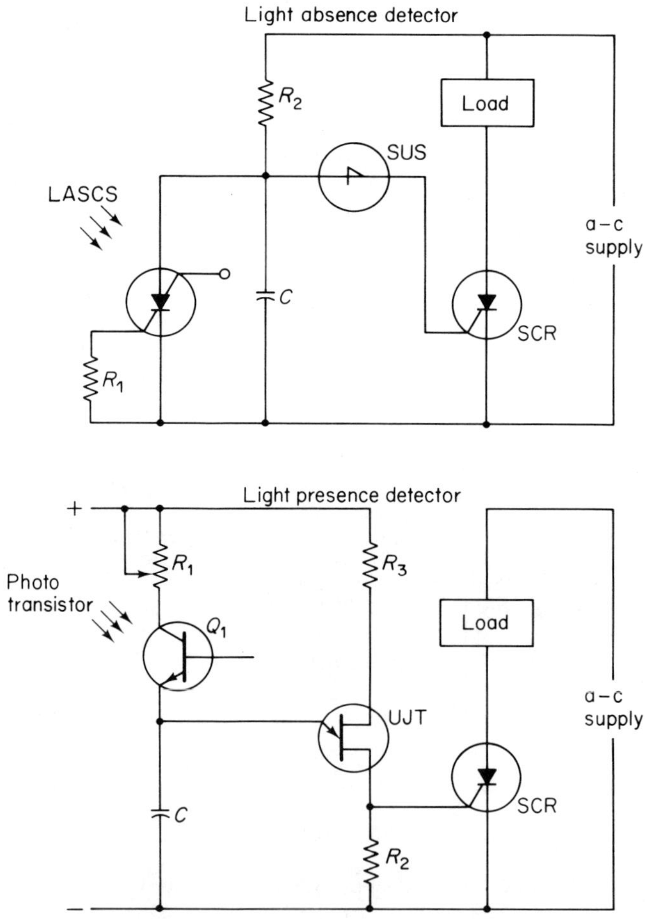


**FIGURE 4-72** Light actuated control circuits

for such a circuit is to activate warning or advertising signs at night by the headlights of approaching vehicles.

In the case of the *light-absence detector*, power to the load is switched on when light is interrupted. As long as there is light on the LASCS, capacitor C is shorted and does not charge. When the light striking the LASCS is removed, capacitor C charges and the SUS anode goes positive on the next positive cycle of the power supply. This triggers the SUS and SCR, applying power to the load. When light is again applied to the LASCS, capacitor C is shorted and prevented from charging. Thus, the SUS and SCR do not trigger and power is removed from the load. A typical application for such a circuit is to switch on house lights in the evening or whenever it starts to get dark (such as on an overcast day).

## EXERCISES

4.1. What is the function of the circuit shown in Fig. 4-1?

4.2. What is the function of C3 in Fig. 4-1?

4.3. In general, what determines the gain, impedance, and stability of the circuit in Fig. 4-1?

4.4. If you find two circuits on a schematic, one identical to Fig. 4-1 and one identical except that C3 is omitted, what would you expect with regard to circuit voltage gain?

4.5. What type of capacitor is often used for C3?

4.6. When one transistor feeds into another transistor as shown in Fig. 4-2, this is known as ____________ .

4.7. What is the advantage of direct coupling such as that shown in Fig. 4-2?

4.8. When an emitter follower feeds directly into another emitter follower, the circuit is often called a ____________ .

4.9. If the base input of a phase splitter goes positive, the emitter also goes positive. True or false?

4.10. In the differential amplifier of Fig. 4-7, if the inputs are both +2 V, what is the approximate output (assuming that the circuit has a gain of 10)?

4.11. Why does the single power supply circuit of Fig. 4-8 have reduced low-frequency response?

4.12. Describe one advantage and one disadvantage of the coupling capacitor in Fig. 4-10.

4.13. Where are the inputs and outputs of the circuit shown in Fig. 4-11?

4.14. In the circuit of Fig. 4-12, what sets the bias on Q2 and Q3?

4.15. In the circuit of Fig. 4-12, Q1 is probably operated as class ______. Q2 and Q3 are probably operated as class ______ or possibly ______.

4.16. Where are the inputs and outputs of the circuit shown in Fig. 4-12?

4.17. In the circuit of Fig. 4-12, R3, R5, and R6 are ______ resistors.

4.18. What is the advantage of class AB operation over class B (in audio amplifiers)?

4.19. Although the terms are often used interchangeably, what is the true difference between a volume control (Fig. 4-14) and a gain control (Fig. 4-15)?

4.20. Operation of most treble and bass controls (Figs. 4-16 and 4-17) is based on the ______ principle.

4.21. Most radio-frequency circuits contain one or more ______ circuits.

4.22. The circuit of Fig. 4-18a is generally used as the ______ stage of an RF transmitter, whereas the circuit of Fig. 4-18b is most likely found as the ______ stage.

4.23. What is the function of C2 in Fig. 4-18?

4.24. What is the function of L1 in Fig. 4-18?

4.25. Where are the inputs and outputs of the circuit shown in Fig. 4-18a?

4.26. The circuit of Fig. 4-18 is probably operated class ______.

4.27. What is the function of C1 in Fig. 4-18?

4.28. What are the purposes of the RFCs in Fig. 4-18?

4.29. If you were not told the function of the circuit in Fig. 4-19, how could you tell that it was an RF voltage amplifier?

4.30. If you were not told the function of the circuit in Fig. 4-20, how could you tell that it had two functions (amplification and mixing)?

4.31. What is the purpose of C1 in Fig. 4-21?

4.32. If the signal input to the circuit of Fig. 4-21 is frequency-modulated by a 3000-Hz audio tone, what output will the circuit produce?

4.33. Name two differences (as shown by the schematic of Fig. 4-22) between the FM discriminator and the FM ratio detector.

4.34. If the signal input to either circuit of Fig. 4-22 is amplitude modulated by a 7000-Hz audio tone, what output will the circuit produce?

4.35. What is the function of CR1 in Fig. 4-23?

4.36. What is the function of CR2 in Fig. 4-23?

4.37. What is the function of C1 in Fig. 4-24?

4.38. All oscillators operate on the ____________ principle.

4.39. Where is the feedback in the circuit of Fig. 4-25?

4.40. Where is the *LC* resonant circuit in Fig. 4-26?

4.41. What are the functions of Y1, L1, and C5 in Fig. 4-28?

4.42. Where is the feedback in the circuit of Fig. 4-28?

4.43. What is the function of C1 in Fig. 4-29?

4.44. What is the probable frequency range of the oscillator shown in Fig. 4-30? What determines this frequency?

4.45. Where is the feedback in the circuits of Fig. 4-31?

4.46. The blocking oscillators shown in Fig. 4-31 operate on the ____________ principle.

4.47. Where is the feedback in the circuits of Fig. 4-32?

4.48. If there is no trigger, what determines the frequency of the circuit shown in Fig. 4-32?

4.49. What type of output is produced when a square-wave signal is applied to a low-pass *RC* filter?

4.50. What type of output is produced when a square-wave signal is applied to a high-pass *RC* filter?

4.51. What determines the frequency of the circuit shown in Fig. 4-40?

4.52. Describe the two types of outputs available from a UJT relaxation oscillator.

4.53. What determines the peak frequency of the circuit shown in Fig. 4-42?

4.54. Name two advantages of the bridge rectifier power supply in Fig. 4-45 over the circuits of Figs. 4-43 and 4-44.

4.55. Resistor R1 in Figs. 4-43 through 4-45 is often called a ____________ resistor.

4.56. If the a-c input voltage to the circuit of Fig. 4-51 is 10 V, what is the approximate a-c output voltage? Will there be any d-c output voltage from the circuit?

4.57. If the value of RS in Fig. 4-52b produced a 10-V drop, what is the circuit input voltage?

4.58. What is the function of R3 in Fig. 4-55?

4.59. If R1 and C1 in Fig. 4-57 were eliminated, the output would then be pure direct current. True or false?

4.60. How does transformer T1 in Fig. 4-58 differ from transformer T2?

4.61. The purpose of the circuits in Figs. 4-57 and 4-58 is to convert ac into dc. True or false?

4.62. What is the difference between the attenuators shown in Fig. 4-59 and those of Fig. 2-48?

4.63. When a high-pass filter is combined with a low-pass filter, a ________ filter is formed.

4.64. When a resistor is connected in series with the line and a capacitor is connected across the line, a ________ filter is formed.

4.65. The amount of attenuation produced by *RC* filters is dependent on the ________ of resistance or reactance to impedance.

4.66. What is the primary advantage of *LC* filters (Fig. 4-61) over *RC* filters (Fig. 4-60)?

4.67. What is the primary advantage of the active filters (Figs. 4-64 through 4-66) over *LC* filters (Figs. 4-62 and 4-63)?

4.68. The control circuits of Fig. 4-67 operate on the principle of ________.

4.69. What determines the amount of power applied to the load in the circuits of Fig. 4-68?

4.70. What is the function of the neon lamp in the circuit of Fig. 4-69?

4.71. In the circuit of Fig. 4-70, if the motor slows down, what happens to CR1, the SCR, and the motor?

4.72. In the circuit of Fig. 4-71, if the motor slows down, what happens to Q1, Q2, the UJT, the SCR, and the motor?

4.73. Describe what happens when light is applied to the LASCS in the light-absence detector of Fig. 4-72.

# 5

# Reading Some Advanced Circuit Schematics

In this chapter, we concentrate on reading the schematics of complete, functioning electronic equipments or sets. You will find that such equipment is made up of the basic circuits discussed in Chapter 4. In turn, the basic circuits are made up of the passive and active devices discussed in Chapters 2 and 3, respectively.

If you have read the book from Chapter 1 through Chapter 4, in order, you should be able to recognize circuits. In effect, you should be able to see a circuit schematic and say "that is an oscillator, and it operates as follows . . . ." By the time you have finished this chapter, you will see how these circuits are combined to form complete electronic equipments. Although the devices described in this chapter are fairly complex, they represent common household electronic equipment, such as a CB set and a TV set. As we go along, you will find some new, specialized schematic symbols, and you will see that there are many variations of basic circuits.

## 5-1 TRANSMISSION AND RECEPTION OF RADIO SIGNALS

The circuits of both CB and TV sets involve the transmission and reception of radio signals. So before we get too far into the circuit details, let us review very briefly the basics of radio.

### *5-1.1 RADIO TRANSMISSION BASICS*

CB involves the transmission of AM (amplitude-modulation) and SSB (single-side-band) signals. TV uses both AM and FM (frequency-modulation) signals. All three of these (AM, FM, and SSB) are based on the transmission of CW (continuous-wave) signals. So let us review the four basic transmission systems, starting with CW.

**CW Transmission.** Figure 5-1 shows the block diagram of a simple radio transmitter system. This transmitter is called a CW system since it produces a continuous radio wave. That is, the radio waves or signals are not modulated in any way, and they are of constant amplitude and frequency. CW transmitters are used primarily to transmit code signals. Generally, the final amplifier stage is keyed to produce the code signals. As shown, the output of the final amplifier is fed to the transmitting antenna and then broadcast at a selected frequency. The transmitter of Fig. 5-1 has three stages or circuits: oscillator, buffer/multiplier, and final amplifier.

The *oscillator* generates the signal at the selected frequency. Most present-day oscillator circuits are crystal-controlled. That is, the frequency of the oscillator circuit is set by a quartz crystal. However, there are variable-frequency oscillators (but not in CB or TV systems).

The *buffer/multiplier* performs either or both of two major functions. One prime function is to amplify the oscillator signal up to a level suitable for driving the final amplifier. (The buffer/multiplier is called the *driver* on some schematics.) The other major function is to multiply the frequency of the oscillator signal. For example, if the oscillator crystal is at 5 MHz and the desired output frequency of the transmitter is 15 MHz, the multiplier stage must triple the signal frequency. (Sometimes, buffer/multipliers are called *doublers*, or *triplers*, on the schematic.) Another possible function of the buffer/multiplier stage is to isolate the oscillator from the final amplifier so that there is no iteration or feedback of signals.

The *final amplifier* (also known as a *power amplifier*, or PA) amplifies the signal from the buffer/multiplier and applies the signal to the transmitting antenna. In very rare cases, the final amplifier

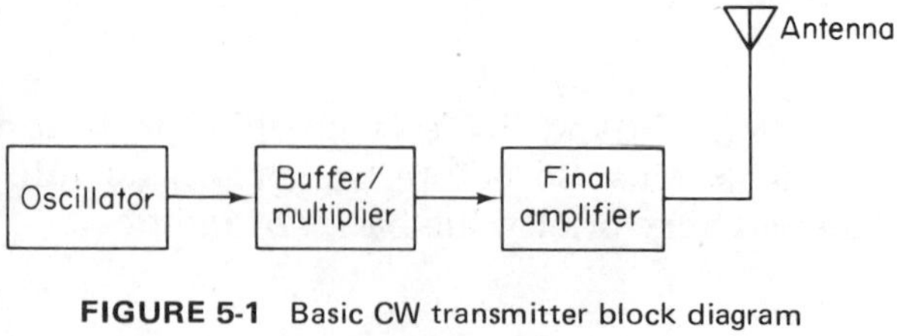


**FIGURE 5-1** Basic CW transmitter block diagram

also multiplies the frequency of the signal. The final amplifier is where the intelligence is inserted into the transmitter of a CW system. The intelligence is in the form of dots and dashes (depending on how long the operator holds the sending key down and lets it up). With the key down (closed), the transmitter produces RF output to the antenna. No output is produced with the key up (open). Generally, the dots and dashes are transmitted in International Morse Code.

**AM Transmitter.** Figure 5-2 shows the block diagram of a basic AM radio transmitter system. This transmitter is called AM since it produces an amplitude-modulated radio wave. That is, the radio waves are amplitude-modulated by audio-frequency signals but are of constant frequency. The AM transmitter has the same three stages as the CW transmitter (oscillator, buffer/multiplier, final amplifier), plus two additional stages: the speech amplifier and the modulator.

Obviously, the *speech amplifier* amplifies the level of the audio signal to be transmitted over the radio waves. In this case, the radio waves are called *carrier waves* (or simply the *carrier*), since they carry the intelligence in the form of amplitude modulation. The *modula-*

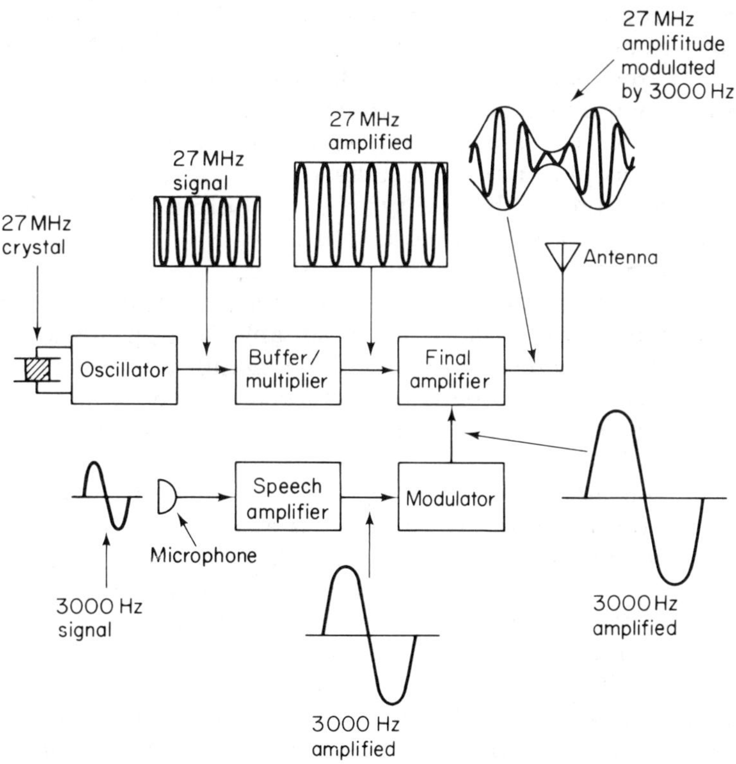


**FIGURE 5-2** Basic AM transmitter block diagram

*tor stage* amplifies the audio information to a level that can be fed to the modulated final amplifier (known as the modulated RF amplifier on some schematics).

The modulated final amplifier, in addition to amplifying the RF signal level as in the CW transmitter, also combines the audio information with the RF carrier to produce a *modulated RF output*, which is fed to the antenna system and transmitted. The inputs to an AM transmitter can be speech (voice), music, or any audio-frequency signal. The AM transmitter processes this and generates an output of modulated RF waves or signals.

Figure 5-2 also shows the waveforms that appear throughout the transmitter. It is assumed that the oscillator crystal is at 27 MHz, that the audio intelligence to be transmitted is a constant 3000-Hz signal or tone, and that there is no multiplication in the buffer stage. Under these conditions, the audio signals through both the speech amplifier and modulator stages are at 3000 Hz, and the RF signals from oscillator up to the final amplifier stage are at 27 MHz. The output of the final amplifier to the antenna is a 27-MHz carrier, amplitude-modulated by a 3000-Hz tone.

**FM Transmitter.** Figure 5-3 shows the block diagram of a basic FM radio transmitter system. This transmitter is called FM since it produces a frequency-modulated radio wave. That is, the radio waves are frequency-modulated by audio-frequency signals but are of constant amplitude. The FM transmitter has the same three stages as the CW transmitter, plus a modulator stage (and possibly a speech amplifier). However, unlike the AM system, the output of the modulator in the FM system is fed to the oscillator.

The oscillator generates an RF signal. The buffer/multiplier amplifies and/or multiplies the oscillator signal. The final amplifier further amplifies the signal to a level suitable for transmission by the antenna. The modulator amplifies the audio intelligence (voice, music, etc.) and applies the amplified signal to the oscillator so that the oscillator frequency is changed in accordance with the audio intelligence.

Figure 5-3 also shows the approximate waveforms that appear throughout the transmitter. Again, it is assumed that the oscillator *center frequency* is 27 MHz and that the audio intelligence to be transmitted is a constant 3000-Hz tone. (The center frequency is the frequency of the signal when there is no modulation applied.) Under these conditions, the audio signals through the modulator stages are at 3000 Hz. The signals from the oscillator vary about the center frequency of 27 MHz at a 3 kHz rate. These signals (which change in frequency above and below the 27-MHz center) are amplified by the

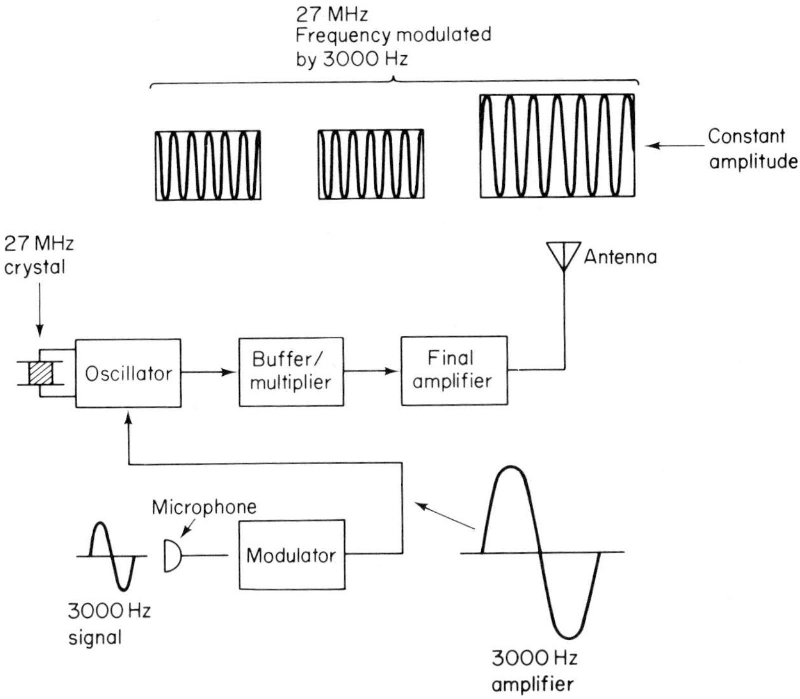


**FIGURE 5-3** Basic FM transmitter block diagram

buffer/multiplier and final amplifier before delivery to the transmitter. The output of the final amplifier is a constant-amplitude 27-MHz carrier, frequency-modulated by a 3000-Hz tone.

**SSB Transmitter.** Figure 5-4 shows the block diagram of a basic SSB radio transmitter. This transmitter uses a special form of AM modulation. As in the case of the basic AM transmitter, the SSB has a speech amplifier to amplify the audio intelligence to be transmitted. The SSB also has an oscillator and a buffer stage. As in the case of AM, these stages generate and amplify an RF signal. The outputs of the buffer (RF) and the speech amplifier (audio) are combined in a *balanced modulator* that removes the carrier signal but leaves the *sidebands.* (These sidebands are the upper and lower amplitude variations of the RF signal. As a matter of interest, most of the power transmitted by an AM transmitter is in the upper and lower sidebands. Most of this power is lost in a conventional AM transmitter.)

The double sideband signals from the balanced modulator are fed to a bandpass filter where one of the sidebands is filtered out. The single-sideband output of the bandpass filter is fed to a mixer, to-

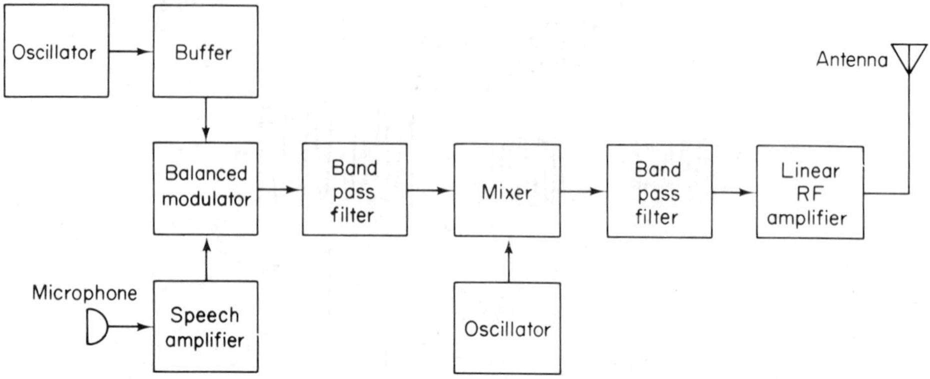


**FIGURE 5-4** Basic SSB transmitter block diagram

gether with signals from another oscillator. The mixer/oscillator circuit shifts the sideband signal (containing the audio intelligence) to a desired transmitting frequency. Some SSB transmitters used several mixers and oscillators to arrive at the desired transmitting frequency.

The bandpass filter following the mixer/oscillator provides additional filtering to remove any *harmonics* or other undesired signals. (Harmonics are multiples of a given frequency.) When two signals are mixed, or *heterodyned*, they produce both a sum and a difference frequency. For example, when a 10-MHz signal is mixed with a 7-MHz signal, a 17-MHz (sum) and a 3-MHz (difference) signal are produced. Similarly, there will be multiples (harmonics) of both the 3- and 17-MHz signals. The bandpass filter passes only the signals of the desired frequency; it rejects all others.

Note that the final amplifier in the SSB transmitter is called a *linear amplifier.* The function and circuit for linear amplifiers are essentially the same as for the RF amplifiers used in other transmitting systems except that a linear amplifier will amplify a wide band of frequencies. A conventional (or nonlinear) RF amplifier is usually designed to tune over a very narrow range of frequencies (and thus automatically reject all harmonics).

### *5-1.2 RADIO RECEPTION BASICS*

Most radio receivers (including CB and TV receivers) use some form of *superheterodyne* principle. The function of any radio receiver is to convert the intelligence in the transmitted RF signals (picked up by the receiving antenna) into usable form. For example, in the case of CB, the receiver converts the audio (voice) intelligence transmitted on the RF carrier back into audio signals that are reproduced on the loudspeaker.

Figure 5-5 is the block diagram of typical superheterodyne (or "superhet") receiver. The superhet receiver takes the incoming signal from the antenna and mixes this signal with another signal developed by an oscillator to produce signals of new frequencies. One band of new frequencies is selected by tuned circuits to be passed along to the remaining receiver circuits for processing. The selected band of frequencies includes a carrier and sidebands (modulation) in which the intelligence is contained. In most cases, the new signals are lower in frequency than the transmitted RF signals but much higher than the audio range (or "super audio"; thus the name "superheterodyne").

The reason for mixing and producing this new frequency (called the *intermediate frequency*, or IF) is because it is easier to build stable, high-gain amplifiers at lower frequencies. Also, if the receiver oscillator is designed so that it can be tuned (changed in frequency) at the same time that the receiver is being tuned to broadcast stations,

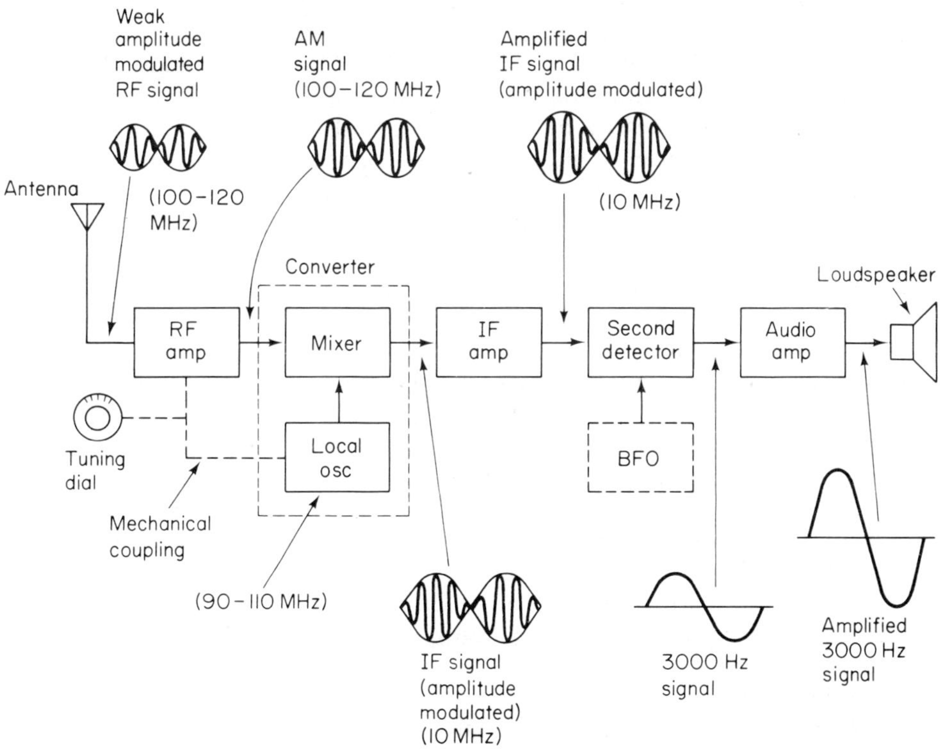


**FIGURE 5-5** Basic superheterodyne receiver block diagram

the difference frequency between the local oscillator and incoming RF will always be the same. That is, the IF stages can be *fixed-tuned*. Thus, only the RF amplifier stage and the local oscillator stage need be tuned or "tracked" simultaneously (usually by mechanical coupling between two variable capacitors connected to the same tuning dial). As an example, if the IF is 10 MHz, the local oscillator frequency must be 10 MHz below the transmitted RF. If the receiver can be tuned to receive signals from 100 to 120 MHz, the local oscillator must tune (simultaneously) from 90 to 110 MHz.

As shown in Fig. 5-5, the incoming RF signals from the antenna are amplified in the *RF amplifier* stage. At this point, the signals are very weak (typically a few microvolts). The output of the RF amplifier is applied to the *converter*, which includes a *mixer* and a *local oscillator*. The RF and local oscillator signals are mixed to produce the IF signals. Tuned circuits (IF transformers) in the IF amplifiers select the desired frequency, usually the difference frequency between the RF and local oscillator.

For example, if the RF is at 100 MHz and the local oscillator is at 90 MHz, there will be signals at 190 MHz (sum) and 10 MHz (difference). In this case, the IF transformers are tuned to 10 MHz, to reject the 190-MHz signals but pass the 10-MHz signals. At the output of the IF amplifiers, the signals are still amplitude-modulated in exactly the same way as the transmitted RF, but at a new frequency and at a much higher amplitude. The signal of a few microvolts at the antenna is typically raised to several hundred millivolts, possibly even 1 V, at the IF output.

The amplified IF signal is then fed to a detector, where the audio intelligence is separated from the carrier. In some superhet receivers, this detector is called the "second detector," since what is now called the converter was at one time called the "first detector." Some literature still refers to the converter as the first detector. Whatever it may be called, the (second) detector *demodulates* the IF signals. If there is no modulation on the transmitted carrier, no signals are produced by the detector. If the transmitted carrier is modulated by a 3000-Hz signal as discussed in Sec. 5-1.1, the output of the detector is the 3000-Hz signal or voltage (typically about ½ V or less if a solid-state diode detector is used).

The detector output is amplified by one or more audio amplifier stages and is reproduced on the loudspeaker. Obviously, the detector must be AM if the transmitted signal is AM, and must be FM to demodulate FM signals. Either way, the output of the final audio amplifier to the loudspeaker is usually of the order of several volts.

Note that there is an additional block shown on Fig. 5-5, called the BFO. This stands for *beat frequency oscillator* and is found on CW and SSB receivers but rarely on AM or FM receivers (unless they are used primarily for communications). The BFO is an oscillator that produces a signal that is mixed (or "beat") with the IF signals to produce an audio signal. This is necessary on CW systems since the CW carrier is not modulated and the detector will not produce an audio tone, even though the carrier is present. (The presence of a carrier will remove background noise.) The BFO is also necessary for SSB systems to replace the lost carrier (the carrier that is removed by the balanced modulator in the SSB transmitter, Sec. 5-1.1).

## 5-2 READING CB CIRCUIT SCHEMATICS

Now that we have rushed through the basics of radio transmission and reception, let us see how these basics can be combined with the discussions of circuits found in Chapter 4 to read some actual schematics. We will take the schematic of a typical hand-held ("walkie-talkie") CB set and go through stage by stage (circuit by circuit) from microphone to antenna, and from antenna to loudspeaker. If you can follow this discussion and answer the questions at the end of the chapter, you should be able to read the schematics of most CB sets. Of course, if you want to get into such refinements as SSB systems, with PLL (phase-locked-loop) control, your attention is invited to the author's *Handbook of Practical CB Service* (Englewood Cliffs, N.J.: Prentice-Hall, Inc., 1978).

### *5-2.1 GENERAL DESCRIPTION OF CB SET*

Figure 5-6 is the block diagram of a Pace CB155 hand-held CB set. Figure 5-7 is the full schematic of the same set. Let us start by noting some of the differences between the two drawings. Both drawings show 14 transistors. However, the transistors are labeled Q on the block and TR on the schematic. Also, the detector and automatic noise limiter (both diodes) are labeled as CR on the block and D on the schematic.

The CB155 is a six-channel, crystal-controlled AM transceiver. (The term "transceiver" is used since most CB sets are a combination *trans*mitter and re*ceiver*.) The set is all solid-state and is operated on a 12-V internal rechargeable battery. The battery is shown at the lower right-hand corner in Fig. 5-7. The battery is controlled by POW CHG (power-charge) slide switch S4-A, and the CHARGE J jack J2. When S4 is in the POW position (normal operation), the positive battery terminal is connected through the ON-OFF switch S3 to the positive distribution line (which runs throughout the schematic). The negative battery terminal is connected through the normally closed contacts of J2 to the negative distribution line.

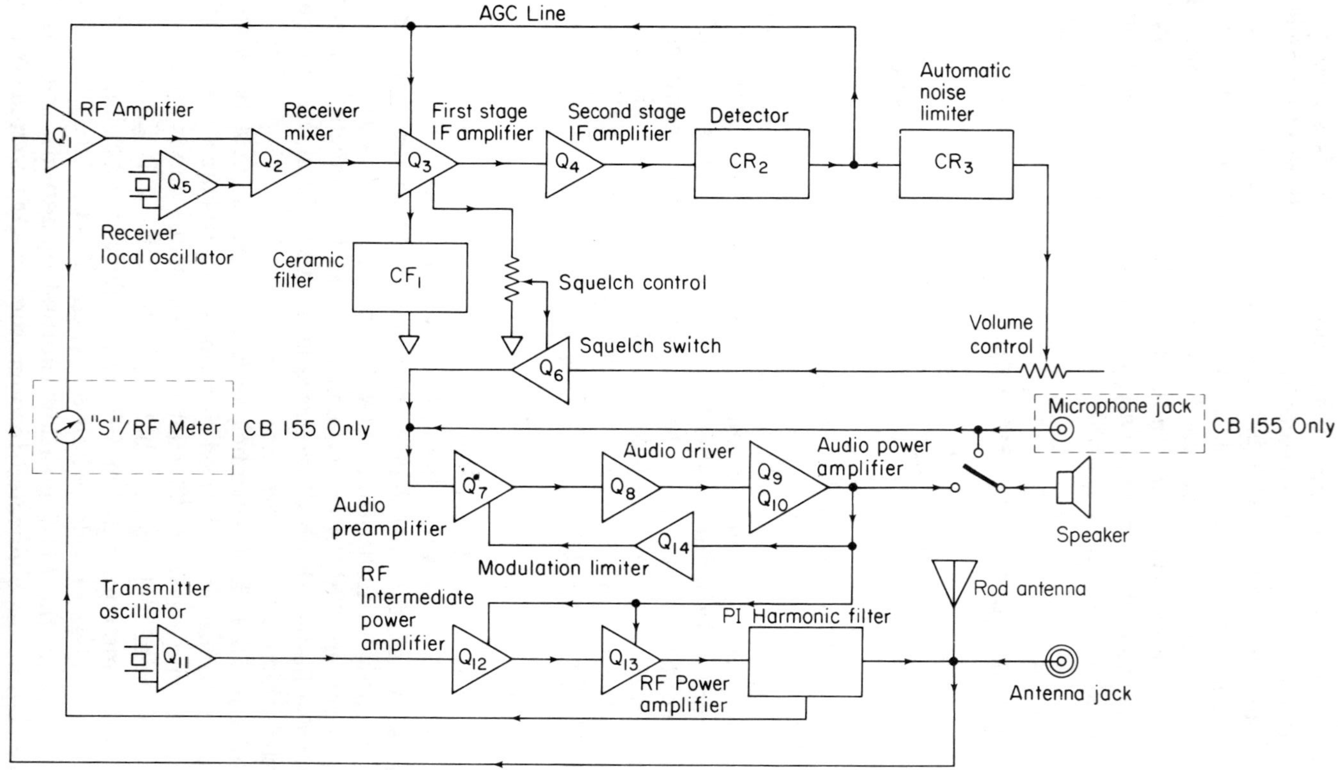


**FIGURE 5-6** Pace CB155 walkie-talkie CB block diagram

When the battery is to be charged, S4 is set to CHG and an external 12-V supply is connected to J2. This applies the external charging voltage to the battery through resistor R43. Although not shown on the schematic, the most obvious function for R43 is to limit charging current to the battery. You can determine the value of R43 from the schematic. It is 240 Ω with a power dissipation of 1 W. If the value is given but is not followed by the letters K or M, the value is presumed to be in ohms. If the value is followed by a K or an M, the value is in kilohms or megohms, respectively. If the wattage is given, the wattage is critical to circuit operation. (The resistor should never be replaced with a resistor of lower wattage rating.) If the wattage is omitted, it is presumed to be some value less than 1 W (possibly ¼ or ½ W) and is not critical.

RF power for the CB155 is switchable to either 1 W or 5 W. A telescoping rod antenna is built in and an external antenna jack is provided. A single device serves as a loudspeaker (during receive) and as a microphone (during transmit). An external microphone and loudspeaker can be connected to the audio section, thus providing a public address (PA) function. The set includes such functions as squelch, volume, and S/RF meter.

### *5-2.2 RECEIVE-TRANSMIT CONTROL*

The receive-transmit function is controlled by an R-T pushbutton switch S1. The various sections of S1 are shown throughout Fig. 5-7. Let us take each section of S1 in turn. S1-A (upper left of Fig. 5-7) connects the antenna to the input of the receiver (during receive) or to the output of the transmitter (during transmit). The external antenna is connected through a jack (with no reference designation) to the arm of S1-A. The rod antenna is connected through L1 and C1 to S1-A. S1-B (above TR 7) connects the input of the audio section to the loudspeaker (acting as a microphone during transmit) or to the receiver output (during receive). S1-C (below T5) connects the output of the audio section transformer T5 to the final amplifier (transmit) so that the audio section acts as a modulator.

S1-D connects the output of the loudspeaker/microphone to the output of the audio section (receive) or the input of the audio section (transmit). (As an exercise, try tracing the loudspeaker/ microphone connections from S1-D to and from the audio section.) S1-E connects the negative terminal of the battery to the transmitter section (transmit) or the receiver section (receive). S1-F connects the positive terminal of the battery to the transmit and receive sections.

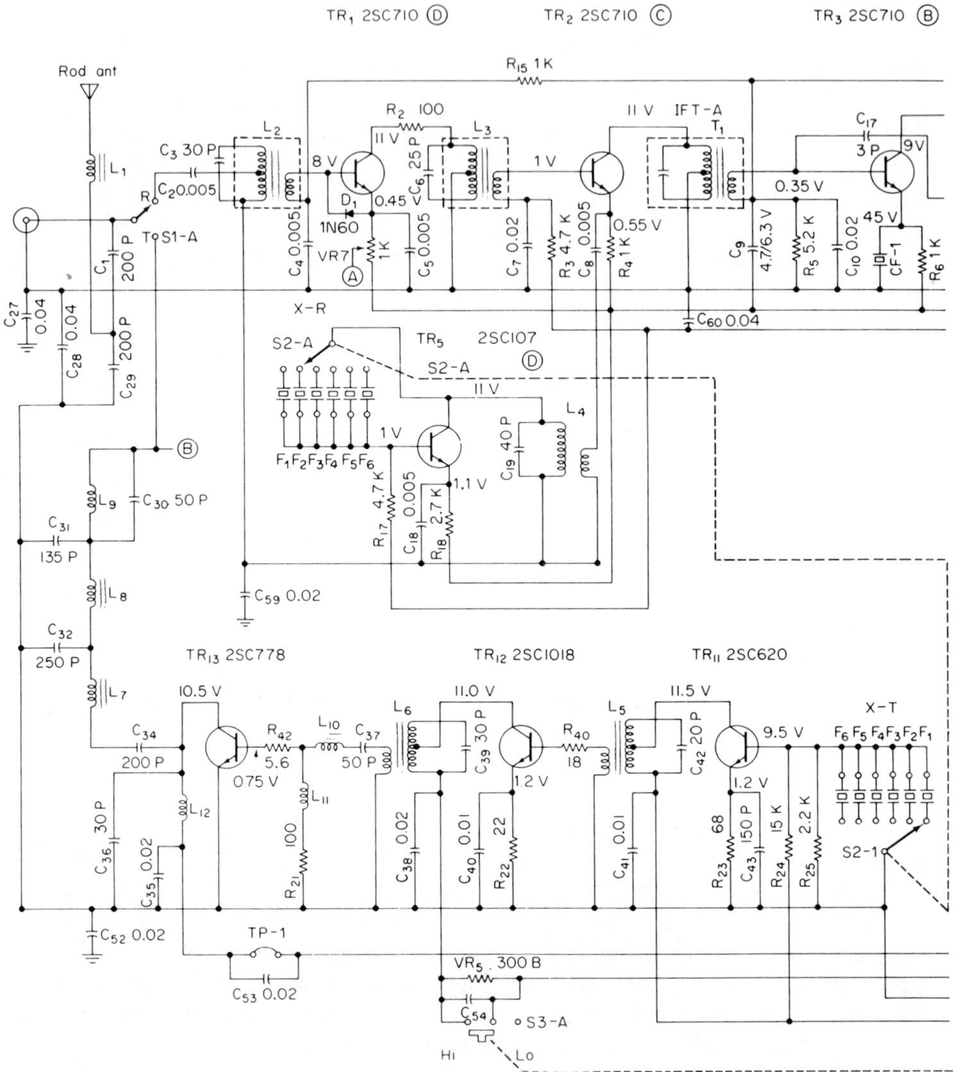

**FIGURE 5-7** Pace CB155 walkie-talkie CB schematic diagram

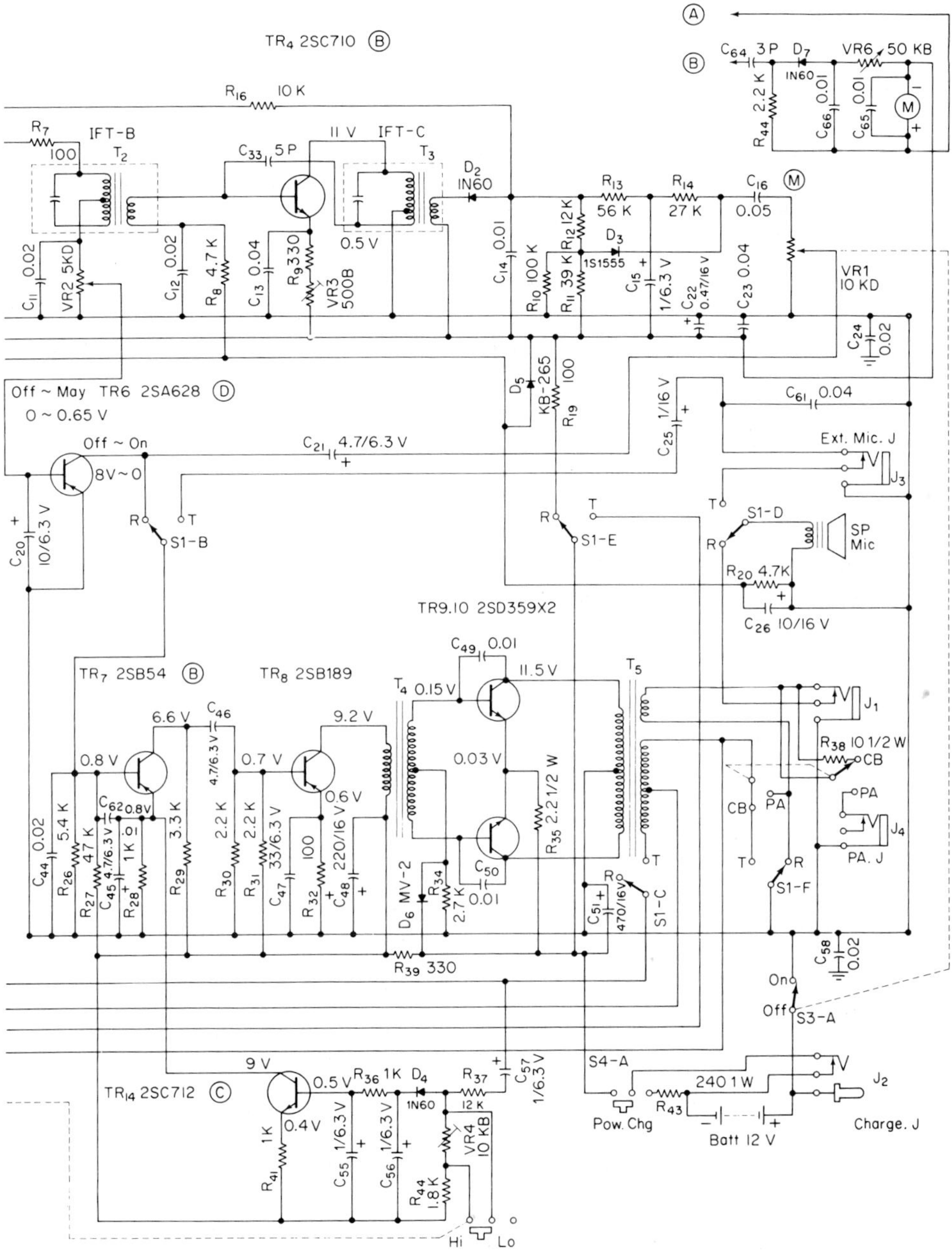
TR4 2SC710 (B)
IFT-B
IFT-C
TR6 2SA628 (D)
TR9.10 2SD359X2
TR7 2SB54 (B)
TR8 2SB189
TR14 2SC712 (C)
Ext. Mic. J
SP Mic
PA. J
Pow. Chg
Batt 12 V
Charge. J

### 5-2.3 HIGH POWER/LOW POWER CONTROL

The transmitter is capable of operating with 1 W of input power or 5 W of input power. This is controlled by the two sections of the HI-LO slide switch shown at the bottom of Fig. 5-7. One section functions to short-out variable resistor VR5 when in the HI position. VR5 limits current (and voltage) to the transmitter section (to TR12 in particular) so as to reduce input to 1 W. With the HI-LO switch in HI, VR5 is shorted out and input power to the transmitter section is raised to 5 W. The other section of the HI-LO switch shorts out VR4 in the audio section (in modulation limiter TR14). As is discussed in Sec. 5-2.4, TR14 functions as a modulation limiter. The amount of modulation must be reduced when the transmitter input is changed from 5 W to 1 W.

### 5-2.4 AUDIO SECTION

The dual-function audio section consists of transistors TR7 through TR10, TR14, and associated circuit parts. During receive, the audio section amplifies the audio output from the receiver detector to a level suitable for reproduction by the loudspeaker. During transmit, the audio section amplifies the loudspeaker (now acting as a microphone) signals to a level suitable for modulation of the transmitter section. The audio section can also be used as the amplifier for a PA system, although this function has no connection with the normal CB function.

Input to the audio section is made through section S1-B of the transmit-receive switch to the base of TR7. During receive, the input is obtained from the receiver detector circuit through coupling capacitor C21 and volume control VR1. During transmit, the input is obtained through coupling capacitor C25, external microphone jack J3, switch S1-D, and the loudspeaker (microphone). When an external microphone is inserted into J3, the contacts open, disconnecting the set's loudspeaker.

Output from the audio section is taken from two secondary windings on transformer T5. The top secondary winding is used during receive and is connected to the loudspeaker through the contacts of J1, S1-D, and S1-F. (Try tracing out the connections on Fig. 5-7.) The bottom secondary winding (which is tapped) is used during transmit. The tapped output is applied to RF intermediate power amplifier TR12 through VR5. Variable resistor VR5 permits adjustment of the modulation level by setting the amount of modulating signal applied to the transmitter section. The untapped output of T5 is applied to the RF power amplifier TR13 through S1-C and TP-1.

Test point TP-1 is actually a jumper wire between two terminals. During test, the jumper is removed and a meter is connected between the two terminals to measure collector current of TR13.

With the possible exception of TR14, the audio-section circuits are quite similar to the examples discussed in Sec. 4-1. Transistors TR7 and TR8 are connected as conventional audio amplifiers with bypassed emitters. C45 and C47 are the emitter bypass capacitors. Both TR7 and TR8 are operated class A (although you cannot tell this from the schematic). Resistors R26 and R27 are the voltage dividers for the base of TR7, and R28 is the emitter feedback resistor (bias stabilization resistor). (Can you identify the corresponding resistors for transistor TR8?) C48 is the bypass capacitor for TR8, and C46 is the coupling capacitor between the TR7 and TR8 amplifier stages.

Transistors TR9 and TR10 are connected as a conventional push/pull audio amplifier. (Note how similar this circuit is to that shown in Fig. 4-12.) The push/pull input is from transformer T4; the push/pull output is at T5. Can you tell which resistor supplies the bias voltage for TR9 and TR10? It is resistor R34. Trace from both bases through the secondary winding to the center tap and then to the positive distribution line. Which resistor supplies the emitter feedback (bias stabilization)? It is resistor R35. Trace from the emitters of TR9 and TR10 to the negative distribution line.

Now can you tell from the schematic if TR9 and TR10 are operated class AB or B? As discussed in Sec. 4-1.6, class AB is where some current flows at all times (with or without a signal passing through the audio amplifier), and true class B is where both transistors of a push/pull amplifier are just cut off. Note that both TR9 and TR10 are NPN transistors (which, as discussed in Chapter 3, require that the base be more positive than the emitter to produce current flow). The schematic shows that the bases of TR9 and TR10 are at 0.15 V, whereas the emitters are at 0.03 V. Since the voltages are not labeled, they are considered to be positive. The bias circuit has made the bases positive in relation to the emitters, and there is current flow (with no signal applied). Thus, TR9 and TR10 are operated class AB.

Transistor TR14 functions as a modulation limiter during transmit. Part of the audio modulating signal fed to TR13 through S1-C is also passed through coupling capacitor C57 to the TR14 circuit. These audio signals are rectified by D4 into pulsating direct current and applied to the base input of TR14. The output of TR14 is applied to the emitter of TR7. If there is a modulation increase above the desired level, the gain of TR7 is decreased by the signal from TR14. Decreases in modulation level increase the gain of TR7.

Thus, the modulation level remains constant (within limits). The amount of modulation can be set by variable resistor VR4, which shorts out some of the audio signals before they arrive at rectifier diode D4. Generally, modulation circuits are adjusted for 100% modulation. Anything more than 100% modulation produces excessive sidebands, which can interfere with transmissions on nearby frequencies. With anything less than 100% modulation, the transmitter is not delivering full power.

### *5-2.5 TRANSMITTER SECTION*

The transmitter section consists of transistors TR11 through TR13 and associated parts. (Note that the transmitter section is drawn opposite to the normal pattern for schematics. That is, the signal flow from oscillator to final amplifier runs from right to left instead of the conventional left to right.) The transmitter section is used only during transmit, and produces an RF carrier signal on one of six frequencies (or CB channels). This carrier is voice-modulated by the audio section, as discussed in Sec. 5-2.4. The transmitter circuits are quite similar to the examples discussed in Secs. 4-2 and 4-3.

Transistor TR11 is connected as a crystal-controlled oscillator. One of six crystals can be selected by means of CHANNEL switch S2-1. This section of S2 is mechanically coupled (indicated by dashed lines) to a corresponding switch S2-A in the receiver section, thus permitting transmission and reception on the same selected channel. The transmitter crystals are labeled as X-T, frequencies F1 through F6. The schematic does not give any hint as to crystal frequencies or whether there is any multiplication of frequencies in the transmitter section. However, the block diagram (Fig. 5-6) calls TR12 and TR13 "amplifiers," so you can assume that there is no frequency multiplication and that the transmitting crystals are at the desired channel frequency (in the 27-MHz CB range).

Resistors R24 and R25 provide base bias for the oscillator transistor TR11. The output of TR11 is coupled by a transformer-like device labeled L5 to the base input of TR12, where the RF signal is amplified. Modulation is also introduced to the collector of TR12, as discussed in Sec. 5-2.4. The amplified and modulated output of TR12 is coupled to the final amplifier TR13 through another transformer L6. The RF signal is further amplified (and modulated) in the TR13 circuit and applied to the transmitting antenna through a low-pass filter (Sec. 4-7) and S1-A (which is set to T during transmit). The function of the filter (consisting of C31, C32, L7, L8, and L9) is to remove any undesired harmonics that might interfere with television or other broadcast signals. The filter passes signals below the 27

MHz range but rejects signals above 27 MHz. Note that the RF output to the antenna is also tapped off (at the junction of L9 and C30) and fed to the S/RF meter (upper right-hand corner of Fig. 5-7). As discussed in Sec. 5-2.7, this permits the output to be monitored.

### *5-2.6 RECEIVER SECTION*

The receiver section consists of transistors TR1 through TR6 and associated circuit parts. The receiver section is used only during receive and removes the audio intelligence (generally voice) from the transmitted RF carrier being broadcast by another CB station. The output of the receiver section is fed to the audio section for reproduction on the loudspeaker.

With the possible exception of TR6, the receiver section circuits are quite similar to the examples discussed in Secs. 4-2 and 4-3. Transistor TR1 is connected as an RF amplifier, which amplifies the signal picked up by the antenna. Note that the emitter resistor VR7 of TR1 is variable. The slider arm of VR7 is connected to the S/RF meter circuit. As discussed in Sec. 5-2.7, this permits the strength of the incoming signal to be monitored.

The output of TR1 is applied to a converter circuit consisting of mixer TR2 and local oscillator TR5. The frequency of the TR5 oscillator is controlled by the X-R crystals, F1 through F6. The output of the local oscillator at L4 is applied to the emitter or mixer TR2, which also receives the incoming RF signal (after being amplified by TR1). Thus, TR2 receives two signals at different frequencies, and both of the frequencies (as well as the sum and difference frequencies) appear at the collector output of TR2. However, since the collector output of TR2 is tuned to only one frequency by IF transformer T1, only that frequency is passed to the next stage. Although you cannot tell it from the schematic, the IF is 455 kHz.

Transistors TR3 and R4 are connected as IF amplifiers, with the inputs and outputs tuned to 455 kHz by IF transformers T1, T2, and T3. Note that a crystal symbol, labeled CF-1, is connected in the emitter circuit of TR3. This crystal acts as a very selective bandpass filter, passing only signals at or very near 455 kHz. This helps improve the selectivity of the IF amplifier stages.

The IF signal is coupled through T3 to the detector D2, which removes audio from the IF signal. The audio signals from D2 are passed through *automatic noise limiter* diode D3 and volume control VR1 to the audio section, as discussed in Sec. 5-1.4. Noise-limiter diode D3 is normally forward-biased so that audio signals pass from the detector to the audio section. However, when there is excessive noise on the incoming signal, the bias changes so that D3 becomes temporarily reverse-biased, to cut off passage of the excessive noise signals.

The squelch-switch transistor TR6 and squelch control VR2 also provide cutoff of undesired noise. (A squelch circuit keeps the receiver audio output cut off, except when signals above a certain amplitude are present.) When a signal above the desired squelch point is present at TR3, the squelch-switch transistor TR6 is biased off (no current flows between emitter and collector). The squelch point is set by squelch control VR2. With TR6 off, the audio signal is not shorted to ground and passes through C21 to TR7 in the audio section for reproduction on the loudspeaker. When there is no signal present at TR3 or the signal is below the desired squelch point, TR6 is biased on (current flows between emitter and collector). This shorts the audio input to transistor TR7 directly to ground, thus squelching the audio.

### *5-2.7 S/RF-METER CIRCUITS*

The purpose of an S meter is to monitor the incoming RF signal strength (from antenna to receiver). S meters are usually calibrated in terms of S1 through S9, with S9 being the highest strength (full-scale). The purpose of an RF meter is to monitor the outgoing RF signal (from transmitter to antenna). The S/RF meter in the upper right-hand corner of Fig. 5-7 performs both these functions.

During receive, the incoming signal forces RF amplifier TR1 toward cutoff. The resultant voltage change across VR7 causes the meter to move up-scale (from S1 to S9). Variable resistor VR7 permits adjustment of the S meter during receive. During transmit, a portion of the transmitted signal is tapped off through coupling capacitor C64 and is rectified and filtered by D7 and C66. An increase in power output causes the meter to move up-scale. Variable resistor VR6 permits adjustment of the RF meter during transmit.

### *5-2.8 TRANSISTOR VOLTAGES*

If you are paying attention, you will note that voltages are given for each element (collector, emitter, base) of all transistors. The fact that all of these voltages are included on the schematic is unusual. However, this schematic was chosen as an example of what information can be found on well-drawn schematics.

The voltages are given on the schematic for troubleshooting, as a starting point for tracing faults in electronic circuits. As an example, assume that you are measuring voltages in the audio section and find that there is no voltage at the collector of TR8, instead of the 9.2 V shown on the schematic. This is probably caused by an open primary winding in transformer T4 (if all other voltages are correct). There could also be other causes for the abnormal voltage condition. However, the fact that the voltage is absent or abnormal provides a good starting point to look for possible trouble. We will not go further into troubleshooting here. If you are truly interested in simplified troubleshooting techniques, your attention is invited to the author's *Handbook of Basic Electronic Troubleshooting* (Englewood Cliffs, N.J.: Prentice-Hall, Inc., 1977).

## 5-3 READING TV SCHEMATICS

If you have survived the CB schematics, you are now ready for the schematics of a TV set. Here, we will take the schematics of a black-and-white TV set and go through circuit by circuit, from antenna to picture tube and loudspeaker. If you can follow this discussion and answer the questions at the end of this chapter, you should be able to read the schematics for most TV receivers. If you want to get into color-TV circuits, your attention is invited to the author's *Handbook of Simplified Television Service* (Englewood Cliffs, N.J.: Prentice-Hall, Inc., 1977).

### *5-3.1 GENERAL BLACK-AND-WHITE TV THEORY*

Before going into circuit details, let us review the basics of TV transmission and reception. It is not intended that this introduction provide a complete course in TV broadcasting, but rather as a summary and a starting point for understanding circuit schematics.

As shown in Fig. 5-8, the basic TV broadcast system consists of a transmitter capable of producing both AM and FM signals, a television camera, and a microphone. TV program sound (or audio) is transmitted through the microphone, which modulates the FM portion of the transmitter. The picture (or video) portion of the program is broadcast by means of the AM transmitter.

The TV broadcast channels are approximately 6 MHz wide, with the sound (FM) carrier at a frequency 4.5 MHz higher than the picture (AM) carrier. For example, on channel 8, the picture (AM) is transmitted at a frequency of 181.25 MHz, whereas the sound (FM) is transmitted at 185.75 MHz. Channel 8 occupies the band of frequencies between 180 and 186 MHz.

The sound portion of the TV signal uses conventional FM broadcast principles, which need not be described here. However, the

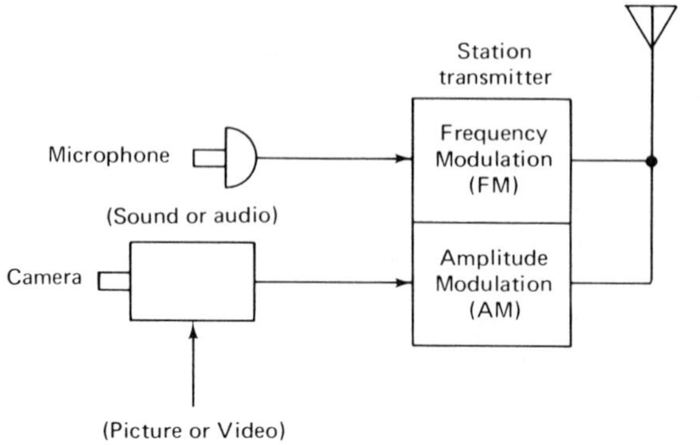


**FIGURE 5-8** Basic television broadcast system

picture portion of the signal is unique to TV in that both *picture information* and *synchronizing pulses* must be transmitted on the AM carrier.

Operation of the TV picture channel can be understood by reference to Fig. 5-9, which shows the relationship between the picture tube in the TV receiver and the station camera picture tube. Both tubes have an electron beam, which is emitted by the tube cathode and strikes the tube surface, as discussed in Sec. 3-1.9. Both tubes have horizontal and vertical sweep systems that deflect the beam so as to produce a rectangular screen (or "raster") on the tube surface. The vertical sweep is at a rate of 60 Hz, whereas the horizontal sweep is at 15,750 Hz.

The electron beam of the station camera picture tube is modulated by the amount of light that strikes the camera tube screen. In turn, the transmitted AM signal is modulated by the electron beam. Thus, the amplitude of the transmitted AM signal is determined by the intensity of the light at any given instant. The position of the electron beam at a given instant is set by the horizontal and vertical sweep circuits, which, in turn, are triggered by pulses in the camera. These same pulses are transmitted on the AM carrier and act as synchronizing pulses ("sync pulses") for the TV receiver horizontal and vertical sweep circuits.

The electron beam of the TV receiver is modulated by the transmitted AM signals so as to "paint" a picture on the tube screen. The amplitude of the AM signal determines the intensity of the light produced on the TV screen at any instant. For example, if the camera sees an increase in light, the electron beams in both tubes (camera and receiver) are increased, and the TV receiver tube shows

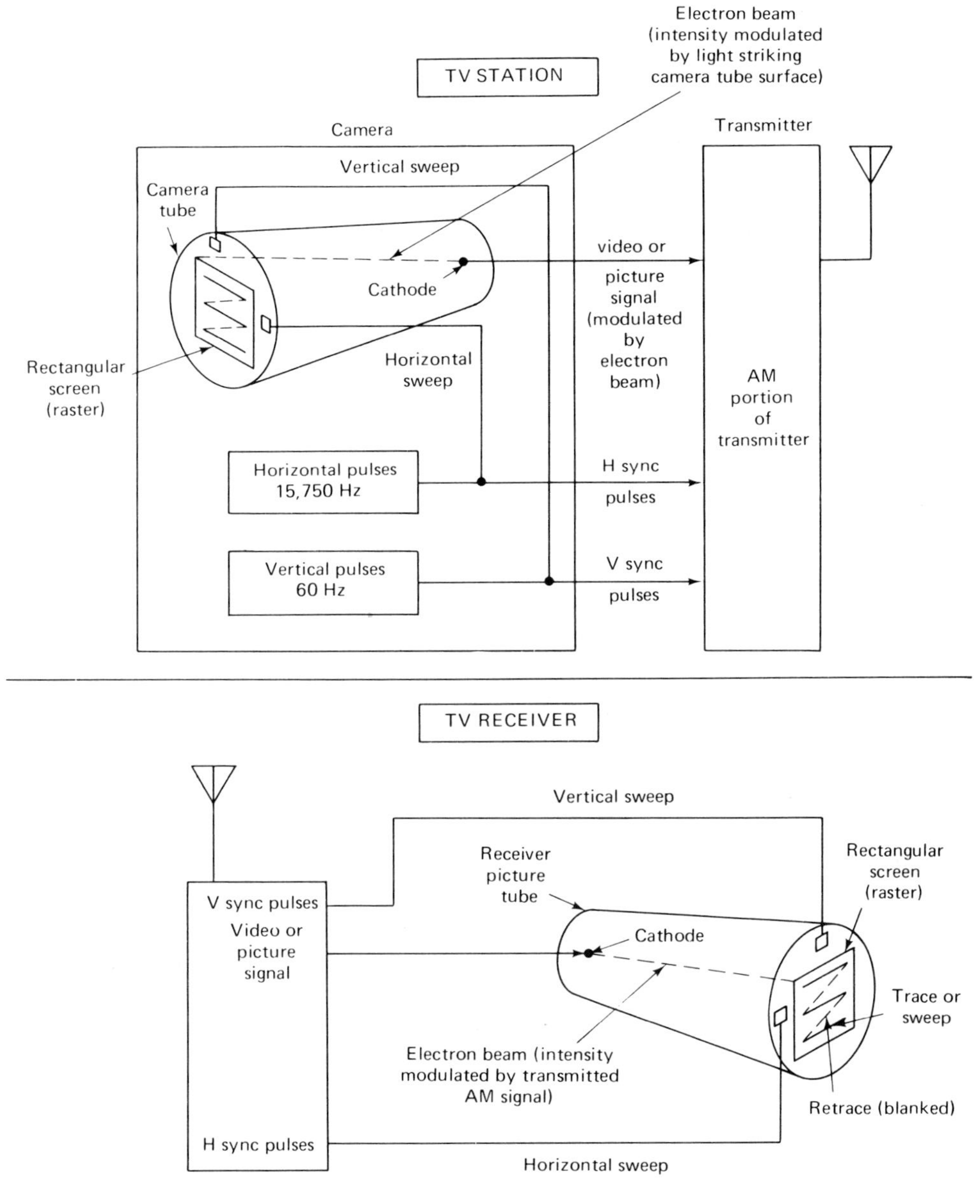


**FIGURE 5-9** Relationship between camera and receiver picture tubes

an increase in light. The TV receiver tube horizontal and vertical deflection systems are triggered by the horizontal and vertical synchronizing pulses transmitted on the AM portion of the TV broadcast signal. Thus, the electron beam of the TV receiver follows the beam in the camera tube. Both beams are at the same corresponding spot, at the same instant.

Assume that the camera is focused on a white card with a black numeral 3 at the center, as shown in Fig. 5-10. As the camera tube electron beam is swept across the surface, light is reflected from the card onto the camera tube surface. When the beam passes across the white-card background portion of the reflected light, the beam intensity is maximum. Beam intensity drops to a minimum when the beam is at any portion of the light reflected from the numeral 3. This varying electron beam modulates the transmitted AM carrier.

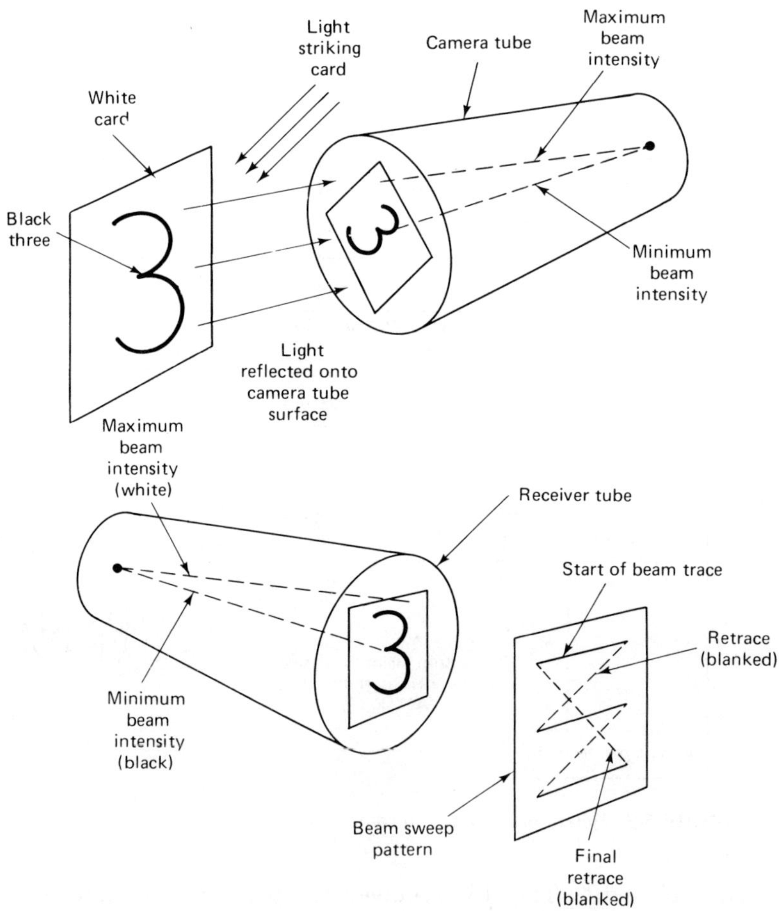


**FIGURE 5-10** Relationship of electron beams in camera and receiver picture tubes

The electron beam starts at the top of the camera tube screen, sweeps across to one side, and is blanked while it returns to the other side (retraces), until the beam finally reaches the bottom of the screen. The beam is then blanked and returned to the top of the screen. Thus, if the camera is focused on the numeral 3 pattern, the camera translates the entire pattern, line by line, into a picture signal (a voltage that varies in amplitude with intensity of the reflected light).

At the TV receiver, the picture-tube electron beam follows the camera beam in both position and intensity. In the example of Fig. 5-10, both electron beams increase when they are swept across any of the white background. Similarly, both beams decrease when they are swept across any portion of the numeral 3. Thus, the TV receiver picture tube reproduces the numeral 3 in black and white (black, or minimum beam intensity on the numeral portion of the sweep; white, or maximum intensity on the card background portion).

### *5-3.2 BASIC BLACK-AND-WHITE RECEIVER CIRCUITS*

Figure 5-11 is the block diagram of a black-and-white TV receiver. The diagram shown is a composite of several types of TV receivers and is presented as a starting point for understanding the circuit schematics that follow.

### *5-3.3 LOW-VOLTAGE POWER SUPPLY*

Figure 5-12 is the schematic of a typical low-voltage power supply circuit. The purpose of the circuit is to provide direct current for all circuits of the receiver, except the high voltages required by the picture tube. The high voltages are supplied by the *flyback circuit* (more properly known as the horizontal output and high-voltage circuit), as described in Sec. 5-3.4. The circuit of Fig. 5-12 consists essentially of a full-wave bridge rectifier CR1, followed by a solid-state (zener diode CR2 and transistors Q1 through Q3) regulator circuit. As with many portable solid-state TV receivers, the set can also be operated with self-contained rechargeable batteries. Thus, the power supply circuit also provides for recharging the batteries. The set can also be operated from an auto battery (12 V) by means of an adapter plug connected to the auto's cigarette lighter output. This arrangement is typical of small portable TV receivers.

**Circuit Operation.** The line voltage is dropped to about 12 V by transformer T1 and is rectified by CR1. The output from CR1 is regulated by Q1 through Q3, as well as by zener CR2, and is dis-

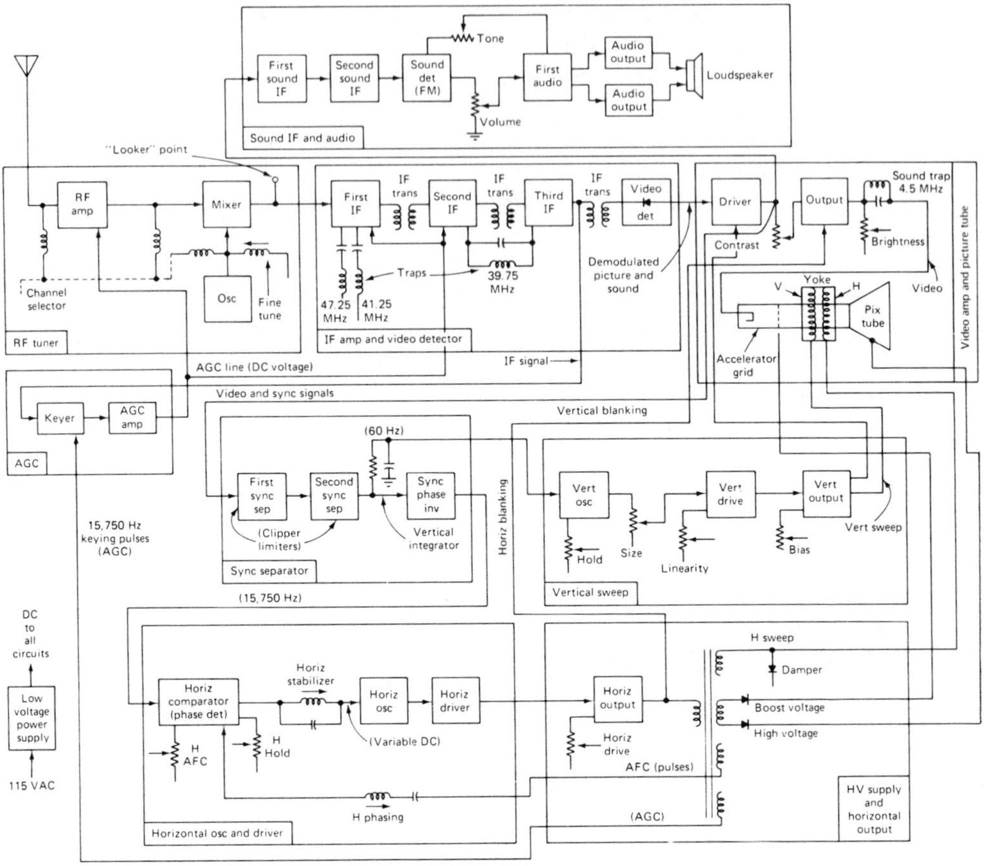


**FIGURE 5-11** Composite black and white television receiver

tributed to three separate circuit branches, each at a slightly different output voltage. Operation of the regulator is similar to the regulator circuits discussed in Sec. 4-5. The emitter of Q1 is held constant by zener CR2, whereas the base of Q1 depends on output voltage. Any variations in output voltage (resulting from changes in input voltage or variations in load) change the base voltage in relation to emitter voltage. These changes appear as a variation in Q1 connector voltage and, consequently, Q2 and Q3 base voltage.

Transistors Q2 and Q3 are connected in parallel or shunt with the output from CR1 and the load. Thus, Q2 and Q3 act as variable resistors to offset any changes in output voltage. For example, if the output voltage increases, the base of Q1 swings more positive, causing a drop in Q1 collector voltage. This causes the bases of Q2

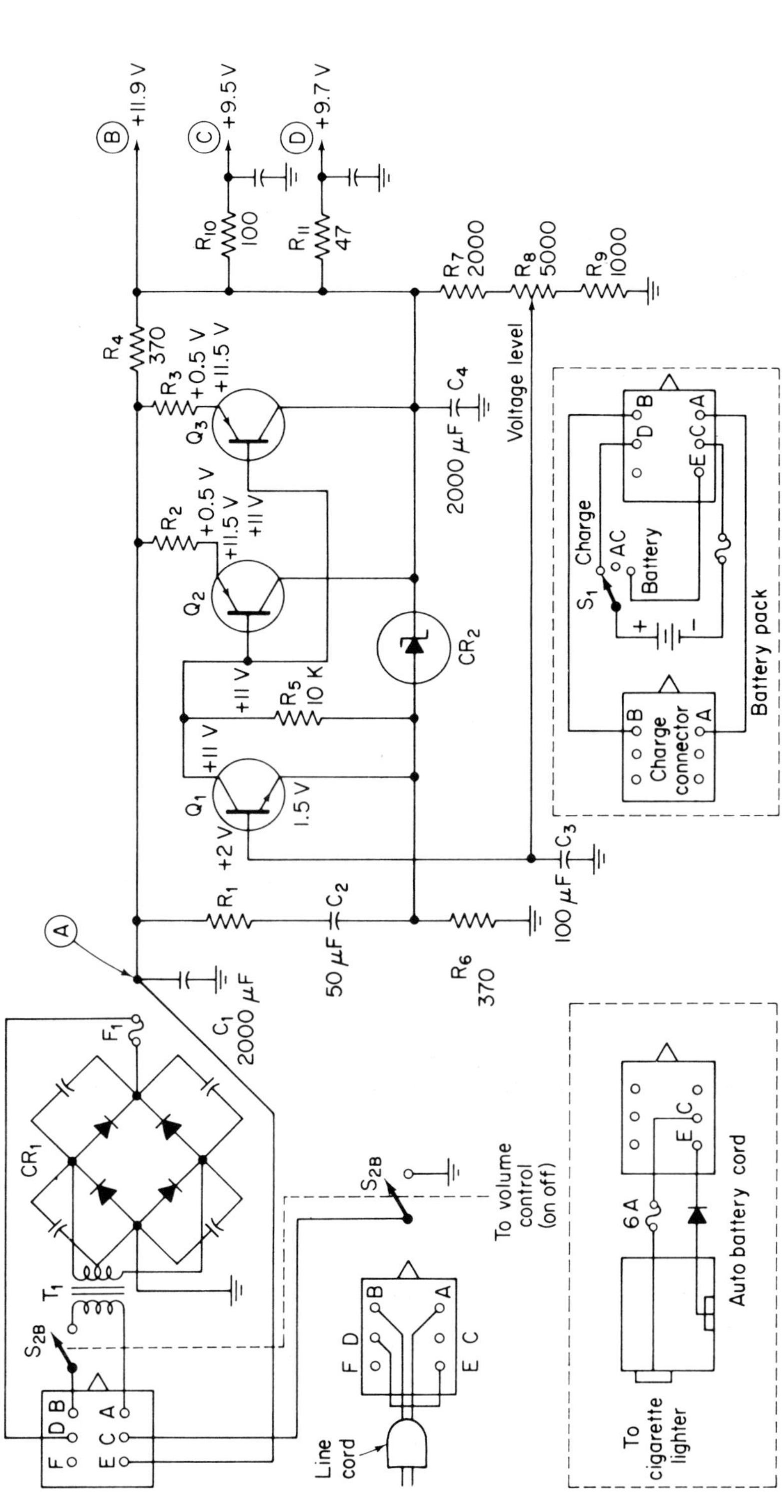


**FIGURE 5-12** Low-voltage power supply

and Q3 to swing negative and results in more collector current flow (since Q2 and Q3 are PNP). Thus, the increase in output voltage is offset. The level of the output voltage is set by R8.

When the set is to be operated on battery power, switch S1 is set to "battery," the a-c power plug is removed, and the battery pack cord is plugged in. When the battery is to be recharged, switch S1 is set to "charge" and the a-c power cord is plugged into a special connector on the battery pack. Although there is no standardization, rechargeable batteries provide about 4 to 6 hours of operation and require about 8 to 12 hours of recharging time. When the set is to be operated from an auto's cigarette lighter, the a-c power plug is removed and the auto battery power cord plugged in.

### *5-3.4 HIGH-VOLTAGE SUPPLY AND HORIZONTAL OUTPUT*

Figure 5-13 is the schematic of a typical high-voltage and horizontal output circuit. The main purposes of the circuit are (1) to provide a high voltage for the picture tube, and (2) to provide a horizontal deflection voltage (horizontal sweep) to the picture-tube deflection coil (yoke). In most receivers, the circuits also supply a boost voltage for the picture-tube focus and accelerating grids. In solid-state sets, the circuits may also supply a voltage for the video output transistor (Sec. 5-3.8), since the output transistor usually requires about 40 to 70 V (instead of the 12 V available from the low-voltage power supply). Often, the circuits also supply an AFC (automatic frequency control) signal to control the frequency of the horizontal oscillator (Sec. 5-3.5) and an AGC (automatic gain control) signal to AGC circuits (Sec. 5-3.11).

The circuits generally have only one input, no matter what combination of outputs are provided. The circuit receives pulses from the horizontal driver (Sec. 5-3.5). These pulses are at a frequency of 15,750 Hz and are synchronized with the TV broadcast picture transmission.

*It is important that you note and remember inputs and outputs when reading any schematic.* This is especially important for troubleshooting since most troubleshooting techniques are based on a comparison or study of inputs and outputs at each circuit (*what signals go in; what signals go out*). For example, in the simplest of terms, assume that the input to the high-voltage supply and horizontal output circuit is normal (15,750 Hz synchronizing pulses of correct amplitude) but that one or more of the outputs (high voltage, horizontal sweep, boost voltage, AGC, AFC, etc.) is absent or abnormal. This pinpoints trouble to some part of the high-voltage supply and horizontal output circuit.

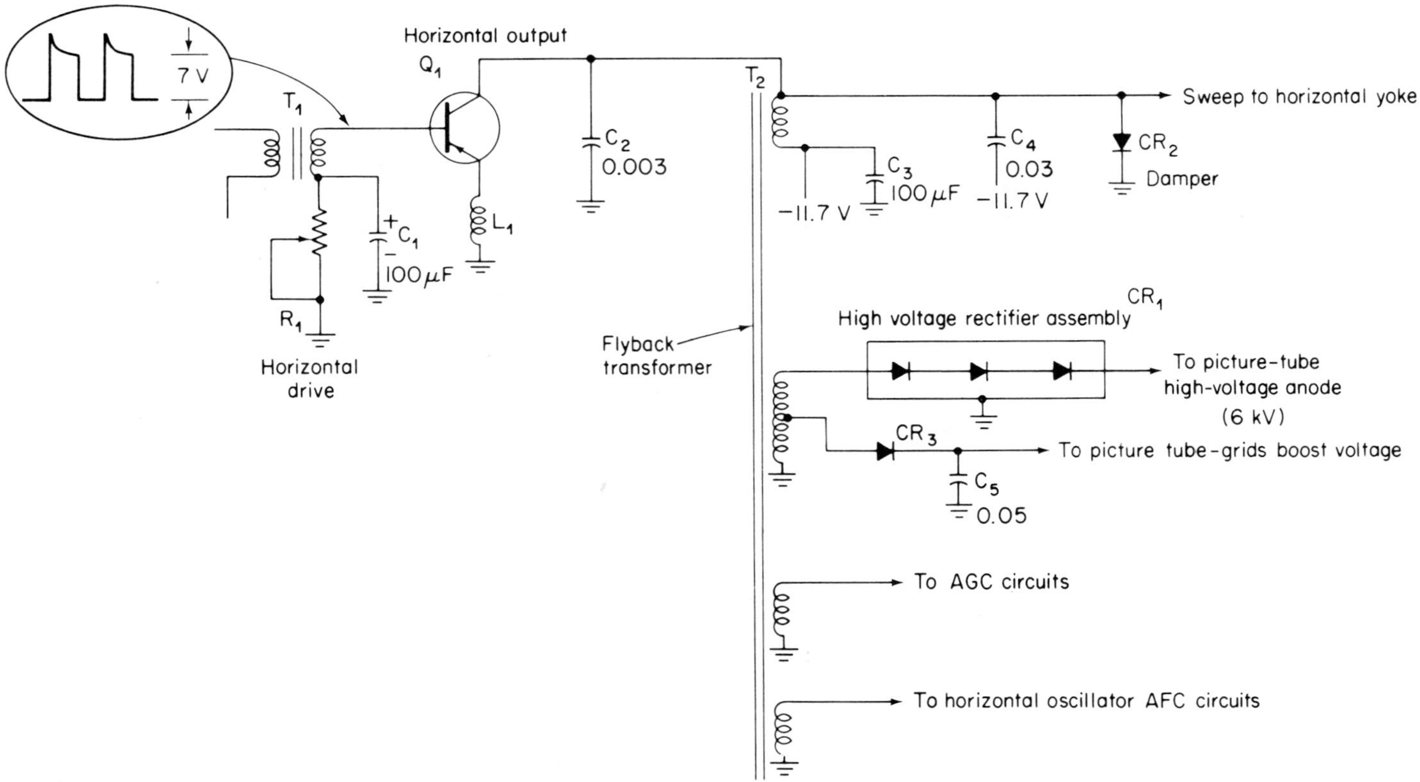


FIGURE 5-13 High-voltage supply and horizontal output circuit

The circuits generally have only one adjustment control, the *horizontal drive* control (R1 in Fig. 5-13). This control sets the amplitude of the horizontal sweep. In some receivers, the horizontal drive also sets the amount of high voltage available to the picture tube. The amount of high voltage required is determined by the size and type of picture tube. Typically, a 23-inch picture tube requires between 25 and 30 kV.

*It is also important that you note and remember adjustment controls when reading any schematic.* Again, this is especially important in troubleshooting, because proper adjustment of controls is a vital part of any troubleshooting technique.

The circuit of Fig. 5-13 receives pulses from the horizontal driver (at 15,750 Hz) synchronized with the picture transmission. These pulses are applied through transformer T1 to horizontal output transistor Q1. Can you tell the approximate amplitude and shape of the pulses from the schematic? The pulses, as they appear on the secondary side of T1 (base of Q1) are rectangular, with a sharp point on the leading edge of the pulse, and are about 7 V in amplitude. It should be remembered that waveforms often appear on the schematics of pulse circuits or of any circuits where waveforms are important. However, do not count on all schematics having waveforms any more that you would on all schematics showing a full set of voltages for all transistors.

The horizontal output transistor is normally biased at or near zero so that one edge of the pulse (negative in this case since Q1 is PNP) drives Q1 into heavy conduction (near saturation); the opposite swing of the pulse cuts Q1 off. In effect, Q1 is operated in the switching mode. The collector current of Q1 is applied through a winding on the flyback transformer T2, resulting in a pulse waveform at all of the other windings. The high voltage is rectified by CR1 and applied to the picture-tube anode. The boost output is rectified by CR3 and applied to the picture-tube focus and accelerator grid anode. On some sets, a separate winding is provided for the horizontal deflection yoke. On the circuit of Fig. 5-13, the horizontal output transistor Q1 current is passed through the deflection coil (yoke). Either way, the horizontal output pulses produce the horizontal sweep voltage.

The AFC and AGC windings (not found on all sets) are applied to the horizontal oscillator AFC circuits (Sec. 5-3.5) and the tuner/IF stage AGC circuits (Sec. 5-3.11), respectively. In some sets, the output of one winding is rectified and applied to the video output transistor as a collector voltage.

During the horizontal forward scan (when the picture is displayed), the damping diode CR2 conducts and Q1 is cut off from the start of the sweep to about the midpoint of the sweep. Then diode

CR2 is cut off and Q1 conducts for the remaining half of the sweep. This sequence is shown in Fig. 5-14. When Q1 is cut off, the picture tube is blanked and the current flows rapidly through the horizontal yoke in the opposite direction, pulling the electron beam back to the left side of the screen. This is known as horizontal retrace or *flyback*. The pulse developed during the *flyback interval* is used by the other windings on the flyback transformer T2.

### *5-3.5 HORIZONTAL OSCILLATOR AND DRIVER*

Figure 5-15 is the schematic of a typical horizontal oscillator and driver circuit. The purpose of the circuit is to provide the drive signal to the horizontal output and high-voltage supply circuits (Sec. 5-3.4). The horizontal oscillator and driver signals are at a frequency of 15,750 Hz and are synchronized with the picture transmission by means of sync signals (taken from the sync separator circuits, as described in Sec. 5-3.7).

Most solid-state horizontal oscillator circuits include some form of AFC system to ensure that the horizontal sweep signals are synchronized (both for frequency and phase) with picture transmission, despite any changes in line voltage and temperature or minor variations in circuit values. The AFC action is accomplished by comparison of the sync signals with horizontal sweep signals, both for frequency and phase.

Deviations of the horizontal sweep signals from the sync signals cause the horizontal oscillator to shift in frequency or phase as necessary, to offset the initial (undesired) deviation. For example, if the horizontal sweep increases in frequency from the sync signals, the horizontal oscillator is shifted in frequency by a corresponding amount but in the opposite direction (a decrease in frequency to offset the undesired change).

**Circuit Operation.** The circuit of Fig. 5-15 is composed of three sections: the horizontal AFC, and horizontal oscillator, and the horizontal driver. The horizontal oscillator Q1 is of the blocking oscillator type (described in Sec. 4-3), operating at a frequency of 15,750 Hz and producing an output of about 1 to 3 V. The output is amplified to about 6 to 8 V by the horizontal driver transistor Q2 and is applied to the high-voltage and horizontal output circuits (Sec. 5-3.4). This output is also returned to the AFC section, through L2 and C9, to serve as a comparison voltage. The voltage is compared with the voltage produced by the sync signals to control frequency of the Q1 horizontal oscillator.

Note that the collector of Q1 is connected to the voltage distribution line (—12 V) through a winding on transformer T1, thermistor R11, and fixed resistor R12. Thermistor R11 provides temperature

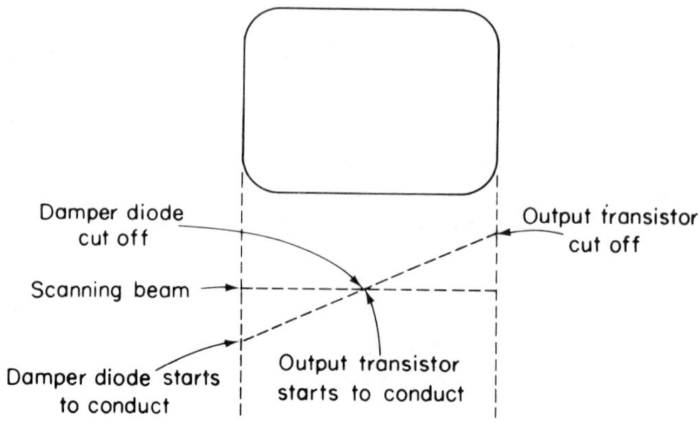


**FIGURE 5-14** Relationship of damper diode and output transistor conduction periods to picture display horizontal sweep

compensation for the Q1 oscillator. If there are temperature variations, the oscillator may tend to speed up or slow down in frequency with the variations. A thermistor is a resistor that varies in resistance value with changes in temperature. (Actually, all resistors tend to change resistance with temperature variations, but a thermistor is made of resistance material that is particularly temperature-sensitive. Generally, resistance increases with temperature decreases, and vice versa.) Thermistor R11 controls the current flow (and voltage) to the collector of Q1. When there is a change in temperature that causes Q1 to increase in frequency, the same temperature change alters the R11 resistance. In turn, this resistance change alters the Q1 collector voltage so as to decrease the oscillator frequency and thus offset the undesired increase in frequency.

The base bias for Q1 is developed by the AFC diodes CR1 and CR2. The sync pulses are applied to opposite ends of CR1 and CR2, whereas the comparison pulses (returning from the horizontal output) are applied to the diode common point (CR1 cathode, CR2 anode) through R5. An adjustable d-c voltage from the *horizontal hold* control R16 is also applied to the diode common (through R15), as is another adjustable d-c voltage from the AFC control R2 (through R3 and R5). In normal operation, the horizontal-hold R16 and AFC R2 controls are adjusted so that the bias on Q1 keeps Q1 locked onto the sync pulse frequency (thus keeping the horizontal sweep synchronized with the incoming picture signals). If the Q1 oscillator deviates from perfect sync for any reason, the comparison voltage will change, changing the bias output voltage from CR1 and CR2 and shifting the Q1 frequency as necessary to restore perfect sync.

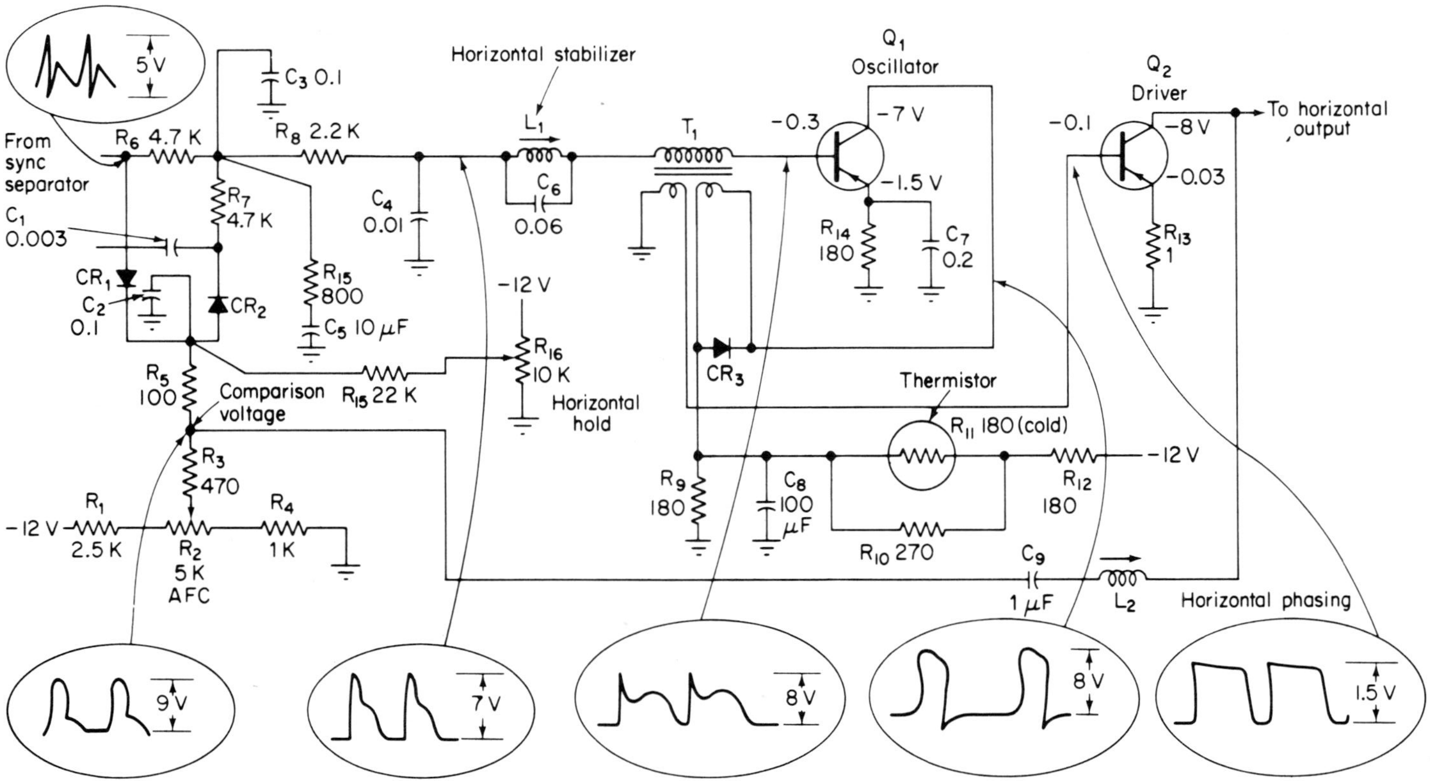


**FIGURE 5-15** Horizontal oscillator and driver with balanced AFC

Note that the bias voltage for Q1 is applied through an adjustable inductance or coil L1 and that the returning comparison pulses are returned through adjustable coil L2. These adjustable coils make it possible to adjust for slight shifts in phase or frequency of the signal voltages. L1 and L2 are not normally adjusted except after extensive repair and/or parts replacement.

### *5-3.6 VERTICAL SWEEP*

Figure 5-16 is the schematic of typical vertical sweep circuits. These circuits provide a vertical deflection voltage (vertical sweep) to the picture tube deflection yoke. The vertical circuits also supply a blanking pulse to the picture tube (usually through the video amplifier, as discussed in Sec. 5-3.10). This blanks the picture tube during retrace of the sweep. The vertical sweep signals are at a frequency of 60 Hz and are synchronized with the picture transmission by means of sync signals (taken from the sync separator section, as described in Sec. 5-3.7).

**Circuit Operation.** The circuit of Fig. 5-16 is composed of three stages: oscillator, driver, and output. In some solid-state TV sets, the driver and output functions are combined into one stage. The vertical oscillator Q1 is of the blocking oscillator type (Sec. 4-3), operating at a frequency of 60 Hz and producing a pulse output of about 1 to 2 V. The Q1 pulse output is modified into a *sawtooth sweep* (as shown at the base of Q2) and is applied through the driver Q2 to the output stage Q3. The final circuit output is a peaked sawtooth waveform (as shown at the vertical coils). The sawtooth portion (about 4 to 5 V) is applied to the picture-tube deflection yoke, whereas the peaked portion (about 50 V) is used as the blanking pulse for the picture tube.

The vertical oscillator Q1 frequency is set by circuit values and by the bias voltage. However, oscillator Q1 is locked in frequency by the sync pulses. Q1 is normally reverse-biased where d-c voltages are concerned. How can you tell this? Q1 is a PNP transistor, with the base a —2.6 V and the emitter at —4.4 V. This is reverse-bias for a PNP (for current flow, forward bias, the base of a PNP must be more negative than the emitter). Feedback for the oscillator is obtained through magnetic coupling of the collector and base windings of transmitter T1. Whenever Q1 conducts, the winding feedback makes the base more negative, increasing the collector current and charging C1 to a high negative value. When the emitter of Q1 goes sufficiently negative, Q1 is cut off and remains so until C1 discharges through R1.

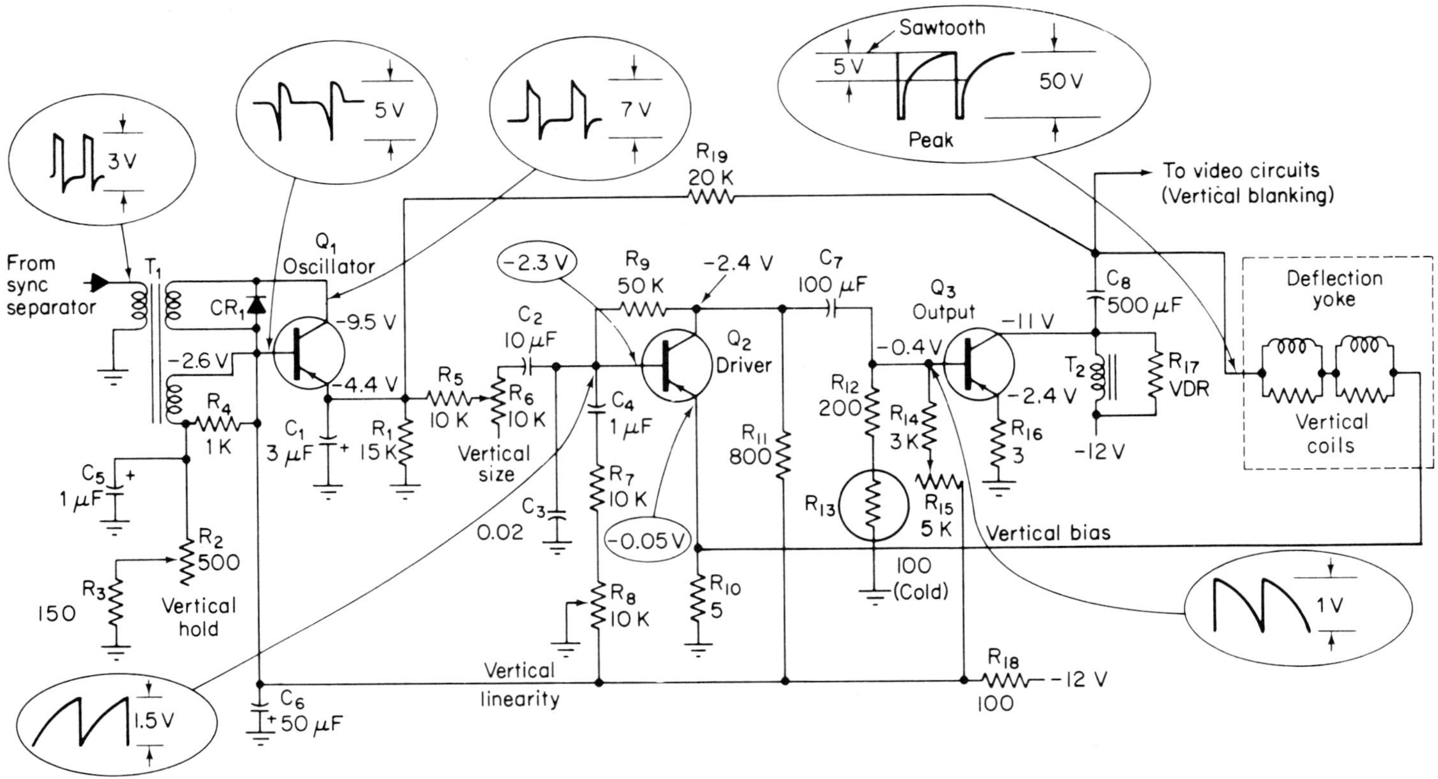




**FIGURE 5-16** Vertical sweep circuits

As with most solid-state blocking oscillators, Q1 is on (conducting) for a very short period of time and is off (nonconducting) for a long period of time during each cycle. The point at which Q1 starts to conduct (and thus frequency of Q1) is set by *vertical hold* control R2, which controls the base bias. Actually, R2 is set so that the frequency of Q1 is just below 60 Hz. In practice, R2 is adjusted until the vertical sweep is locked with the picture transmission. That is, the sync pulses trigger Q1 into conduction earlier in the cycle than Q1 would otherwise start to conduct. Diode CR1 acts as a blocking diode to prevent the Q1 signal from going back into the sync separator and to prevent excessive peak voltages on the base of Q1.

The Q1 pulse output is converted to a sawtooth sweep by means of emitter network R1C1. The discharge of C1 through R1 produces the basic sweep waverform. The sweep is made linear through feedback of the output by R19. The waveform is also shaped by the network C3-C4-R7-R8. *Vertical linearity* control R8 provides a manual control for the sawtooth waveform. The amplitude of the sawtooth waveform to driver Q2 (and thus the vertical size of the picture raster) is set by *vertical size* control R6. Note that the emitter of Q2 is connected to the deflection yoke. This provides negative feedback to make a more linear output.

The deflection yoke of a solid-state TV requires heavy current even though the sweep voltage is low. For this reason, the output transistor Q3 is of the power type (often mounted on a *heat sink* to dissipate excess heat). Since there is always the danger of *thermal runaway* (Sec. 3-3.7) in power transistors, thermistor R13 is included in the base circuit of Q3. If collector current increases, heating Q3 and causing more current flow, the resistance of R13 drops, lowering the emitter-base voltage differential. In turn, this lowers the emitter-base and emitter-collector current flows. Transistor Q3 is further stabilized by unbypassed emitter resistor R16.

If it becomes necessary to replace transistor Q2, it may be necessary to adjust vertical bias control R15. Most solid-state vertical output circuits have a bias control, since the characteristics of replacement transistors can be different from the original transistor. The peaked portion of the output waveform is obtained by inductive "kickback" of T2. Typically, the peak of the waveform (shown at the top of Fig. 5-16) is about 50 V, even though the sawtooth portion is usually about 5 V. Resistor R17 is a *voltage-dependent resistor* (VDR) found on many late-model sets. Resistor R17 keeps the amplitude of the output waveform within narrow limits. The resistance of the VDR is varied by the output voltage to the deflection yoke. If the output tends to increase, the resistance of R17 decreases, keeping the output constant.

Figure 5-17 is the schematic of typical sync separator circuits. These circuits function to remove the vertical and horizontal sync pulses from the video circuits (Sec. 5-3.10) and apply the pulses to the vertical (Sec. 5-3.6) and horizontal (Sec. 5-3.5) sweep circuits, respectively. Thus, the sync separator circuits have one input and two outputs. The sync separator circuits also function as clippers and/or limiters to remove the video (picture) signal and any noise. Thus, the horizontal and vertical sweep circuits receive sync pulses only and are free of noise and video signal (in a properly functioning receiver).

**Circuit Operation.** As shown in Fig. 5-17, the video input is negative-going and input transistor Q1 is PNP. Thus, both the sync pulses and the signal/noise turn Q1 on. However, with Q1 biased near zero (the base at 9.3 V and the emitter at 9.5 V) the large sync pulses drive Q1 into saturation at a level never reached by the signal and/or

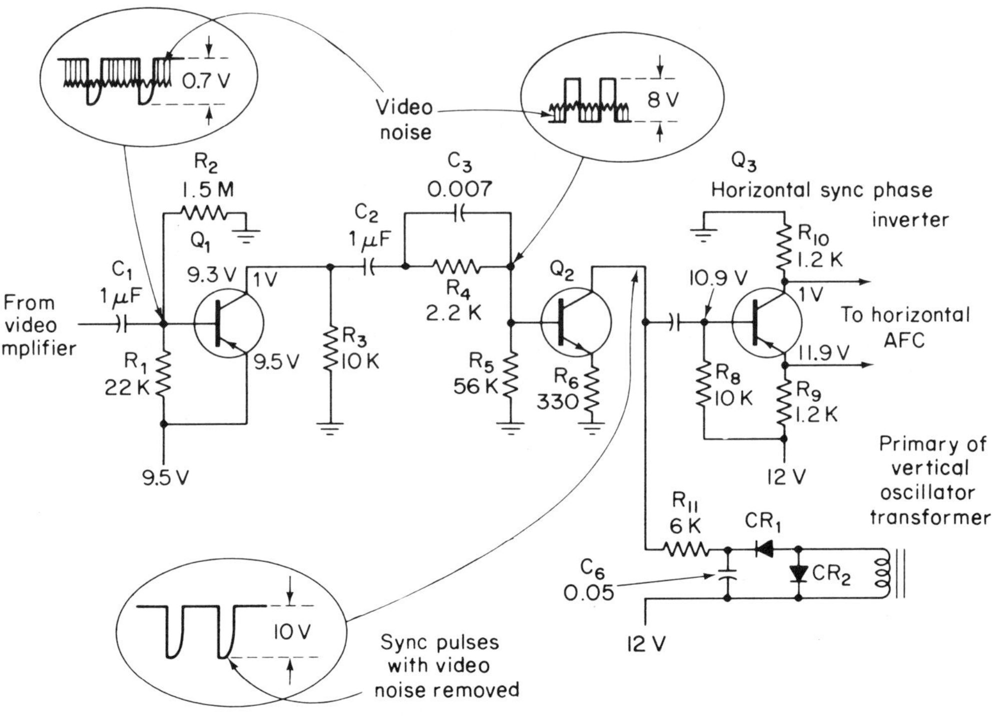


**FIGURE 5-17** Sync separator circuits

noise. The second sync separator transistor Q2 is reverse-biased, so that the first portion (about 0.3 typically) of the Q1 output is clipped. This further removes any signal and/or noise.

The output of Q2 is applied to the low-pass filter (also known as a *vertical integrator*) of C6 and R11. The 60-Hz vertical signals are applied to the vertical sweep circuit input (the primary of the vertical oscillator transformer) through series and parallel diodes CR1 and CR2. These diodes are arranged to pass the sync pulses to the vertical sweep input but prevent the vertical oscillator pulses from passing back to the sync circuits. The output of Q2 is also applied to Q3, which acts as a phase splitter or phase inverter. The output from Q3 is applied to the horizontal AFC circuits.

### *5-3.8 RF TUNER*

Figure 5-18 is the schematic of a typical RF tuner. The functions of an RF tuner in a TV receiver are essentially the same as the RF sections in other AM and FM receivers. The typical tuner in a TV receiver consists of an RF amplifier, mixer, and oscillator. Each TV channel consists of two carriers: the FM sound carrier and the AM picture carrier. Both carriers are amplified by the RF amplifier. Both carriers are mixed with signals from the oscillator to produce two IF signals in the mixer. These two signals are amplified by the IF amplifiers (Sec. 5-3.9). Thus, the RF tuner has one input and one output, even though there are two carriers.

The RF tuner also receives automatic gain control (AGC) signals from the AGC circuits (Sec. 5-3.11). These AGC signals are generally applied only to the RF amplifier (not the mixer and oscillator). The AGC signals function to control gain of the RF amplifier in the presence of carrier signal variations. For example, if the carrier signal strength increases, the AGC signals decrease the RF amplifier gain and thus offset the initial increase in carrier strength.

**Circuit Operation.** As shown in Fig. 5-18, the tuning coils are of the turret type, where a separate set of drum-mounted coils is used for each channel, and the entire drum rotates when the channel is selected. Some tuners are of the switch type, where series-connected coils are mounted on wafer switches. All three stages of the tuner (RF amplifier, oscillator, and mixer) are connected in the common-emitter configuration. This is typical for transistor tuners, although a few tuners use the common-base configuration. The RF amplifier Q1 is *neutralized* by C308. Neutralization is often required in RF amplifiers to prevent oscillation. The collector and base of a transistor act somewhat like a capacitor. If circuit conditions are just right, part of

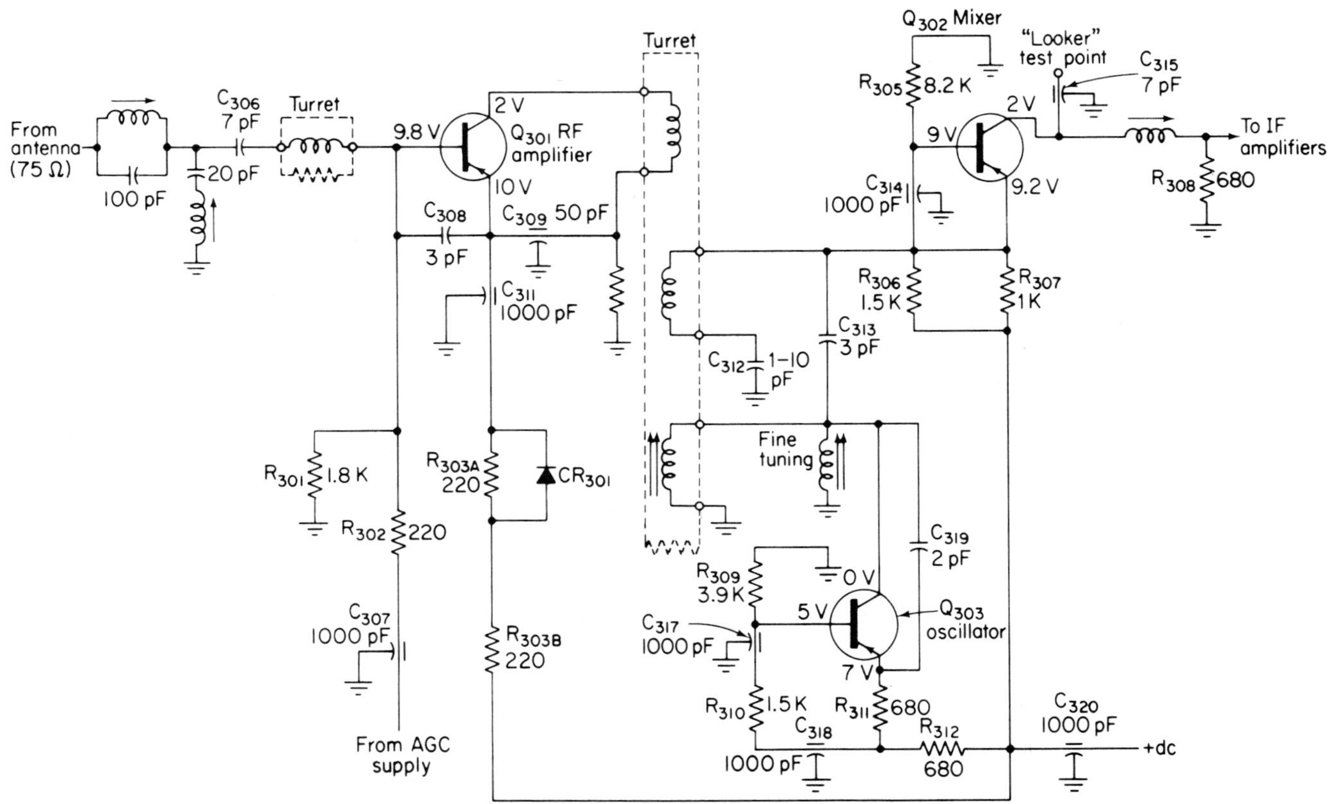


**FIGURE 5-18** RF tuner circuits

the signal at the collector can be fed back to the base in sufficient strength to produce oscillation. This feedback can be offset by a neutralizing capacitor connected between the emitter (or the opposite end of the resonant tuning coil) and the base. The signal through the neutralizing capacitor is equal to (but of opposite phase) that of the base-collector feedback and thus cancels the undesired base-collector feedback.

The RF tuner has provisions for tuning each channel. Tuning is accomplished by three slug-tuned coils (one for the RF amplifier, one for the mixer, and one for the oscillator) at each channel. There is also a *fine-tuning* control (slug-tuned coil) connected in parallel across the oscillator tuning coil. The fine-tuning control is a front-panel (user) adjustment. The RF tuner also includes two traps at the antenna input. One trap is series-resonant, whereas the other trap is parallel-resonant. The traps function to bypass (series) and reject (parallel) undesired signals before they can enter the tuner circuits, and thus prevent the undesired signals from appearing on the voice or video channel signals.

The RF amplifier base is connected to the AGC network. Most solid-state TV AGC networks operate on a different basis from vacuum-tube systems. With solid-state, a strong signal increases the forward bias and drives the transistor into saturation, thus reducing gain. With vacuum-tube systems, strong signals increase the negative bias on the vacuum-tube grids to reduce gain. A few solid-state TV AGC systems also use increased negative bias to desensitize the circuits.

Most tuners are provided with at least one test point (often called the *looker point*). In the circuit of Fig. 5-18, the test point is at the collector of Q303. This test point is used primarily to monitor the tuner output with an oscilloscope. The same point can also be used to inject a signal into the IF amplifiers.

### *5-3.9 IF AND VIDEO DETECTOR*

Figure 5-19 is the schematic of typical IF and video detector circuits. These circuits amplify both picture and sound signals from the RF tuner and mixer, demodulate both signals for application to the video amplifier (Sec. 5-3.10) and sound IF amplifier (Sec. 5-3.12), provide signals to the AGC circuits (Sec. 5-3.11), and trap (or reject) signals from adjacent channels.

In most receivers, both picture and sound outputs from the video detector are applied to the video amplifier circuits, where both signals are amplified by at least one stage of the video amplifier. At this point, the sound signals are applied to the input of the sound IF

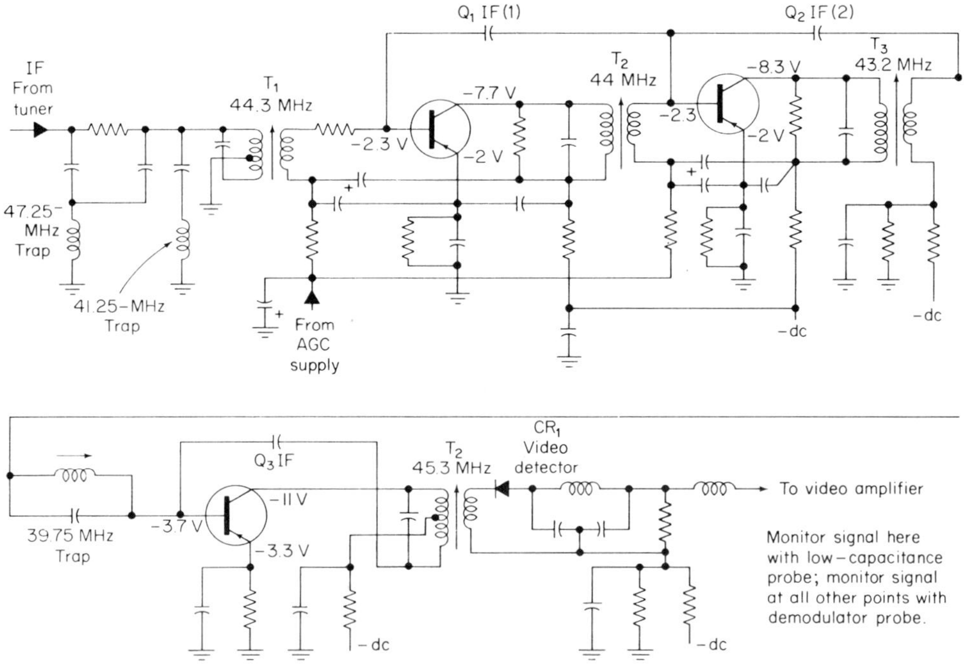


**FIGURE 5-19** IF and video detector circuits

amplifier, and the picture signals are applied to the remaining stages in the video amplifier (and to the sync separator). In a few receivers, the sound signals from the video detector are applied directly to the sound IF amplifiers. In addition to the signal input and output, most stages of the IF amplifier receive an AGC signal from the AGC circuits. These signals, which are in the form of a varying d-c bias voltage, are the same as those applied to the RF tuner (Sec. 5-3.8).

**Circuit Operation.** As shown in Fig. 5-19, three stages of IF amplification are used, with each stage in the common-emitter configuration. All stages are forward-biased. The first two stages, Q1 and Q2, receive their forward bias from the AGC circuit. This same AGC line is connected to the RF amplifier in the tuner. On strong signals, the forward bias is increased, driving Q1 and Q2 into saturation and reducing gain. The third IF stage, Q3, does not have any AGC control. All three stages are neutralized to prevent oscillation in the IF circuits.

Each IF stage is tuned at the input and output by corresponding slug-tuned transformers. As is the case with most FM receivers (and some wide-band AM receivers), the IF stages are *stagger-tuned*. That is, each IF transformer is tuned to a different frequency (one tuned to the high end, one to the low end, and the remainder tuned to various points in between). This gives the IF amplifiers sufficient *bandwidth* (about 3.25 MHz) to pass both the picture and sound signals.

Three traps are used. The 41.25- and 47.25-MHz traps are series-resonant, whereas the 39.75-MHz trap is parallel-resonant. The traps are tuned to reject and remove signals from adjacent TV channels. Traps are necessary because TV signals may be broadcast simultaneously on adjacent channels, and the combination of wide-band broadcast channels together with wide-band IF circuits may result in overlap of signals.

The output of the video detector CR1 to the video amplifier is typically about 1 V. That is, the peaks of the sync pulses are about 1 V, with the picture signals about 0.25 to 0.5 V. In many new solid-state TV sets, the entire IF function, including the video detector, is performed by an IC.

### *5-3.10 VIDEO AMPLIFIER AND PICTURE-TUBE CIRCUITS*

Figure 5-20 is the schematic of typical video amplifier and picture-tube (video) circuits. These circuits, which are composites of many typical TV circuits, have three inputs and three outputs. Two stages, driver Q1 and Q2, are used. Transistor Q1 is connected as an emitter follower. Typically, the signal input and output at Q1 are about 1 V. This signal consists of the sound, video, and sync pulses, as taken from the video detector (Sec. 5-3.9).

The output of Q1 is applied to the input of video output transistor Q2, to the input of the sync separator circuits (Sec. 5-3.7), and to the input of the sound IF circuits (Sec. 5-2.12). The output of Q1 is applied to Q2 through *contrast* control R3. Transistor Q2 amplifies the 1-V output to a level of about 25 to 50 V, depending on the picture-tube screen size. The contrast control R3 determines the amount of drive signal applied to the picture tube. An increase in drive signal produces an increase in contrast. R3 is, in effect, a volume or gain control between the video driver and video output stages.

*Brightness* control R12 sets the voltage level on the collector of Q2. This determines the amount of fixed bias applied to the picture-tube cathode. In turn, this sets the intensity of the electron beam (with or without a signal present). The output of Q2 is applied through a 4.5-MHz sound trap to the cathode of the picture tube. This trap, when properly adjusted, prevents any sound information from entering the video circuits. The Q2 output should contain only the video information (picture signal) plus the horizontal (15,750-Hz)

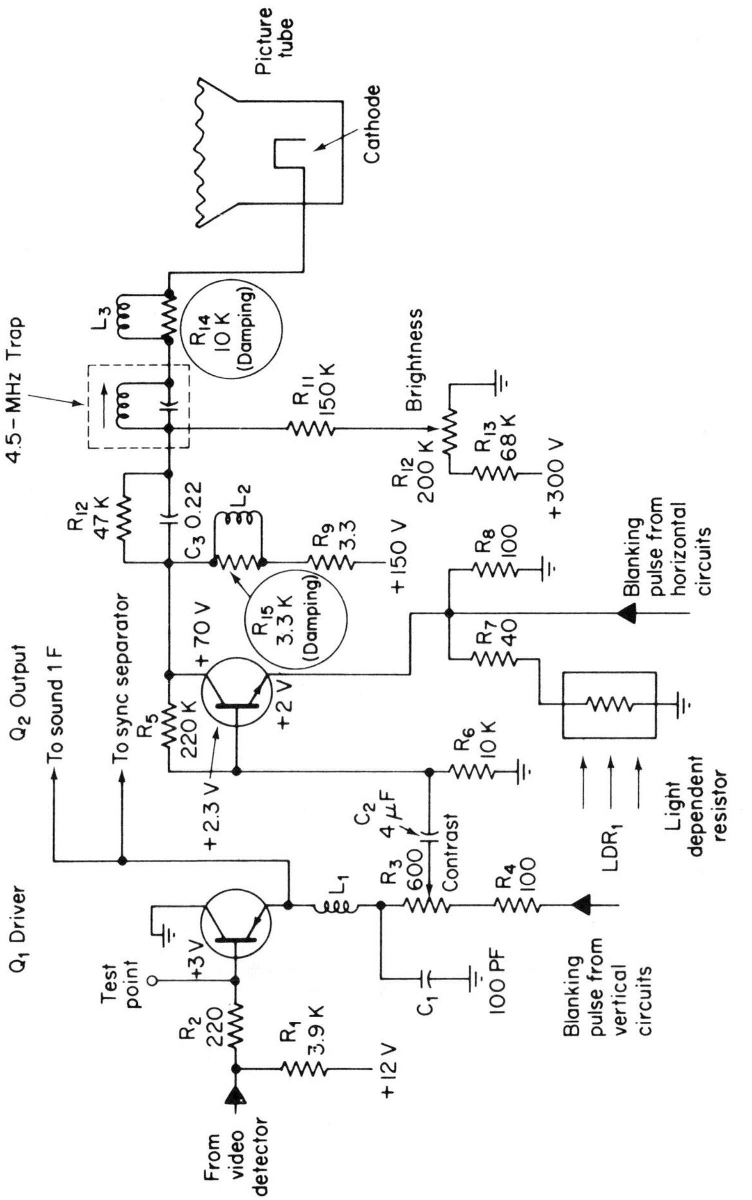


**FIGURE 5-20** Video amplifier and picture-tube (video) circuits

and vertical (60-Hz) retrace blanking pulses. Since all of this information is applied to the cathode of the picture tube, the video (picture) signal is negative, whereas the blanking pulses are positive. (A positive pulse applied to the picture-tube cathode decreases brightness, whereas a negative pulse increases brightness.)

The vertical retrace blanking pulses from the vertical sweep circuits (Sec. 5-3.6) are applied to the emitter of Q1. Horizontal blanking pulses are applied to the emitter of Q2. Note that both Q1 and Q2 are forward-biased during normal operation. The bias on Q2 is partially determined by light-dependent resistor LDR1 connected between the emitter and ground. As the ambient lighting around the set varies, the bias on Q2 (and thus the gain of the video amplifier) varies. This variable-gain feature provides the necessary changes in picture contrast to accommodate changing conditions of ambient light.

The horizontal and vertical blanking pulses blank the picture tube (cut off the electron beam) during the retrace period of the horizontal and vertical sweeps. During the trace periods of both sweeps, the picture signal varies the intensity of the electron beam so as to "paint" a picture on the picture-tube screen. The electron beam of the picture tube "follows" the broadcast camera tube beam, in both intensity and position, so as to reproduce the picture focused in the camera lens.

### *5-3.11 AGC CIRCUITS*

Figure 5-21 is the schematic of typical AGC circuits. The circuits shown are of the keyed, saturation type. The RF tuner and IF stage transistors connected to the AGC line are forward-biased at all times. On strong signals, the AGC circuits increase the forward bias, driving the transistors into saturation, thus reducing gain. Under no-signal conditions, the forward bias remains fixed.

Keyed AGC circuits have two inputs. One input is from the IF circuit (Sec. 5-3.9) output at the IF center frequency. The other input is from the horizontal output circuit (Sec. 5-3.4) at a frequency of 15,750 Hz. The AGC output is a varying d-c bias voltage (controlled by bursts of IF signals). A portion of the IF signal is taken from the IF amplifiers and is pulsed, or keyed, at the horizontal sweep frequency. The resultant keyed bursts of signal control the amount of d-c voltage produced on the AGC line. The AGC output is applied to both the IF amplifiers and the RF tuner.

In the circuit of Fig. 5-21, keyer transistor Q1 is an IF amplifier, with its collector tuned to the IF center frequency of 42 MHz by transformer T1. No d-c voltage (as such) is supplied to the elements of Q1. The keying pulses from the horizontal flyback transformer are applied to the collector through diode CR1. This produces an average collector voltage of about 1 V. When Q1 is keyed on, the bursts of IF signals pass through T1 and are rectified by CR2. A corresponding

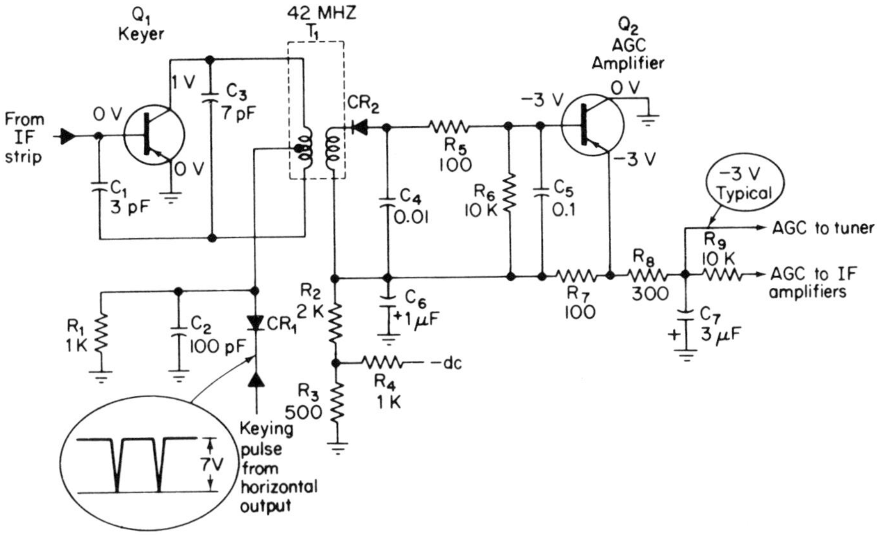


**FIGURE 5-21** Keyed AGC circuits

d-c voltage is developed across C4 and acts as a bias for AGC amplifier Q2. Transistor Q2 is connected as an emitter follower, with the AGC line being returned to the emitter. Variations in IF signal strength cause corresponding variations in Q2 bias, Q2 emitter voltage, and the AGC line voltage.

### *5-3.12 SOUND IF AND AUDIO*

Figure 5-22 is the schematic of typical sound IF and audio circuits. These circuits amplify the 4.5-MHz sound carrier, demodulate the FM sound (remove the audio signals), and amplify the audio signals to a level suitable for reproduction on the loudspeaker. The sound carrier is at a frequency of 4.5 MHz, since the picture and sound broadcast carriers are 4.5 MHz apart on all channels. After the AM carrier is demodulated by the video detector (Sec. 5-3.9), only the FM carrier remains. In most TV receivers, the FM carrier is demodulated or detected by means of a ratio detector as shown (following two stages of IF amplification). However, some receivers use a discriminator-type sound detector after the IF amplifiers. The audio portion of the circuits is similar to the audio circuits of AM or FM broadcast receivers. As shown, the audio circuits include a *volume* control R12 and a *tone* control R13. These are front-panel controls, readily available to the user.

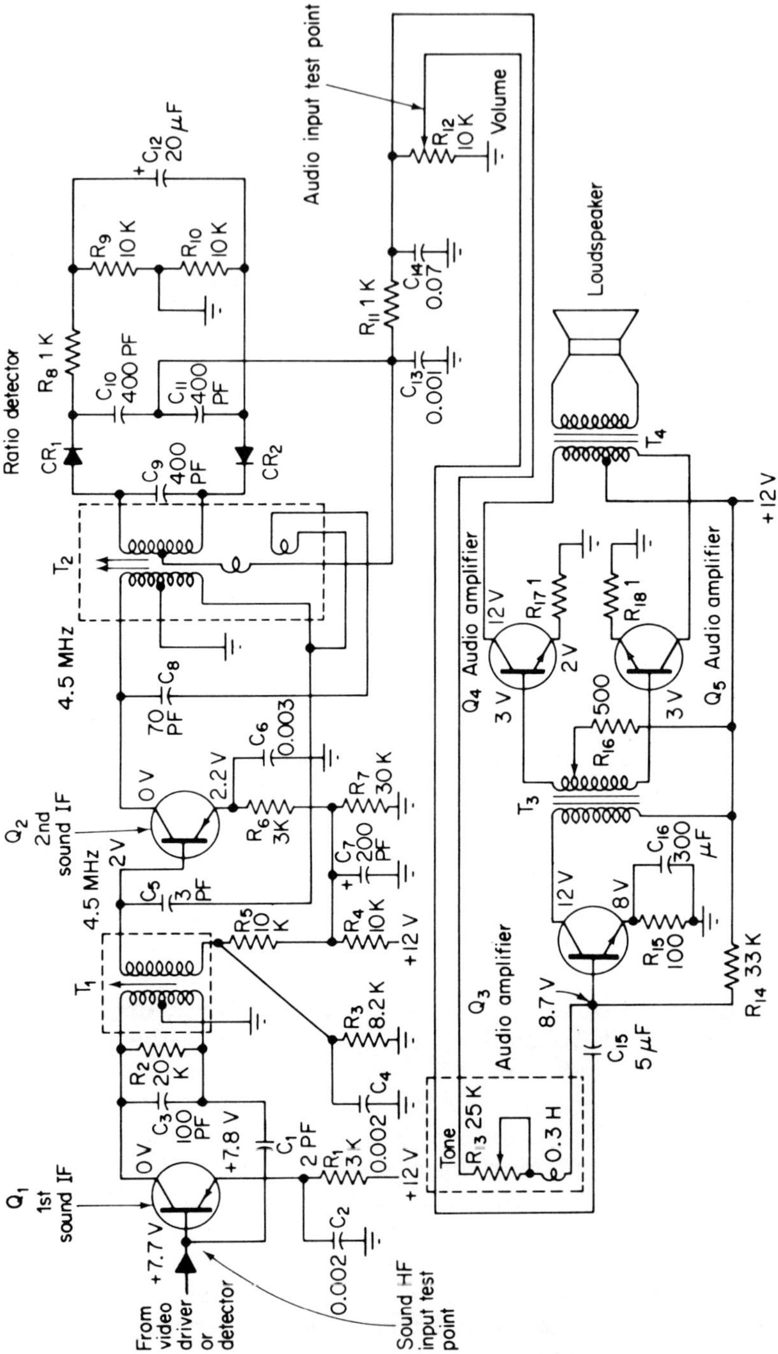


**FIGURE 5-22** Sound IF and audio circuits

**Circuit Operation.** As shown in Fig. 5-22, two stages of IF amplification and two stages of audio amplification are used, with all stages in the common-emitter configuration. All stages are forward-biased. The final audio stage is push/pull and thus operates class B (with slight forward bias) or possibly class AB (with slightly increased forward bias). Unlike the video IF stages, the sound IF stages do not have any AGC circuits, and the forward bias is provided by fixed-resistance networks.

Both sound IF stages Q1 and Q2 are neutralized by capacitors C1 and C5 to prevent oscillation. Each stage is tuned at the output by corresponding transformers T1 and T2 to 4.5 MHz. The overall bandwidth of the sound IF amplifiers is typically 50 to 60 kHz in solid-state circuits.

## EXERCISES

5.1. Generally, where is intelligence inserted into a CW transmitter, and in what form (Fig. 5-1)?

5.2. In the block diagram of Fig. 5-2, if the input to the speech amplifier (from the microphone) is changed to 7000 Hz, what will the carrier signal frequency be?

5.3. In the block diagram of Fig. 5-3, if the input to the modulator (from the microphone) is changed to 7000 Hz, what will the carrier signal frequency be?

5.4. By what percentage will the amplitude change in the transmitter of Fig. 5-3 when there is a 3000- to 7000-Hz modulating signal change (as in Exercise 5.3)?

5.5. Is the SSB transmitter of Fig. 5-4 AM or FM?

5.6. What is the advantage of the SSB transmitter of Fig. 5-4?

5.7. If the receiver of Fig. 5-5 is tuned to 103 MHz and receives an unmodulated signal at that frequency, what will be the output from the second detector? From the loudspeaker?

5.8. Name the functions of the following devices shown in Fig. 5-7: TR1, TR2, TR3, TR4, TR5, TR6, TR7, TR8, TR9, TR10, TR11, TR12, TR13.

5.9. During receive, where are the inputs and outputs of the circuit shown in Fig. 5-7?

5.10. During transmit, where are the inputs and outputs of the circuit shown in Fig. 5-7?

5.11. During transmit, how is the loudspeaker connected to the audio section (Fig. 5-7)?

5.12. During receive, what determines the channel or operating frequency (Fig. 5-7)?

5.13. During transmit, what determines the channel or operating frequency (Fig. 5-7)?

5.14. During receive, where is the output of the audio section applied (Fig. 5-7)?

5.15. During transmit, where is the output of the audio section applied (Fig. 5-7)?

5.16. What are the purposes of C45 and C47 (Fig. 5-7)?

5.17. Are TR9 and TR10 operated class AB or B? How can you tell (Fig. 5-7)?

5.18. What are the functions of L7, L8, L9, C31, and C32 (Fig. 5-7)?

5.19. How would you tune or adjust the stages of the transmitter section (Fig. 5-7)?

5.20. What is the purpose of VR7 (Fig. 5-7)?

5.21. Through what component is the output of the local oscillator applied to the IF stage (Fig. 5-7)?

5.22. What is the purpose of CF1 (Fig. 5-7)?

5.23. Describe the operation of D3 (Fig. 5-7).

5.24. What is the purpose of VR2 (Fig. 5-7)?

5.25. What is the purpose of VR1 (Fig. 5-7)? How does VR1 operate during transmit?

5.26. During receive, how is the audio output from VR1 applied to the output of the audio section (Fig. 5-7)?

5.27. During transmit, how is the audio output from VR1 applied to the input of the audio section (Fig. 5-7)?

5.28. What is the main purpose of the transistor voltages shown in Fig. 5-7?

5.29. The sound or audio portion of a TV program is broadcast with ____________ modulation, and the picture or video portion is broadcast with ____________ modulation.

5.30. The picture or video portion of TV broadcast signal contains both ____________ information and ____________ pulses.

5.31. Referring to Fig. 5-11, name the inputs, outputs, adjustment controls, and stages of the vertical sweep circuits.

5.32. Identify (by reference designation) each of the items called for in Exercise 5-31 on the schematic of Fig. 5-16.

5.33. Referring to Fig. 5-11, name the inputs, outputs, adjustment controls, and stages of the sync separator circuits.

5.34. Identify (by reference designation) each of the items called for in Exercise 5-33 on the schematic of Fig. 5-17.

5.35. If you monitored the AGC line shown in Fig. 5-11, what type of signal would you expect?

5.36. What type of regulator is used in the circuit of Fig. 5-12?

5.37. What part sets the output voltage in the circuit of Fig. 5-12?

5.38. What is the purpose of CR2 (Fig. 5-12)?

5.39. What are the circled letters on the schematic of Fig. 5-12?

5.40. Name five functions for the circuit of Fig. 5-13.

5.41. What is the purpose of R1 (Fig. 5-13)?

5.42. What is the frequency of the signals applied to the circuit of Fig. 5-13?

5.43. Where are the inputs and outputs for the circuit of Fig. 5-15?

5.44. What kind of oscillator is used in the circuit of Fig. 5-15?

5.45. What is the purpose of R11 (Fig. 5-15)?

5.46. What are the purposes of L1 and L2 (Fig. 5-15)?

5.47. What is the frequency of oscillator Q1 in Fig. 5-16?

5.48. What converts the pulse output from Q1 into the sawtooth voltage applied to Q2 (Fig. 5-16)? What keeps the sawtooth voltage linear?

5.49. What part (in the circuit of Fig. 5-16) sets the height of the picture on the TV screen?

5.50. Transistor Q3 in Fig. 5-16 is probably mounted on a ____________.

5.51. What is the frequency of the signal applied to the circuit of Fig. 5-17? What is the output frequency?

5.52. What is the purpose of Q3 (Fig. 5-17)?

5.53. What parts in Fig. 5-17 form the vertical integrator?

5.54. Name the inputs to the circuit of Fig. 5-18.

5.55. What is the purpose of the test point at C315 (Fig. 5-18)?

5.56. What is the purpose of C308 (Fig. 5-18)?

5.57. Through what part is the local oscillator signal injected into the mixer (Fig. 5-18)?

5.58. Name the inputs to the circuit of Fig. 5-19.

5.59. Are the transistors in Fig. 5-19 forward- or reverse-biased?

5.60. Why are the IF transformers in Fig. 5-19 tuned to different frequencies?

5.61. Name the inputs and outputs of the circuit in Fig. 5-20.

5.62. Describe the three adjustment controls for the circuit of Fig. 5-20.

5.63. What is the purpose of LDR1 (Fig. 5-20)?

5.64. What is the purpose of R12 (Fig. 5-20)?

5.65. What are the frequencies of the input signals to the circuit of Fig. 5-21?

5.66. On strong signals from the IF circuits, does the AGC output from the circuit of Fig. 5-21 increase or decrease? How does this affect the RF tuner, IF amplifier, and audio amplifier?

5.67. What is the frequency of the input signals to the circuit of Fig. 5-22?

5.68. Are transistors Q1 and Q3 in Fig. 5-22 forward- or reverse-biased?

5.69. Are transistors Q3, Q4, and Q5 in Fig. 5-22 forward- or reverse-biased?

5.70. What is the purpose of C1 (Fig. 5-22)?

5.71. What are the purposes of C6 and C16 (Fig. 5-22)?

5.72. Name four adjustment controls in the circuit of Fig. 5-22.

5.73. C13, C14, and R11 (Fig. 5-22) form a ______________ filter.

5.74. Q4 and Q5 are operated class ______________ . What on the schematic of Fig. 5-22 tells you this?

5.75. The FM detector on Fig. 5-22 is identified as a ratio detector. If the circuit were not so identified, how would you know that it was a ratio detector?

# Answers to Exercises

*CHAPTER 1*

1.1. Symbols.

1.2. Circuits.

1.3. Theoretically, never! Current flows from *cathode to anode* when the anode is more positive than the cathode. As a practical matter, there may be some *leakage current* or *reverse current* from anode to cathode, but this should not be a concern in reading schematics.

1.4. Alternating current (a-c) power source.

1.5. The resistor symbol.

1.6. Because schematics are normally read from left to right, with signal or function flow shown from left to right.

1.7. Because the power is turned off.

1.8. Because the diode is reverse-biased or "turned off" since the anode is negative with respect to the cathode.

1.9. Half-wave rectifier.

1.10. You know that CR2 is a rectifier (converts ac into dc) and that it is called a "detector" in Fig. 1-3, but you do not know the detector function until you have read further (Chapters 3 through 5).

1.11. Q1.

1.12. Because it is improperly labeled. The capacitor should be C12.

1.13. That C2A and C2B are mechanically coupled or "ganged."

1.14. The second IF is not shown in Fig. 1-3.

1.15. Because Q1 and the associated parts have two functions.

1.16. 10 kΩ (10,000 Ω).

1.17. The value is not given.

1.18. −8.5 V.

1.19. Also −8.5 V.

1.20. The voltage varies between 0 and −1 V, depending on signal conditions.

1.21. Transistor Q1.

1.22. −3 V. How did you know which was the emitter? (We have not talked about transistor elements yet.)

1.23. A sawtooth signal of about 0.1 V in amplitude.

1.24. Minus (−) or negative.

1.25. Resistor R3 is the *horizontal hold* control (probably for a television set), but you must read further to understand the function of such a control (Chapter 5).

1.26. The −2.5-V collector voltage of Q1 should change when R9 is adjusted.

1.27. Test points.

1.28. Smaller.

1.29. 4, 4.7, and 6 V (at emitter, base, and collector, respectively).

1.30. Positive. Since no polarity symbol is given, the voltages are assumed to be positive (+) with respect to ground.

1.31. Four diodes.

1.32. None. Fig. 1-6 shows physical relationships.

1.33. Signal flow is not shown on Fig. 1-7. Physical relationships and point-to-point wiring are shown on practical wiring diagrams.

1.34. Black.

1.35. The starred numbers are *major test points.* The circled letters are secondary test points.

1.36. The signals start at the antenna (left) and go through mixer Q1, IF amplifier Q2, detector CR2, audio Q3, transformer T5, audio power Q4 and Q5, transformer T6 to the loudspeaker (right).

1.37. Detector CR2.

1.38. Three IF stages, operating at a frequency of 455 kHz.

1.39. The tuning capacitors, which are part of transformers T2 and T3.

1.40. The schematic. (That is why you are studying this book!)

## CHAPTER 2

2.1. Connected.

2.2. To carry electrical current (or voltages, signals, pulses, etc.).

2.3. Possibly, but the current (if any) will be very small through the lamp (and the lamp will not light). Most of the current will flow through the wire short.

2.4. There is no way of telling from the information given in Fig. 2-2.

2.5. 1000 Ω: $R = E/I = 3/0.003 = 1000$.

2.6. A rheostat uses only two of the three potentiometer terminals.

2.7. Ohms (or multiples and submultiples of ohms: kilohms, megohms, milliohms, etc.).

2.8. Capitol omega (Ω).

2.9. Wattage and tolerance.

2.10. The resistance will not be less than that given on the schematic (but the resistance could be greater that the value indicated).

2.11. Resistors control the flow of electrical current, and divide voltages.

2.12. Series, parallel, and series-parallel.

2.13. One-half the value of each resistor. (For example, two 10-Ω resistors in parallel produce 5 Ω.)

2.14. Something slightly less than 10 Ω, probably about 9 Ω.

2.15. 1.5 V.

2.16. 1200 Ω: $R = E/I = 1.5/0.001 = 1500$ Ω; 100(R1) + 200(R2) = 300; $1500 - 300 = 1200$.

2.17. In no way. Battery voltage is not affected by a change in circuit resistance.

2.18. Electrolytic.

2.19. Air.

2.20. You have no way of knowing. It is a fixed capacitor and not electrolytic. The dielectric could be mica, paper, ceramic, or even air.

2.21. Trimmer.

2.22. 90 to 105 pF.

2.23. Pass alternating, block direct.

2.24. 4000 μs.

2.25. The coil has a magnetic core (probably iron).

2.26. Powdered iron.

2.27. Adjustable.

2.28. A resonant circuit.

2.29. The coil tends to impede the flow of alternating current.

2.30. A coil has very little effect on a d-c circuit (except to possibly drop the voltage because of the coil resistance).

2.31. False. Signals at the resonant frequency will pass through a series-resonant circuit.

2.32. True. Signals at frequencies other than the resonant frequency will pass through a parallel-resonant circuit.

2.33. Approximately 0.5 $\mu$H.

2.34. To oppose alternating current (to provide impedance to ac).

2.35. To oppose alternating currents at radio frequencies (to prevent RF signals from passing through a circuit).

2.36. Pass direct, oppose alternating.

2.37. To tune equipment to a particular frequency (either to pass or reject signals at certain frequencies).

2.38. 2.65 MHz.

2.39. The resonant frequency is still 2.65 MHz.

2.40. CR1 rectifies the RF signals into direct current so that they can be indicated by meter M1. C1 tunes the parallel-resonant circuit to the desired frequency.

2.41. C1 tunes the series-resonant circuit to the desired frequency (frequency to be rejected). Signals at the resonant frequency pass directly to ground (are trapped), and signals not at the resonant frequency pass to the TV set.

2.42. The resistor is probably a swamping resistor used to broaden the tuning of the resonant circuit.

2.43. Anything below about 20,000 Hz (20 kHz).

2.44. Anything above about 20,000 Hz (20 kHz).

2.45. About 120 V across L1, and about 208 V from L1 to L2.

2.46. Star or wye (Y) connections.

2.47. The delta connection.

2.48. True.

2.49. None! Direct currents are not stepped up or down by transformers, since the magnetic field produced by a direct current remains stationary.

2.50. Increased.

2.51. The IF transformers in Fig. 2-45 provide tuned coupling between stages.

2.52. 80 V.

2.53. You have no way of knowing for sure, since the schematic symbols for transformers rarely show actual turn ratios of primary to secondary. However, it would be reasonable to assume that there is a voltage step-down and that the secondary voltage is something less than 7 V.

2.54. The longer line or bar indicates positive.

2.55. The crystals shown in Fig. 2-50 are generally used to control the frequency of oscillators used in transmitters (or possibly the oscillators used in receivers).

2.56. Electrical signals into mechanical sound waves.

2.57. Mechanical sound waves into electrical signals.

2.58. False. The value given is probably the impedance.

2.59. NC contacts open, and NO contacts close when coil is energized.

2.60. Probably direct current (although there are a-c relays).

2.61. Any kind of current (ac, dc, RF, AF, pulse, etc.) can be used on relay contacts, provided that the contacts are capable of handling the voltage and current.

2.62. False. NC pushbutton contacts are always closed unless you push the button. On some pushbutton switches, the NC contacts will remain open if you hold the pushbutton.

2.63. Voltage, in the millivolt range.

*CHAPTER 3*

3.1. Cathode ray tube (CRT or KRT).

3.2. Electrostatic and electromagnetic.

3.3. None.

3.4. Whenever the anode is positive with respect to the cathode.

3.5. A zener will also pass current in the reverse direction when the avalanche or breakdown condition occurs.

3.6. Orange, yellow and brown, as shown in Fig. 3-5A. Neither the color code nor the schematic identification of 1N34A tells you the current capability of the diode. You must consult the diode data sheet.

3.7. Emitter, collector, and base.

3.8. The arrowhead points away from the base for NPN and toward the base for PNP.

3.9. The letter Q.

3.10. Nothing.

3.11. The emitter.

3.12. C1 and C2 are coupling capacitors. R1 and R2 are connected as a voltage divider to provide bias, RL is the collector or load resistor. RE is the emitter feedback resistor.

3.13. The power supply would be reversed (positive connected to $V_{CC}$ or ground).

3.14. Because a common-emitter circuit inverts the signal (output swings negative when input swings positive).

3.15. The common-base circuit has very low input impedance and high output impedance.

3.16. The common-collector circuit has high input impedance and low output impedance.

3.17. There is no voltage gain in a common collector (emitter-follower) circuit.

3.18. Bias networks set the operating point, stabilize the circuit, and set the input/output impedance of the circuit.

3.19. RE provides negative feedback.

3.20. The circuit values (particularly RL, RE, and RB).

3.21. The gate of a MOSFET is insulated from the substrate.

3.22. Normally positive. However, the polarity of the voltage applied to the drain and source of a FET can be interchanged.

3.23. Base, collector, and emitter.

3.24. N-channel.

3.25. The designation 3N153 does not identify JFET or MOSFET, nor does it tell the gain. The type of FET can be identified by the symbol. The data sheet must be consulted for the gain.

3.26. C1 and C2 are coupling capacitors. R1 and R2 are connected as a voltage divider to provide bias. RL is the drain or load resistor. RS is the source feedback resistor.

3.27. Input and output.

3.28. A relaxation oscillator.

3.29. The controlled rectifier must be triggered or turned on.

3.30. The SCS can be turned on and off by gate or trigger signals.

3.31. The triac conducts in both directions and can be used in controlling a-c power.

3.32. There is no anode or cathode on a triac symbol.

3.33. A diac.

3.34. UJT.

3.35. SUS devices, alternate positive and negative.

3.36. The photovoltaic cell acts like a battery and produces current when exposed to light. The photoconductive cell changes resistance when exposed to light.

3.37. Solor batteries or solar cells (or possibly sun batteries).

3.38. Light-dependent resistors (LDR) or light-sensitive resistors.

3.39. Arrows pointing toward the symbol.

3.40. The arrows used to show a phototransistor.

3.41. The junctions of a phototransistor are exposed to light, whereas the junctions of a conventional transistor are sealed from light.

3.42. A photoconductive cell.

3.43. The LASCS can be turned on and off by gate or trigger signals. The LASCS symbol has gates on both anode and cathode.

3.44. Capacitor value is controlled by a variable voltage applied to the VVC.

3.45. The saturable reactor has a slanted line (with parallel ends) passing through the windings and core.

3.46. The notch is an index to locate terminal 1.

3.47. Because IC25 is an amplifier.

3.48. Because it is assumed that parts within an IC array will be used together in some common circuits (since an array is an assortment of parts on a common base).

3.49. Key internal parts of an IC are shown on a pinout diagram (usually in block form).

3.50. Internal parts of an IC are shown on a pinout diagram to help the user understand operation of the circuit and to show the relationship to external parts.

## *CHAPTER 4*

4.1. To amplify audio-frequency signals.

4.2. To bypass emitter resistor R3 and increase voltage gain (without changing the d-c characteristics of the circuit).

4.3. Circuit values.

4.4. The circuit with C3 will have higher voltage gain but with a limited low-frequency response.

4.5. An electrolytic capacitor.

4.6. Cascading.

4.7. Direct coupling eliminates the interstage coupling capacitor. This makes it possible to amplify changes in d-c levels (as well as a-c signals).

4.8. Darlington compound.

4.9. True.

4.10. Theoretically, there will be no output since a differential amplifier produces an output only when one input differs from the other input.

4.11. Because of the coupling capacitor to the load.

4.12. The advantage is that only one power supply is needed. The disadvantage is that the capacitor produces limited low-frequency response.

4.13. The input is at C1. The output is at the loudspeaker.

4.14. Resistor R4 (and possibly R5 and R6).

4.15. Q1, class A; Q2 and Q3, class AB or B.

4.16. The input is at T1 (or possibly C1 if the alternate circuit is used). The output is at T3 to the load.

4.17. Emitter-feedback resistors.

4.18. Class AB minimizes crossover distortion (which can be a problem in class B operation of audio amplifiers).

4.19. A volume control is usually found between stages and sets the level of the signal passing through the amplifier. A gain control is usually part of a stage and sets the gain of one stage (thus setting overall gain).

4.20. Feedback.

4.21. Resonant.

4.22 Figure 4-18a is used as the final or power amplifier. Figure 4-18b is used as a buffer, multiplier, or driver.

4.23. C2 tunes the amplifier to the desired frequency.

4.24. L1 is part of the resonant LC circuit (the other part is C2).

4.25. The input is at the base of the transistor (across the RFC). The output is at C2 (to the antenna).

4.26. Class C.

4.27. C1 matches the circuit output impedance to the antenna (loading function).

4.28. The RFCs isolate the RF signals from the power supply.

4.29. The fact that T1 and T2 have powered-iron cores rather than the laminated iron cores found in audio circuits means that the circuit is probably operated at radio frequencies. The fact that C1 is used indicates that the circuit is to provide high voltage gain. Thus, the circuit is probably an RF voltage amplifier.

4.30. The fact that there are two tuned circuits (T1 and T2) coupled together, and a third circuit (T3) tuned separately, indicates that two signals are being mixed to produce signals at a third frequency.

4.31. C1 provides a path for RF signals around R1.

4.32. If there is no amplitude modulation of the input signal, the output will be a direct current.

4.33. In the ratio detector, the audio output is taken from the junction of R1 and R2. Also, the diodes CR1 and CR2 are reversed in the ratio detector. In the discriminator, the audio is taken from across R1 and R2. Also, the anodes of both CR1 and CR2 are connected to the input transformer T1.

4.34. If there is no frequency modulation of the input signal, there will be no output from either circuit.

4.35. CR1 is a variable shunt across the input used to reduce the gain of large signals.

4.36. CR2 is a combination AM detector and AGC bias source.

4.37. C1 tunes the resonant circuit to set the operating frequency of the oscillator.

4.38. Feedback.

4.39. Between the junction of C2-C3 and the emitter of Q1.

4.40. There is no resonant *LC* circuit in Fig. 4-26.

4.41. Y1 sets the operating frequency of the oscillator. L1 tunes the oscillator over a narrow range of frequencies. C5 matches the impedance of the circuit to the antenna.

4.42. Between the junction of C1 and C2 and the emitter of Q1.

4.43. C1 tunes the resonant circuit to set the operating frequency of the oscillator.

4.44. RC oscillators are used for audio frequencies, so the probable frequency range is below 20 kHz. The exact frequency is set by the values of C1, C2, C3, R3, R4, R5, and R6.

4.45. Through capacitor C1.

4.46. Relaxation.

4.47. Through capacitors C1 and C2 and resistors R3 and R4.

4.48. The *RC* time constants.

4.49. An integrated output (ramp waveform).

4.50. A differentiated output (spiked or pulse waveform).

4.51. The values of R1, R2, and C1.

4.52. Sawtooth or ramp output from the emitter. Spiked or pulse output from base 1.

4.53. The values of L1 and C1.

4.54. The bridge is more efficient than the half-wave supply and does not need the center-tapped transformer of the full-wave supply.

4.55. Bleeder.

4.56. There is no a-c voltage output. There will be a d-c output voltage of approximately 30 V.

4.57. 40 V.

4.58. R3 sets the output voltage of the regulator circuit.

4.59. False. The output would then be pulsating dc, or dc with considerable ripple.

4.60. T1 is a saturable reactor, whereas T2 is a conventional transformer.

4.61. False. The overall purpose of the circuits is to convert dc into ac, step up the ac, and then convert the stepped-up ac into stepped-up dc.

4.62. The attenuators of Fig. 4-59 provide fixed attenuation, and all the resistors are shown. The attenuators of Fig. 2-48 provide variable attenuation, and the symbol represents many resistors.

4.63. Bandpass.

4.64. Low-pass filter.

4.65. Ratio.

4.66. *LC* filters are more precise than *RC* filters.

4.67. Active filters provide the equivalent of an inductive reactance without an actual heavy and bulky inductance. Active filters also eliminate signal loss normally associated with passive filters.

4.68. Phase control.

4.69. The phase angle of the a-c waveforms at which the control device is triggered.

4.70. The neon lamp triggers Q1 (when the voltage across C1 exceeds the breakdown voltage of the neon lamp).

4.71. If the motor slows down, CR1 is forward-biased, the SCR is triggered, and power is applied to speed up the motor.

4.72. If the motor slows down, the tachometer produces a lower output, diodes CR1 through CR4 produce a lower voltage (negative swing), Q1 inverts the negative to a positive, C1 is charged to a higher value, the SCR triggers at an earlier point, and power is applied to speed up the motor.

4.73. Capacitor C2 is shorted and does not charge. No power is applied to the load.

## *CHAPTER 5*

5.1. Generally, the final amplifier stage is keyed to produce code (International Morse Code) signals (dots and dashes).

5.2. The carrier signal frequency will remain at 27 MHz, but the carrier will be amplitude-modulated by 7000 Hz.

5.3. The carrier signal center frequency will remain at 27 MHz, but the carrier will be modulated about this center frequency by 7000 Hz.

5.4. There will be no change in amplitude (if the transmitter is operating properly).

5.5. AM, but a special form of AM using only one sideband (instead of both sidebands and the carrier as in conventional AM).

5.6. Most of the power transmitted by an AM transmitter is in the sidebands and is lost by conventional AM transmission. SSB eliminates this problem.

5.7. The output from the second detector will be a d-c voltage (with no modulation). This will not be reproduced on the loudspeaker. However, the presence of an unmodulated carrier signal may remove the background noise normally heard from the loudspeaker.

5.8. TR1 RF amplifier, TR2 mixer, TR3 first IF amplifier, TR4 second IF amplifier, TR5 local oscillator, TR6 squelch switch, TR7 audio preamplifier, TR8 audio driver, TR9 and TR10 audio power amplifiers, TR11 transmitter oscillator, TR12 RF intermediate power amplifier, and TR13 RF power amplifier.

5.9. Input is at the antenna; output is at the loudspeaker.

5.10. Input is at the loudspeaker (acting as a microphone); output is at the antenna.

5.11. During transmit, the loudspeaker is connected to the input of the audio section through S1-D, J3, C25, and S1-B.

5.12. The receiver local oscillator TR5 crystals X-R, F1 through F6.

5.13. The transmitter oscillator TR11 crystals X-T, F1 through F6.

5.14. To the loudspeaker through J1 and S1-D.

5.15. To the transmitter amplifier transistors TR12 and TR13.

5.16. Both are emitter bypass capacitors used to increase the voltage gain of the audio amplifier section.

5.17. Class AB, because the transistors are slightly forward-biased.

5.18. These parts form a filter to remove undesired harmonics that might interfere with TV or other broadcast signals.

5.19. The transmitter stages are not tuned or adjusted. The frequency is set by the oscillator TR11 crystals.

5.20. VR7 is the emitter resistor for TR1. VR7 also permits adjustment of the S/RF meter during receive.

5.21. From L4 through C8 to the emitter of TR2.

5.22. CF1 acts as a very selective bandpass filter, passing only signals at or very near 455 kHz.

5.23. Noise-limiter diode D3 is normally forward-biased so that audio signals pass from detector D2 to the audio section. When there is excessive noise on the incoming signal, the bias changes so that D3 is reverse-biased to cut off passage of the noise signals.

5.24. To set the squelch point or level (the point below which no signals will pass to the audio section).

5.25. Volume control. VR1 is not operative during transmit.

5.26. The output of VR1 is not applied to the output of the audio section under any conditions. VR1 is applied to the input of the audio section during receive.

5.27. The output of VR1 is not applied to the input of the audio section during transmit.

5.28. To aid in troubleshooting.

5.29. Sound broadcast with frequency modulation; picture broadcast with amplitude modulation.

5.30. Picture information and synchronizing pulses.

5.31. Input is from the sync separator; output is to the yoke (vertical deflection coil and vertical blanking) of the picture tube; adjustment controls include vertical hold, size linearity, and bias; stages include vertical oscillator, driver, and output.

5.32. The input is at T1 primary; output is at the collector of Q3 and emitter of Q2, vertical hold R2, vertical size R6, vertical linearity R8, vertical bias R15, vertical oscillator Q1, vertical driver Q2, and vertical output Q3.

5.33. Input is from the video and picture-tube driver; output is to the vertical sweep circuits and to the horizontal oscillator; there are no adjustment controls; stages include first sync separator, second sync separator, and sync phase inverter.

5.34. Input is at the base of Q1 through C1; output to the vertical sweep circuits is from Q2 through R11, C6, CR1, and CR2; output to the horizontal oscillator is from the collector and emitter of Q3.

5.35. A d-c voltage that varied with changes in signal strength (a negative d-c voltage that becomes more negative when signal strength increases).

5.36. Shunt with feedback.

5.37. R8.

5.38. Zener CR2 keeps the emitter voltage of Q1 constant (as a reference for the output voltage fed back to the Q1 base).

5.39. Test points.

5.40. (1) Provides high voltage for the picture tube, (2) provides horizontal deflection voltage (horizontal sweep), (3) provides boost voltage for the picture tube, (4) provides AFC signal to horizontal oscillator, and (5) provides AGC signal to AGC circuits.

5.41. R1 sets the amplitude of the horizontal sweep and the amount of high voltage available to the picture tube.

5.42. 15,750 Hz.

5.43. The input is at C1; the output is at the collector of Q2.

5.44. Blocking oscillator.

5.45. Thermistor R11 provides temperature compensation for the Q1 oscillator.

5.46. L1 and L2 make it possible to adjust for slight shifts in phase or frequency of the signal voltages.

5.47. 60 Hz.

5.48. The discharge of C1 through R1 produces the basic sweep waveform. The sweep is made linear through feedback of the output by R19.

5.49. Vertical size control R6.

5.50. Heat sink.

5.51. The input has signals of two frequencies (15,750 Hz horizontal, 60 Hz vertical). The output to the horizontal AFC is 15,750 Hz; the output to the vertical oscillator is 60 Hz.

5.52. Q3 is a phase inverter to provide two signals ($180^\circ$ out of phase with each other) to the horizontal circuits.

5.53. R11 and C6.

5.54. One input is the RF signal from the antenna to the base of Q301. The other input is an AGC signal (varying d-c voltage), also applied to the base of Q301.

5.55. The test point at C315, called the "looker point," is used mostly to monitor the tuner output with an oscilloscope but can also be used to inject a signal into the IF amplifiers.

5.56. To neutralize C301.

5.57. C313.

5.58. One input is the RF signal from the tuner to transformer T1. The other input is an AGC signal (varying d-c voltage) applied to the bases of Q1 and Q2.

5.59. Forward-biased, since all transistors are PNP and all bases are more negative than the corresponding emitters.

5.60. The IF transformers are stagger-tuned to give the IF amplifiers sufficient bandwidth to pass both the picture and sound signals.

5.61. The inputs include video (picture) and audio (sound) signals from the video detector, blanking pulses from the vertical circuits, and blanking pulses from the horizontal circuits; the outputs include signals to the sound IF, signals to the sync separator, and picture signals to the picture-tube cathode.

5.62. R3 adjusts the contrast of the picture. R12 adjusts the brightness of the picture. The 4.5-MHz trap is adjusted to prevent any sound information from entering the video circuits.

5.63. LDR1 provides variable picture contrast to accommodate changing conditions of ambient light.

5.64. R12 sets the brightness of the picture.

5.65. The IF signals applied to the base of Q1 are in the 42-MHz range. The keying pulses from the horizontal circuits (applied to the collector of Q1 through CR1) are at 15,750 Hz.

5.66. Strong signals cause the negative AGC voltage to increase (become more negative), increasing the forward bias and driving the tuner and IF amplifier transistors into saturation (thus reducing gain to compensate for the strong signals).

5.67. The input signals are frequency-modulated (in the audio range) about a center frequency of 4.5 MHz.

5.68. Forward-biased. (Q1 is PNP with base more negative, less positive; Q3 is NPN with base more positive.)

5.69. Forward-biased. (All are NPN with base more positive.)

5.70. Neutralization for Q1.

5.71. Emitter bypass capacitors to increase voltage gain.

5.72. T1, T2, R12, and R13.

5.73. A low-pass filter for the audio output of the FM detector.

5.74. Class AB. The bases are more positive than the emitters, and both Q4 and Q5 are NPN.

5.75. Diodes CR1 and CR2 are reversed (the anode of CR1 and the cathode of CR2 are connected to the secondary of T2).

# Index